CENTRALLY ACTING PEPTIDES

BIOLOGICAL COUNCIL
The Co-ordinating Committee for Symposia
on Drug Action

CENTRALLY ACTING PEPTIDES

Edited by

J. HUGHES, Ph.D.

Department of Biochemistry,
Imperial College of Science and Technology,
University of London

First published 1978 by
THE MACMILLAN PRESS LTD
London and Basingstoke
Associated companies in Delhi Dublin
Hong Kong Johannesburg Lagos Melbourne
New York Singapore and Tokyo

Typeset by
Reproduction Drawings Ltd
Sutton, Surrey

British Library Cataloguing in Publication Data

Centrally acting peptides.
    1. Neurotransmitters – Congresses 2. Peptides
    – Congresses
    I. Hughes, John II. Biological Council.
    Co-ordinating Committee for Symposia on
    Drug Action
    612'.82   QP364.7

ISBN 978-1-349-03670-7   ISBN 978-1-349-03668-4 (eBook)
DOI 10.1007/978-1-349-03668-4

Biological Council Co-ordinating Committee for Symposia on Drug Action

*Report of a symposium held on 4 and 5 April 1977 in London
at The Middlesex Hospital Medical School*

Sponsored by:

Biochemical Society
British Pharmacological Society
British Society for Antimicrobial Chemotherapy
British Society for Immunology
Chemical Society
Nutrition Society
Pharmaceutical Society of Great Britain
Physiological Society
Royal Society of Medicine
Society for Applied Bacteriology
Society for Drug Research
Society for Endocrinology
Society of Chemical Industry, Fine Chemicals Group

*The organisers are grateful to the following for the generous financial support
which made the meeting possible:*

Boehringer–Ingelheim Limited
Fisons Limited
Imperial Chemical Industries Limited
Janssen Pharmaceuticals Limited
Lepetit Pharmaceuticals Limited
Reckitt and Colman Limited
Riker Laboratories
Roche Products Limited
Sandoz Products Limited
Stag Instruments
The Wellcome Trust

Organised by a symposium committee consisting of:

J. Hughes (Chairman and Hon. Secretary)
L. L. Iversen
D. W. Straughan
J. R. Vane

Symposium Contributors

Professor P. B. Bradley, The University of Birmingham, Department of Pharmacology (Preclinical), The Medical School, Vincent Drive, Birmingham, B15 2TJ

Dr Michael J. Brownstein, National Institute of Mental Health, Laboratory of Clinical Science, Bethesda, Maryland 20014, U.S.A.

Dr R. P. Elde, Department of Anatomy, University of Minnesota, 262 Jackson Hall, Minneapolis, Minnesota, U.S.A.

Dr J. Hughes, Unit for Research on Addictive Drugs, University of Aberdeen, Marischal College, Aberdeen, AB9 1AS

Dr L. L. Iversen, MRC Neurochemical Pharmacology Unit, Department of Pharmacology, Medical School, Hills Road, Cambridge, CB2 2QD

Professor H. W. Kosterlistz, Unit for Research on Addictive Drugs, University of Aberdeen, Marischal College, Aberdeen, AB9 1AS

Professor F. Lembeck, M.D., Institute of Pharmacology, University of Graz, Universitätsplatz 4, Graz, Austria

Dr R. Miller, The Wellcome Research Laboratories, Wellcome Foundation, Research Triangle Park, North Carolina, U.S.A.

Dr H. R. Morris, Department of Biochemistry, Imperial College, London, SW7 2AZ

Professor A. G. E. Pearse, Royal Postgraduate Medical School, Department of Histochemistry, Ducane Road, Hammersmith, London

Dr Arthur Prange, University of North Carolina, Chapel Hill, North Carolina, U.S.A.

Dr L. P. Renaud, Division of Neurology, Montreal General Hospital, McGill University, Montreal, Canada

Dr D. G. Smyth, Peptide Research Group, National Institute for Medical Research, The Ridgeway, London, NW7 1AA

Dr Solomon H. Snyder, The Johns Hopkins University, School of Medicine, Department of Pharmacology and Experimental Therapeutics, 725 North Wolfe Street, Baltimore, Maryland 21205, U.S.A.

Dr Lars Terenius, Department of Medical Pharmacology, University of Uppsala, Uppsala, Sweden

Professor David de Wied, Rudolf Magnus Institute for Pharmacology, University of Utrecht, Medical Faculty, Vondellaan, 6, Utrecht, The Netherlands

Foreword

This volume contains papers by sixteen of the eighteen participants to a symposium held on 4 and 5 April, 1977 at the Middlesex Hospital Medical School and organised by the Co-ordinating Committee for Symposia on Drug Action of the Biological Council. The authors have bravely attempted to review their particular areas of interest as well as presenting up to date research data.

I should like to express my heartfelt appreciation for the advice and help given by the many people who made this Symposium possible. In particular I should like to thank Miss G. M. Blunt for her indefatigable enthusiasm and hard work and Professor D. W. Straughan for his invaluable advice. The financial support for the meeting depended on the generosity of the supporting Societies and the Pharmaceutical Industry and this is gratefully acknowledged.

Less than ten years ago it was possible to confine a discussion of central neurotransmitter agents to a few amino acids and biogenic amines such as acetylcholine, the catecholamines and indolealkylamines and one peptide, Substance P. It now seems that the peptides must be considered as a major group of neurotransmitter agents with roles distinct from their involvement in neuroendocrine processes. Current interest in the central role of peptides has been given added impetus by the discovery of new peptides such as neurotensin, the enkephalins and endorphins. However, at the present time major advances are being recorded by the increasing use of radioimmunoassay and immunofluorescent techniques to localise and quantitate brain peptides. In addition the interdisciplinary nature of this field is well illustrated in this volume, thus electrophysiological, biochemical and behavioural techniques are all being used to delineate and explore peptide action.

One great gap remains and that is the correlation between the presence and activity of a particular peptide and its relevance to physiological processes. This problem is being urgently addressed but at present most correlates of peptide action are speculative and must therefore be viewed with caution. The infinite diversity, information content and biological potency of the peptides makes them obvious candidates for a variety of central functions. No doubt many will be found to meet the criteria for neurotransmitters but many more may not fit this role and we may have to seek alternative designations. The term 'neuromodulator' although controversial may have value in this respect. This term may be used in a narrow or wide sense but it may be particularly applicable to peptides that are released from neurons, perhaps in concert with a neurotransmitter, to produce subtle or long-term effects (Barker (1976) *Physiol. Rev.*, **56**, 435-51; Barker and Smith (1976) *Brain Res.*, **103**, 167-70) not normally associated with neurotransmitters. Subtle changes in membrane potential or neuronal metabolism may have profound effects on overall neuronal activity (Schmitt *et al.* (1976) *Science*, **193**, 114-20) and it seems likely that these concepts may aid our understanding of the behavioural effects of many centrally active peptides.

The potential therapeutic benefits of neuropeptide research are as yet unknown although several possibilities are discussed by authors in this volume. It may be reasonable to assume that dysfunction of peptidergic neurons may have undesirable sequelae and that new approaches to the treatment of cerebral dysfunctions may soon be revealed. Future research is likely to lead to a deeper understanding of neuronal physiology with corresponding therapeutic benefits.

Aberdeen, August 1977   John Hughes

Contents

1
Isolation and identification of biologically active peptides

Howard R. Morris (Department of Biochemistry,
Imperial College, South Kensington, London, U.K.)

INTRODUCTION

Following the discovery of some unidentified biologically active substance as a constituent of a living system, a full understanding of its role within that system must necessarily await its purification and the elucidation of a chemical structure. Without such detailed characteristisation we are limited to working with and observing the biological activity of a crude substance; this may lead to vague and speculative theories on interactions with other components in the system. Here, we are limiting our discussion to biologically active peptides, but unfortunately, because of the special factors discussed below, these are among the most difficult substances to evaluate structurally. For example, almost a decade of effort went into the structural determination of the hypothalamic hormone responsible for the release of thyroid stimulating hormone, despite the simplicity of the structure ultimately assigned to thyrotropin releasing hormone (TRH) (Burgus *et al.*, 1969; Schally *et al.*, 1966). Nevertheless, the work and the studies which this effort in turn stimulated has led to a revolution in our understanding of hypothalamic and pituitary function, and to a new appreciation of small peptides, not just as inactive fragments resulting from the enzymic digestion of proteins, but as a special class of compound exerting profound effects on the central nervous system.

There are four main reasons why the characterisation of biologically active peptides may prove more difficult than conventional peptide or protein structure determination.

(1)   Biologically active peptides are normally present in the tissue of origin in very small amounts, often just nanomoles or even picomoles.

(2)   The biological activity, often the only means of following or locating the peptide, is lost or diminishes rapidly during purification.

(3)   Once purified, the quantity of peptide available poses severe problems, even for modern structural techniques.

1

(4)   Since the peptides are not obviously of protein origin, that is not produced from a protein by the investigator, it follows that extreme caution must be exercised in structural assignment; sequence methodologies were developed for the study of protein-derived peptides having free (unblocked) N- and C-termini, 'normal' amino acids and α-linked peptide bonds. There is no reason to believe that a new substance isolated from, say, brain tissue will conform to all or indeed any of the above structural features.

In this paper the problems associated with the above four points are investigated, and solutions demonstrated or suggested by reference to work in the literature or to work in progress in the author's laboratory.

PEPTIDE ISOLATION AND PURIFICATION

Most work involves an initial extraction of active material from the tissue of origin, using for example ice-cold acidified solvents, such as HCl/acetone (Kitabgi *et al.*, 1976, Schally *et al.*, 1966). After several extractions of the tissue homogenate, the soluble fraction may be dried to yield a powder which should be stored in the deep-freeze ready for purification. The combination of a mineral acid and organic solvent serves the dual purpose of conveying solubility to peptides rather than proteins, and inactivating many of the proteolytic enzymes that may be present either by denaturation, precipitation or simply swinging the pH away from the optimum for proteolytic catalysis. Having said that, it is still a common phenomenon for biological activity of the crude extract to diminish rapidly during storage; it is particularly dangerous to store extracts of some tissues in solution, even below $-10\,^{\circ}C$.

Tissue extraction, which has been the major if not the only technique of hormone isolation, will almost inevitably result in the co-extraction of biological precursors and intermediates which may themselves show some activity. An attempt to isolate a true hormone, that is one *secreted* by the tissue in which it is synthesised, has been made for corticotropin releasing hormone (CRH) (Gillham *et al.*, 1976; Jones *et al.*, 1976), and purification of this material is now underway (Morris, Gillham and Jones, unpublished).

Peptide purification is a major problem when dealing with microgram or submicrogram amounts of active peptides contaminated by milligram or even gram quantities of other substances. Much of this is due to non-specific adsorption of the peptide to the contaminants, glassware, or other materials (for example column packings) introduced during purification. Some solutions are obvious, for example silanising glassware; other not so obvious tricks will emerge during the course of this discussion.

Purification
The following methods can be used for further purification of the crude extract.

(a)   gel filtration
(b)   adsorption to a stationary phase
(c)   ion exchange chromatography
(d)   high voltage paper electrophoresis

(e)   paper or thin layer chromatography
(f)   counter-current distribution
(g)   high pressure liquid chromatography (h.p.l.c.)

This shows basic choices rather than an exhaustive list, and some options overlap. Gel filtration or adsorption chromatography are excellent first steps in any purification. In (b) the ideal example would be an affinity column, for example the purification of neurotensin from bovine intestinal tissue (Kitabgi *et al.*, 1976). In most work such a possibility would not exist, and one may for example be using XAD-2 resin to bind enkephalins. Washing with water removes salts and many other contaminants and the active compound is eluted using a relatively non-polar solvent such as alcohol. Gel filtration on, say, Sephadex G-25 will not only give an indication of the molecular weight of the unknown peptide, but will also remove any remaining enzyme or protein in the void volume of the column. It should be noted that molecular weight estimation using gel columns (even in the presence of standards) is very crude and deviation from the true value may be as much as 100 per cent. The method usually gives rather poor resolution, and hydrophobic peptides are absorbed and retarded on the resin matrix at low ionic strengths. Peptide yields on this type of column can be as good as 90 per cent, and volatile solvents can be used to afford concomitant desalting of most peptides of molecular weight greater than 500.

Ion exchange chromatography
This is probably the most powerful method for high resolution purification of complex peptide mixtures. Resins include the sulphonated polystyrenes (for example Dowex 50) and the carboxymethyl and diethylaminoethyl celluloses and dextrans.

The choice of resin to suit a particular purpose is important; generalised statements are usually unreliable in protein chemistry because of the diversity of structure and unpredictable properties of many peptides. However, the following hints may be useful. In the polystyrene resins it is often desirable to use a highly crosslinked bead (× 12) to prevent peptide penetration of the resin. Columns can be monitored at D_{230} for peptides provided that salt buffer systems are used for elution. Desalting of small peptides can prove difficult, and excellent chromatography can be achieved using volatile pyridine/acetic acid based buffers as seen in figure 1.1. Here the column effluent is best monitored by analytical electrophoretic analysis of each fraction (see later). A disadvantage of resins of the Dowex 50 type is that hydrophobic or very basic peptides are often lost. Indeed, when dealing with very small quantities of peptide using new resins, experience shows that virtually everything may be irreversibly adsorbed, despite pretreatment of the resin to leach out impurities. In these cases we have found it useful to 'block' the active sites of a new resin by pre-chromatographing a relatively large quantity of a variety of peptides sufficient to encompass the type of structure which the unknown active peptide may belong to, either very acidic, basic or hydrophobic. This is best done by eluting a protein digest, for example ribonuclease or chymotrypsinogen. The main precaution is that the protein must be of known structure so that chance co-elution of any remaining protein-derived peptide with the unknown will not lead to an erroneous identification of structure!

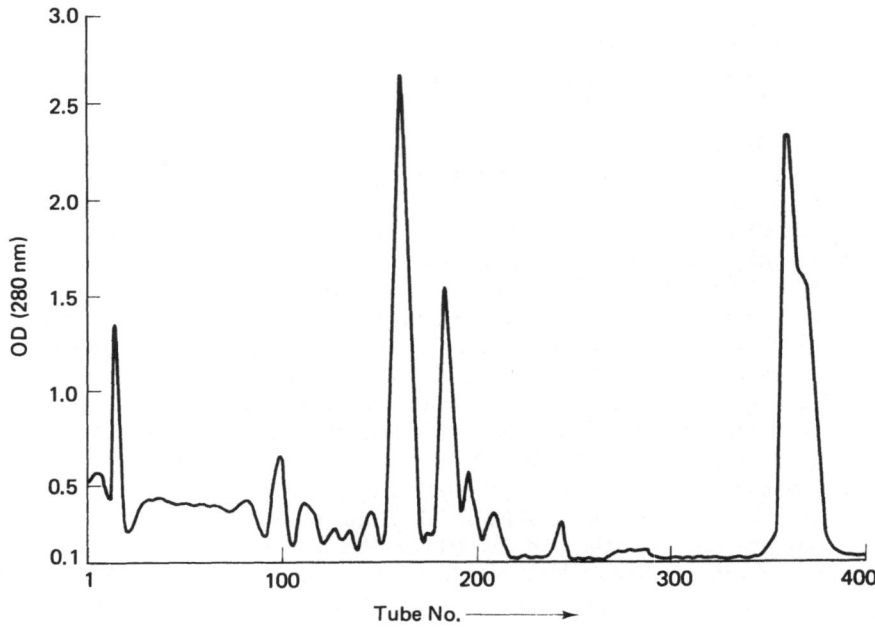

Figure 1.1 Typical absorption profile at 280 nm of the effluent from Dowex 50 chromatography of a mixture of peptides eluted using volatile pyridine/acetic acid buffers.

The point about blocking active sites on columns or other surfaces for high sensitivity work cannot be overemphasised, since this type of adsorption is the major reason for low yields on purification. It should also be noted that as the purification proceeds losses may increase because of the removal of contaminants which acted as 'carriers' by competing for surface active sites. The extent of purification should therefore be as low as the structural methods to be used will allow.

Carboxymethyl and DEAE cellulose or Sephadex resins have excellent properties for separating large peptides, and volatile buffer systems based on ammonium bicarbonate or acetate molarity gradients can often be used, as was the case in an early purification of TRH (Schally *et al.*, 1966).

Electrophoresis
High voltage paper electrophoresis is an excellent method for the separation and purification of small quantities of polar peptides. Common buffer systems are pH 6.5 pyridine/acetic acid/water and pH 2.1 acetic acid/formic acid/water. The pH values are chosen to give fully protonated or deprotonated amino and carboxyl functions. For example, acid or basic peptides will carry a net charge at pH 6.5 and will migrate towards the anode or cathode respectively. The electrophoretic method is clean, and applicable to surprisingly small quantities of peptide if proper precautions are taken (see below). For example figure 1.2 shows an autoradiograph of a pH 2.1 electrophoretogram of a hormone whose purification and structure we are carrying out in this laboratory. The sample was treated with [^{14}C]

Figure 1.2 Autoradiograph of pH 2.1 high voltage paper electrophoretogram showing 15 pmol [^{14}C] carboxymethylated vasopressin.

iodoacetic acid, and the lower band in figure 1.2 represents a 'contaminant' in the preparation; it corresponds to only 15 pmol of carboxymethyl vasopressin. The radioactive marker shows how well even this small quantity of material will migrate in the presence of larger amounts of other substances, provided the right conditions can be found.

Good results can only be obtained if certain precautions are taken; these have included any or all of the following.

(a)   pre-washing of paper in the buffer or formic acid to remove contaminants
(b)   cold-air drying or even incomplete drying of spotted out material before buffering up
(c)   the choice of paper may need to be changed depending on the properties of the peptide (see below)
(d)   incomplete drying after run, and avoidance of overheating during run; most electrophoresis systems are fitted with cooling coils or blocks
(e)   elution of activity in cold room or cabinet

It must also be remembered that extremes of pH (such as pH 2.1) may destroy biological activity in exceptional circumstances, and pilot runs must be made to ascertain this.

Figure 1.3 illustrates point (c) above and shows work on the purification of human vasoactive intestinal peptide (VIP) carried out in the author's laboratory in collaboration with Drs Albuquerque and Bloom. Monitoring after elution by radio-immunoassay shows that Whatman No. 1, a popular paper for sensitive peptide work is of little value for a good yield after purification of VIP. In contrast the use of cellulose acetate gives virtually full recovery of activity after electrophoresis. Figure 1.3 represents the loading of 1 mg of a crude product containing 1 μg of VIP immunoassayable activity, and the elution from cellulose of all the activity. This does not necessarily mean 100 per cent yield since other factors, that is impurities, may have influenced the immunoassay of the crude material. Nevertheless, good activity has been recovered by the correct choice of support.

Human   Vasoactive   Intestinal   Peptide

250 picomole: HVPE pH 6.5 pyridine acetate

80V/cm, 1.5 h : ELUTION 0.01M acetic

PAPER   YIELD by RADIOIMMUNOASSAY

Whatman No. 1   5%

Cellulose Acetate   ~100%

Figure 1.3   High voltage paper electrophoresis data on the purification of vasoactive intestinal peptide (Morris, Albuquerque and Bloom, unpublished). See the text for details.

Thin layer chromatography, paper chromatography counter-current distribution and h.p.l.c. are also valuable techniques where peptides cannot be separated by mass or charge alone. Solvent systems made up from organic solvent/acid or base/water can separate very similar molecules. For example on ethyl acetate/pyridine/acetic acid/water system has been used by Morgan and co-workers to separate leucine-and methionine-enkephalin, two very similar structures which are difficult to separate by other methods. Similarly, h.p.l.c. is becoming a valuable tool in protein chemistry and has been used in the purification of endorphins from bovine hypo-thalami (Ling *et al.*, 1976).

Concluding this section, the problems inherent in the purification of small quantities of biologically active peptides can be overcome, providing the sorts of precautions described are taken. For those workers fortunate enough to be able to work on hundreds of thousands of animals, the problems are minimised; for the rest of us extra skill and ingenuity must be brought to bear to obtain meaningful results.

STRUCTURAL STUDIES

We will assume that the active compound being studied has been identified as a peptide from experiments monitoring loss of activity on treatment with proteases. A point worth noting here is that loss of activity after protease treatment does not necessarily mean that the active compound is a peptide or that it contains a bond susceptible to that particular protease. If the activity is derived from only picomolar levels of sample, loss of activity may be due to adsorption to the protease, unless a correspondingly small amount has been used, normally a 50 : 1 substrate–enzyme ratio.

A first step in structure determination is usually amino acid analysis. Most modern analysers are capable of good results at nanomolar levels. Several amino acids are prone to oxidation or destruction on acid hydrolysis, notably cysteine and tryptophan. If tryptophan is suspected it may be protected by hydrolysis in *p*-*p*-toluene sulphonic acid; cysteine can sometimes be estimated as cysteic acid or carboxymethyl cysteine but its absence or presence is best proven during sequencing or with a radiolabelling reaction (see earlier).

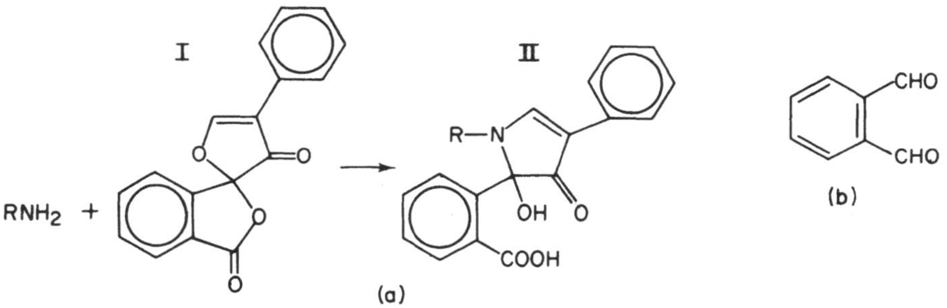

Figure 1.4　Chemical structures of fluorescamine, orthophthalaldehyde and the product of reaction of fluorescamine with a primary amino group.

Today's problems in peptide research are demanding higher and higher levels of sensitivity in amino acid detection. This can be achieved by the production of fluorescent derivates and monitoring by spectrofluorimetry (Roth, 1971; Udenfriend *et al.*, 1972). Figure 1.4 shows the reagents fluorescamine and orthophthalaldehyde, and the product of the former reacting with a primary amino group. With these reagents and microbore columns it is possible to obtain good amino acid analyses at the 10 pmol level (Bensen, 1975). Disadvantages are that proline does not react with either reagent, and some problems also exist with cysteine and lysine. However, even partial information at this level of sensitivity is valuable.

Sequencing

Two sequencing methods will be discussed here, the first an ultrasensitive method based on the Edman degradation reaction, and the second a mass spectrometric method applicable to unusual structures and structures of uncertain assignment.

Figure 1.5 shows the coupling, cleavage and conversion steps involved in the

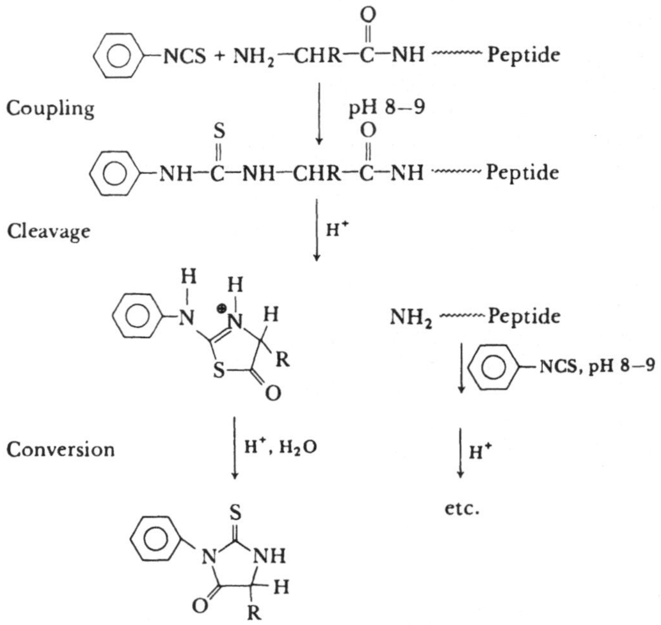

Figure 1.5   Chemical reactions taking place during the Edman degradation.

standard Edman procedure, leading to a phenyl thiohydantoin (PTH) derivative and a new shorter peptide. Sequencing is achieved either by identification of the PTH derivative or of the new amino terminus by the dansyl technique (Hartley, 1970). A modification of this method involves coupling of the peptide to an inactive support resin. Some of the disadvantages of the normal method, associated with reagents, solubility and impurities, may then be removed by adequate washing of the resin-bound peptide. The disadvantage lies in the inefficiency of most coupling procedures. One of the most successful is shown in figure 1.6, which involves coupling of the peptide to a bi-functional reagent, phenyl diisothiocyanate followed by coupling to an amino-resin support or to an activated glass. Acid treatment starts the first step of the Edman procedure, and 'holes' are left in the sequence wherever lysine or ornithine side chains are bound to the resin (glass). A high sensitivity procedure of potential value in the active peptide area is based on the use of radio-labelled phenylisothiocyanate for the formation of [^{35}S] phenyl thiohydantoin derivates (Bridgen, 1975), and we may see applications of this method in the near future. One problem with the method is the normal one associated with the identification of spots on t.l.c. plates in the presence of high background.

Many of the biologically active peptides found to date have been shown to possess unusual or blocked structural features. Examples are TRH, PCA–His–Pro–NH$_2$, melanocyte stimulating hormone release inhibiting factor (MSH-IF), Pro-Leu–Gly–NH$_2$ and the Acetyl-Ser N-terminus of α-MSH. These structures were extremely difficult to establish classically (Celi, *et al.*, 1973; Harris, 1960, Schally *et al.*, 1966) and the methods such as those described above would be quite inapplicable.

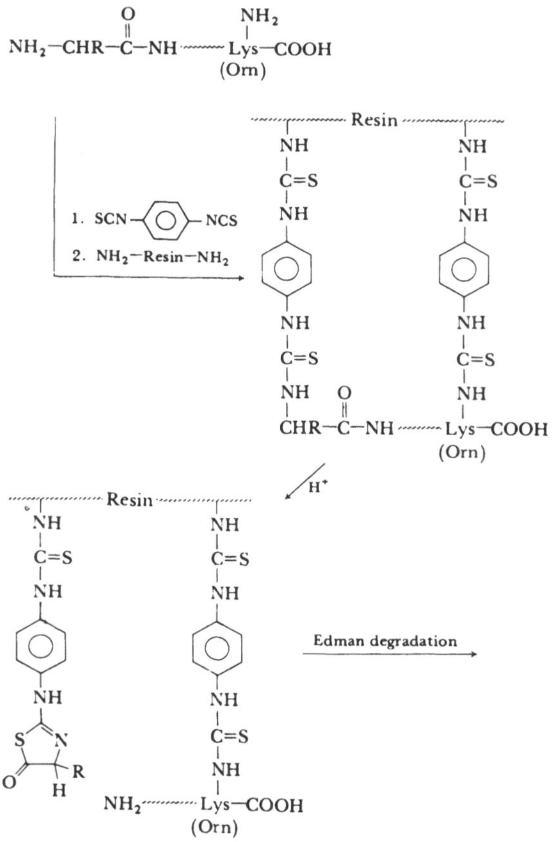

Figure 1.6 A method of coupling peptides to resins using phenyl diisothiocyanate. Peptides are coupled via their amino groups.

One modern technique which can overcome some of these problems is mass spectrometry, and the structures of three of the most recently identified biologically active peptides, the enkephalins and adipokinetic hormone, were deduced using mass spectrometric methods (Hughes *et al.*, 1975; Stone *et al.*, 1976). In both cases the structures were examined by classical methods, but because of problems peculiar to each of the studies, complete structures could not be assigned. Mass spectrometric investigation led to definitive structures which, although unexpected from the initial interpretations of classical data, have since proved to be correct.

The mass spectrometric method referred to above is based on the examination of samples that have first been treated to acetylate free amino groups (Thomas *et al.*, 1968) and then to block all other polar functions by permethylation under special conditions (Morris, 1972). Since it is the practice of the method to block functional groups, peptides which are blocked in nature pose little problem, and the structure of the blocking group(s) can normally be determined along with the sequence. Figure 1.7 shows the structure determined for adipokinetic hormone, the first peptide hor-

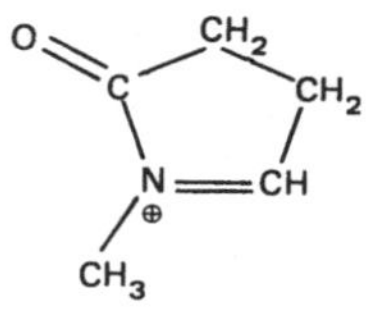

(a)

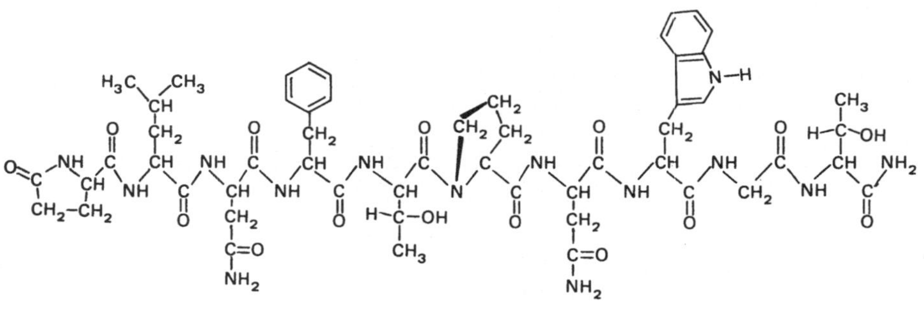

(b)

Figure 1.7 Structure of adipokinetic hormone.

mone of insect neuroendocrine origin to be fully characterised (Stone *et al.*, 1976). The identity of the N-terminal PCA-group was fixed by the presence of a signal in the mass spectrum at *m/e* 98 (Morris *et al.*, 1971), corresponding to the structure in figure 1.7a. Similarly a C-terminal amide was assigned via a signal at *m/e* 428 which would have been at *m/e* 415 had the hormone possessed a free carboxyl group prior to derivative formation. Moreover, the mass spectrum unexpectedly showed the presence of a tryptophan residue; this can be missed, because of destruction, in classical sequence approaches. The overall sequence was then assigned from the spectra to give an N- and C-terminally blocked decapeptide structure for adipokinetic hormone (figure 1.7b).

Figure 1.8 shows the first mass spectrum of enkephalin obtained from approximately 25 nmol of the preparation of Hughes and co-workers. Classical studies had indicated a partial structure of Tyr–Gly–Gly–Phe for the peptide. Both amino acid analysis and chromatographic data suggested a peptide length of 8–10 residues, and an ultraviolet fluorescence measurement had indicated the presence of a possible tryptophan-like amino acid in a non-equimolar amount, and a high glycine value plus traces of other amino acids could have been due to impurities or to real reflections of some structural feature in enkephalin. Referring to point (4) under *Introduction*, such anomalies do exist in real problems at this level of sensitivity, and when as here the sample is not protein-derived it is wise to adopt a technique that can give a definitive answer on the question of structural problems. A more easily understood example of this reasoning concerns the recent discovery of a new amino acid, γ-carboxyglutamic acid, in the sequences of some

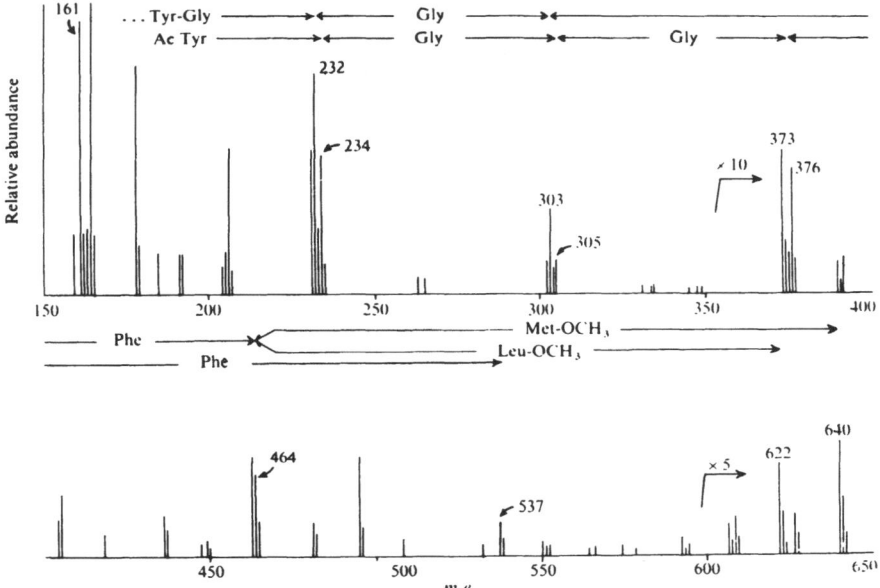

Figure 1.8 The first mass spectrum on which the structure of enkephalin was based. The figure shows 25 nmol of an acetyl permethyl derivative.

blood proteins. Classical methodology leads to the facile decarboxylation of this amino acid, and to the assignment of glutamic acid at the positions under discussion. Mass spectrometry played a major role in the discovery of this new substance (Magnusson *et al.*, 1974; Morris *et al.*, 1976) which in turn has led to a new understanding of vitamin K and Ca^{2+} involvement in the blood clotting process. The moral is to beware when using standard techniques in situations demanding an unbiased approach to structure analysis.

Returning to figure 1.8, an initial conclusion is that no tryptophan, oxidised or other simply modified tryptophan is present in enkephalin since all the low mass signals from aromatically stabilised species can be assigned. The spectrum suggests a major component of either (*A*) Tyr–Gly–Gly–Phe–Met or (*B*) Tyr–Gly–Gly–Phe–Met–Phe–Gly–Gly–Phe and (*C*) a minor component of Tyr–Gly–Gly–Phe–Leu.

This was the first indication, via the signals at *m/e* 622 and 640 that enkephalin could be a mixture of two substances, but clearly the ambiguity of the major component had to be resolved for a full interpretation leading to synthesis. Deuterium labelling experiments confirmed the above assignments, but did not resolve the ambiguity.

It was decided to use a protein chemistry trick to solve the problem. Note that both possible structures contain methionine; cyanogen bromide will react with a methionine-containing peptide to produce chain cleavage and conversion of methionine to homoserine (which has a different mass). Note also that alternative (*B*) (which incidentally would fit the amino acid analysis very well) has a mid-chain methionine, and cleavage would have generated a new peptide Phe–Gly–

Gly–Phe. The remaining 20 nmol of the enkephalin sample was therefore treated
with CNBr and the product examined in the mass spectrometer. The spectrum
(figure 1.9) clearly shows a shift of the methionine signal at m/e 640 to homoserine at m/e 624, but no new peptide. Furthermore m/e 622 assigned to leucine
remained. This structure of enkephalin was thus unequivocally assigned as a
mixture of Tyr–Gly–Gly–Phe–Met and Tyr–Gly–Gly–Phe–Leu.

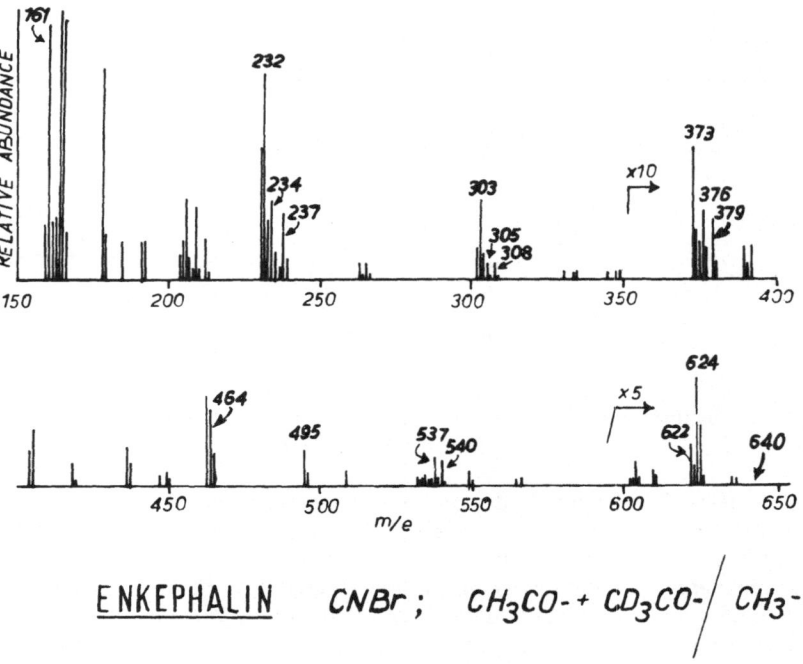

Figure 1.9   Mass spectrum of an acetyl permethyl derivative of 20 nmol of a CNBr digest of
enkephalin.

Following this sequence assignment, those of us associated with the work
(Hughes *et al.*, 1975) began to search the protein chemistry literature for a possible
precursor of such excitingly active peptides. One could find several Tyr–Gly
sequences in proteins but nothing approaching the overall structure. Soon after
this, Dr Smyth was invited to this laboratory to speak on aspects of secondary
and tertiary structure of β-lipotropin. During his talk, when commenting on a
trypsin-like enzymatic cleavage for the generation of β-MSH, he showed a slide
containing –Lys–Arg–Tyr–Gly–. Having seen several Tyr–Gly sequences previously
I was not particularly excited but decided to follow up the remaining sequence of
β-lipotropin after the seminar. To my astonishment, there, buried in the middle
of β-lipotropin was the newly discovered sequence of methionine-enkephalin, in
an ideal position to be released at the N-terminus (figure 1.10). The sequel to
this story is now obvious and exciting revelations continue to be made almost
weekly. No putative precursor has yet been found for leucine-enkephalin, but it

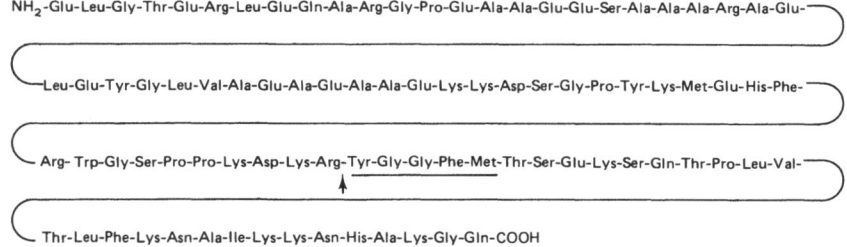

Figure 1.10   Structure of β-lipotropin; the methionine-enkephalin sequence is underlined.

seems likely that this sequence may well prove to be a fragment of 'pro-ACTH' which we may presume to exist, and which is now being looked for in several laboratories.

QUANTIFICATION

One final aspect of identification worthy of mention here is the quantification and localisation of a substance once its structure has been determined. Radio-immunoassay procedures have proved to be very powerful in this respect, but problems do exist either in antibody generation or cross-reactivity. The latter phenomenon could give rise to misleading data if the specificity of the assay is questionable.

A capability of the mass spectrometric methodology is, in certain cases, to examine signals which are absolutely specific for a particular structure and, in the presence of a suitable internal standard, to quantify the substance accurately.

Quantification of the enkephalins is a good example of this. Suitable internal standards can be made by deuterium exchange to give tetradeutero-derivatives which are stable to the isolation conditions. m/e 622 and 640 in figure 1.8 are specific for leucine- and methionine-enkephalin respectively. It follows that by tuning into these signals or scanning slowly across them we can monitor the presence or absence of either peptide. Crude unpublished work of this nature on rat brain preparations by Hughes *et al.* was among the first data to suggest a differential location of the two enkephalins in the brain. We have now developed (Dell and Morris, unpublished) a more sensitive and facile method based on rapid switching between relevant signals using a multipeak monitor device (Costa and Holmstedt, 1973). Figure 1.11 shows the output from only 25 pmol of leucine-enkephalin demonstrating that very small quantities of enkephalin are recognisable. Note that channel 5 was tuned to methionine-enkephalin which was not present in this sample. Work is now progressing in this laboratory in collaboration with Drs A. Crow and J. Edwardson, to identify and quantify brain peptides in general, as part of a study designed to uncover their putative role in the genesis or expression of mental illness.

CONCLUSION

Some of the more powerful methods for the isolation and purification of biologically active peptides have been examined, and where possible suggestions for their proper use have been made. The diversity of potential peptide structures is

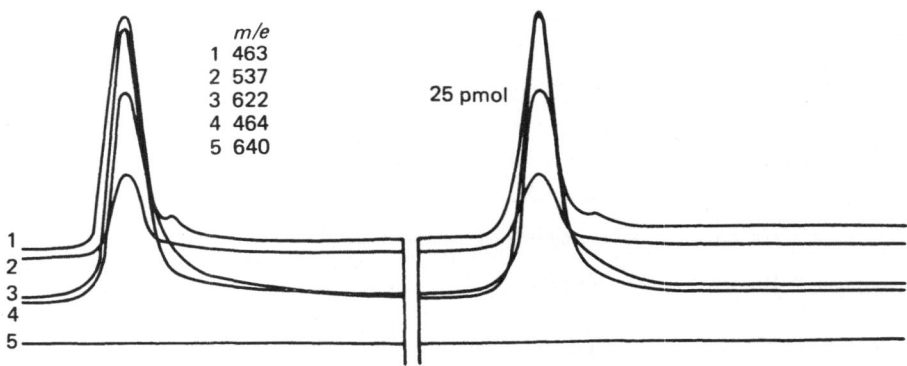

Figure 1.11 Multiple peak monitoring of leucine-enkephalin showing specificity (zero signal for methionine-enkephalin).

such that accurate predictions cannot be made for their behaviour in many chromatographic systems. However, an awareness of the problems, particularly when working with small quantities of peptide, should make it easier to decide on the correct purification step. Two modern extensions of analytical procedures for structure determination have been described, covering ultrasensitive analysis of 'normal' peptides, and a definitive method for normal or unusual structures. The quantitative power of one of these methods has been demonstrated but remains to be fully exploited in this area.

REFERENCES

Benson, J. R. (1975). In *Instrumentation in Amino Acid Sequence Analysis*, (ed. R. N. Perham). Academic Press, New York

Bridgen, J. (1975). *FEBS Lett.*, **50**, 159–62

Burgus, R., Dunn, T., Desiderio, D. and Guillemin, R. (1969). *C.r. hebd Séanc Acad. Sci., Paris*, **269**, 1870–73

Celi, M. E., Macagno, R. and Taleisnik, S. (1973). *Endocrinology*, **93**, 1229–33

Costa, D. and Holmstedt, B. (eds) (1973). In *Gas Chromatography–Mass Spectrometry in Neurobiology*, Raven Press, New York

Gillham, B., Hillhouse, E. W. and Jones, M. T. (1976). *J. Endocr.*, **71**, 60–61

Harris, I. (1960). *Br. med. Bull.*, **16**, 189–95

Hartley, B. S. (1970). *Biochem. J.*, **119**, 805–22

Hughes, J., Smith, T. W., Kosterlitz, H. W., Fothergill, L., Morgan, B. A. and Morris, H. R. (1975). *Nature*, **258**, 577–79.

Jones, M. T., Hillhouse, E. W. and Burden, J. L. (1976). *J. Endocr.*, **69**, 1–10.

Kitabgi, P., Carraway, R. and Leeman, S. E. (1976). *J. biol. Chem.*, **251**, 7053–58

Ling, N., Burgus, R. and Guillemin, R. (1976). *Proc. natn. Acad. Sci. U.S.A.*, **73**, 3942–46

Magnusson, S., Sottrup-Jensen, L., Petersen, T. E., Morris, H. R. and Dell, A. (1974). *FEBS Lett.*, **44**, 189–93

Morris, H. R. (1972). *FEBS Lett.*, **22**, 257–60

Morris, H. R. Williams, D. H. and Ambler, R. P. (1971). *Biochem. J.*, **125**, 189–201

Morris, H. R., Dell, A., Petersen, T. E., Sottrup-Jensen, L. and Magnusson, S. (1976). *Biochem. J.*, **153**, 663–79

Roth, M. (1971). *Analyt. Chem.*, **43**, 880–83

Schally, A. V., Bowers, C. Y., Redding, T. W. and Barrett, J. F. (1966). *Biochem. biophys. Res. Commun.*, **25**, 165–69

Stone, J. V., Mordue, W., Batley, K. E. and Morris, H. R. (1976). *Nature,* **263,** 207–9

Thomas, D. W., Das, B. C., Gero, S. D. and Lederer, E. (1968). *Biochem. biophys. Res. Commun.,* **33,** 519–27

Udenfriend, S., Stein, S., Bohlen, P., Dairman, W., Leimgraber, W. and Weigele, M. (1972). *Science,* **178,** 871–73

2
Immunohistochemical localisation of peptides in the nervous system

Robert Elde*, Tomas Hökfelt, Olle Johansson, Åke Ljungdahl,
Göran Nilsson and S. L. Jeffcoate (Departments of Histology and
Pharmacology, Karolinska Institute, Stockholm,
Sweden and National Institute for Biological Standards and Control,
Holly Hill, Hampstead, London U.K.)

INTRODUCTION

Peptides from the nervous system that act as extracellular messengers have been studied for many years by neuroendocrinologists. The activity of these hypothalamic hormones, such as vasopressin, oxytocin and the releasing factors, was thought to be directed only to target tissues outside of the nervous system. However, in recent years, it has become apparent that some of these neuroendocrine peptides as well as other peptides, are widely distributed in the nervous system and are capable of influencing neuronal activity. Therefore, certain peptides may act as interneuronal messengers, a possibility now being tested in many neurobiological laboratories. In this paper we present and discuss some findings from our laboratory on the regional distribution of neuropeptides as revealed by an immunohisto-chemical approach. Some of these findings have been reviewed (Hökfelt *et al.*, 1977*b, d*; Elde and Hökfelt, 1978) from somewhat different perspectives.

METHODOLOGICAL CONSIDERATIONS

The gross distribution of certain neuropeptides has been determined using bio-assays of extracts of brain regions (Pernow, 1953; Lembeck and Zetler, 1962; McCann *et al.*, 1975; Vale *et al.*, 1975), but the sensitivity of bioassay procedures is generally less than with immunochemical techniques. In addition, bioassays are susceptible to false negative results if antagonists of the activity being tested are present in the extract. For example, Vale *et al.*, (1975) found no somatostatin in the hypothalamic ventromedial nucleus, whereas radioimmunoassay (Brownstein *et al.*, 1975) and immunohistochemical studies (Hökfelt *et al.*, 1975*a*) clearly

*Present address: Department of Anatomy, University of Minnesota, Minneapolis, Minnesota 55455, U.S.A.

17

demonstrated somatostatin in this nucleus. The reason for this discrepancy is probably due to the presence of a yet uncharacterised growth hormone releasing factor in the ventromedial nucleus whose activity neutralised the effect of f somatostatin.

It is, however, clear that immunochemical approaches to the localisation of peptides are not without limitation. Although antibodies may be used as highly sensitive and specific probes of antigens in solution or within a tissue section, there remains the same crucial problem of establishing the specificity of a reaction for both radioimmunoassay and immunohistochemistry. In principle, this can be obtained only by demonstrating chemical identity between the substance bound by the antibody and the peptide used as the immunogen. The small amounts of peptides present in discrete areas of the brain make such chemical characterisation difficult. For this reason two immunochemical approaches have been used to test for specificity.

First, potential cross-reacting peptides may be identified in a competitive peptide binding study. Although this will determine potential cross-reactivity with characterised and readily available peptides, such testing is unable to determine cross-reactivity with unknown and as yet uncharacterised peptides that may be found in the tissues being studied.

Secondly, in radioimmunoassay, different dilutions of an extract can be compared. Lack of parallelism between the standard and sample curve indicates a molecular dissimilarity. However, even parallel curves are not sufficient criteria to establish identity (see Rodbard, 1974). Thus, it is still possible that a precursor molecule, peptide fragment or biosynthetically unrelated peptides with a partial amino acid sequence homology may cross-react with the antibody and, therefore, be localised with the immunohistochemical technique. It should be stressed that wherever reference is made to 'nerve terminals containing peptide A', this is actually 'peptide A *immunoreactivity*' and further characterisation of this immunoreactivity has not been accomplished.

FLUORESCENCE IMMUNOHISTOCHEMISTRY

The indirect immunofluorescence technique of Coons (1958) has been used for most of our studies. This technique is especially suitable for detection of peptides in small structures such as axons and nerve terminals (Johansson *et al.*, in preparation). Although the newer unlabelled peroxidase, anti-peroxidase method of Sternberger (1974) uses a much more dilute solution of the primary antiserum, we have found it to be less 'sensitive'.

The use of fluorescence immunohistochemistry in the nervous system has been described previously (Hökfelt *et al.*, 1975c). Briefly, tissues are prepared by trans-cardiac perfusion of the whole animal with ice-cold formalin in a phosphate buffer (Pease, 1962). After perfusion for 30 min, the tissues are dissected from the animal and immersed in the fixative for an additional 90 min. Thereafter, the tissues are transferred to 5 per cent sucrose in phosphate buffer. The tissue is then sectioned at $10\,\mu$m in a Dittes cryostat (Heidelberg).

The primary antibody (table 2.1) and the fluoresceinisothiocyanate-labelled second antibody were diluted in phosphate-buffered saline (PBS) containing 0.3 per cent Triton X-100 (Hartman, 1973). In addition, antibodies in the primary antiserum directed against the carrier protein were removed by incubating the

Table 2.1 Structural characterisation of peptides and sources of antisera

Peptide	Abbreviation*	Structural characterisation	Carrier protein	Antisera Code†	Antisera Source
Vasopressin	VP	Du Vigneaud (1956)	Thyroglobulin	GP24a	Elde (1974)
Oxytocin	OXY	Du Vigneaud (1956)	Hemocyanin	R111g	Elde¶
Neurophysins	NP	Walter et al. (1971)	—	RNF9‡	Seybold and Elde¶
				R81aφ	Elde¶
Thyrotropin releasing hormone	TRH	Burgus et al. (1970): Nair et al. (1970)	Albumin	R79	Jeffcoate et al. (1973)
Luteinising hormone releasing hormone	LHRH	Amoss et al. (1971): Schally et al. (1971)	Albumin	R-N	Jeffcoate et al. (1974)
			Haemocyanin	R122a	Elde¶
Somatostatin	SOM	Brazeau et al. (1973)	Thyroglobulin	GP2d	Elde and Parsons (1975)
			Haemocyanin	R141c	Elde et al. (submitted)
Substance P	SP	Chang et al. (1971)	Albumin	R16	Nilsson et al. (1975)
Leucine-enkephalin	ENK	Hughes et al. (1975)	Thyroglobulin	GP14b	Elde et al. (1976b)
Methionine-enkephalin	ENK	Hughes et al. (1975)	Haemocyanin	R6c	Elde et al. (1976b)

*Refers to immunoreactivity identified by antibodies to the named peptide
†GP, antisera raised in guinea pigs; R, antisera raised in rabbits
‡Antisera specific for bovine neurophysin II
φAntisera directed against all porcine neurophysins
¶Unpublished

diluted serum with an excess of the appropriate carrier protein (table 2.1). The primary antiserum diluted $\frac{1}{10}$ to $\frac{1}{100}$ was applied to the mounted tissue section and incubated in a humid atmosphere at 37 °C for 30 min. After incubation, the tissue was rinsed with PBS to remove unbound antisera. The second, labelled antibody (Statens Bakteriologiska Laboratorium, Stockholm) was diluted $\frac{1}{4}$ and incubated on the tissue as before. After rinsing, the tissue was mounted with a coverslip using glycerine: PBS (3 : 1), examined and photographed with a Zeiss transmission fluorescence microscope.

Immunohistochemical controls were performed using blocked primary antisera made by pretreatment of the antiserum with an excess of the antigen. In some cases where regions of the nervous system contained overlapping terminals or cell bodies of several different peptide systems, additional controls were performed. In such cases, an antiserum was also pretreated with an excess of the other peptides found in that area.

Tissues studied were obtained from normal male Sprague-Dawley rats (150-200 g) and normal, male albino guinea pigs (200-300 g). In some cases, rats were treated with an intraventricular, intracisternal or intraspinal injection of colchicine (5-25 μg; 1-10 μg/μl saline) 24-48 hours before they were killed in order to inhibit axonal transport and thereby accumulate larger quantities of peptides in neuronal perikarya (Hökfelt *et al.*, 1977*e*). In other cases neuronal perikarya were destroyed by stereotaxically-placed thermal lesions, or the axons of certain perikarya were interrupted in order to study the pathways of peptidergic systems. Terminals arising from the perikarya or axons were allowed to degenerate over a period of 5-10 days before the rats were killed. The placement of these lesions is discussed below.

The antisera used in these studies were produced in our laboratories (table 2.1). We shall describe, on a regional basis, immunoreactivity similar to vasopressin (VP), oxytocin (OXY), neurophysins (NP), thyrotropin releasing hormone (TRH), luteinising hormone releasing hormone (LHRH), somatostatin (SOM), substance P (SP), and the enkephalins (ENK). The following sections contain our own immunohistochemical findings as well as those of others. In view of space limitations, we have not repeatedly cited our own work, but refer the reader to the following: Elde (1973; 1974), Hökfelt *et al.* (1974; 1975*a-j*; 1976*a, b*; 1977*a-e*, Elde and Parsons (1975), Elde *et al.* (1976*a, b*; 1977) and Elde and Hökfelt (1978).

DISTRIBUTION OF PEPTIDES

The identification of biological activities attributable to peptides, the structural characterisation of these factors and their synthesis are necessary before immuno-chemical techniques can be used to study their distribution. Remarkable advances in peptide chemistry have made available relatively large quantities of synthetic neuropeptides and this has enabled production of specific antibodies. The 'state of the art' in peptide identification is discussed in this volume by Morris, Snyder, Terenius and Hughes (chapters 1, 6, 11 and 12, respectively). This field has also been reviewed by Guillemin (1977).

Table 2.1 gives a summary of the work leading to the characterisation of the peptides presented in this study. The distribution of some of these peptides as determined by radioimmunoassay of extracts of 'punches' of discrete nuclei (Palkovits, 1975) of the diencephalon and of the central nervous system (CNS) has been reviewed by Brownstein *et al.* (1976 and this volume, chapter 3).

A great deal of progress has been made in immunohistochemical localisations of peptides in hypothalamic neuroendocrine systems during the past several years, and the results with various peptides have been reviewed quite recently (Barry, 1976; Dubois, 1976; Parsons *et al.*, 1976; Pelletier *et al.*, 1976; Watkins, 1975; Zimmerman, 1976; Hökfelt *et al.*, 1977*b*; Knigge, 1977). We will present our neuroendocrine data without extensive citation and refer the reader to the above reviews. We will instead concentrate, for the most part, on recent work on localisations of neuropeptides not directly related to the hypothalamus–pituitary axis.

Diencephalon

Hypothalamus
Virtually all of the recognised peptides of neuronal origin occur in hypothalamic neurons or terminals. This may reflect the phylogenetically primitive nature of the hypothalamus as a repository of nearly all neuroeffector systems, some of which have been 'selected' for use in more specialised and phylogenetically recent structures. On the other hand, the apparent richness of peptides in hypothalamic structures may simply reflect the fact that hypothalamic extracts have been the starting material for characterisation of many biologically active peptides (see Guillemin, 1977).

Prechiasmatic zone
From the lamina terminalis to the rostral border of the optic chiasma, elements of several peptidergic systems are found. Throughout this region, a diffuse, somewhat medially disposed group of LHRH cell bodies can be identified. They are not restricted to a single nucleus (Barry, 1976). Some fibres from these cells project within this zone to terminate on the vascular channels of the organum vasculosum laminae terminalis (OVLT). In addition, terminals containing SOM, and a few terminals containing TRH are located in this enigmatic organ. Weindl and Sofroniew (1976) reported NP perikarya in this area.

Some SP-positive perikarya have been found in the lateral preoptic area. The medial preoptic nucleus is densely innervated with SP fibres and terminals. A number of ENK perikarya are located in this nucleus as are ENK fibres and terminals. ENK cell bodies also extend along the preoptic periventricular nucleus. A moderate component of TRH fibres is found throughout this region.

Suprachiasmatic zone
Elements of most peptidergic systems are present in this zone above the optic chiasma. The magnocellular neurons of the supraoptic nucleus contain VP or OXY and their respective neurophysins (figure 2.1). Accessory supraoptic nuclei are scattered along the fibres coursing from the paraventricular nucleus to the supraoptic nucleus. These perikarya are usually associated with blood vessels and stain intensely with antibodies to VP or OXY and their respective neurophysins. The magnocellular elements of the paraventricular nucleus also contain VP or OXY and the appropriate neurophysin. Most of their axons pass laterally and then ventrally, passing the accessory supraoptic nuclei. However, some fibres diverge and arch dorsally. Brownfield and Kozlowski (1977) have followed these fibres and note that they innervate the choroid plexus of the lateral ventricles.

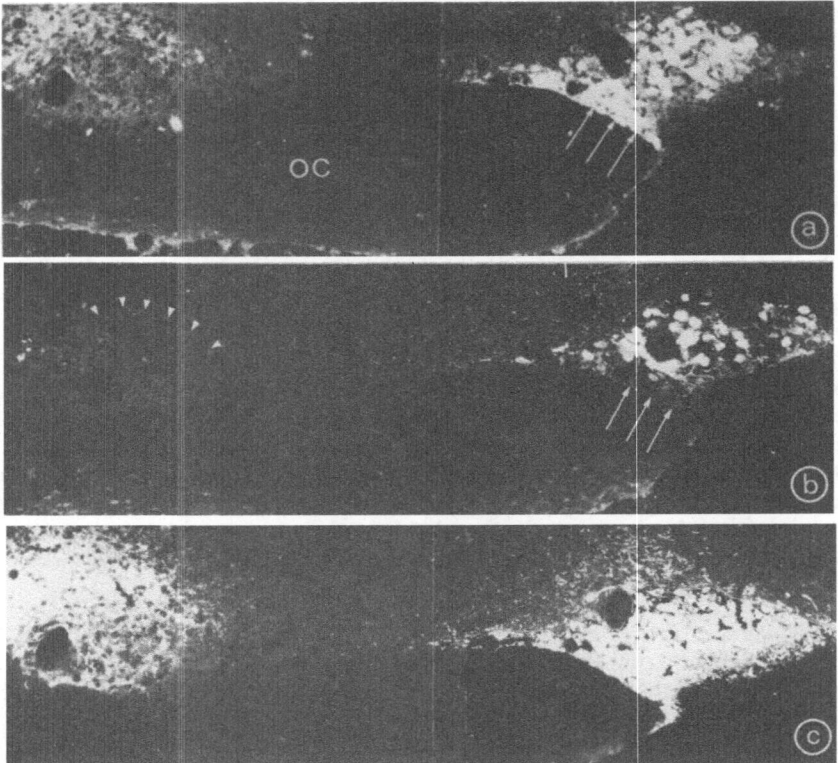

Figure 2.1 Immunofluorescence micrographs (mounts) of coronal serial sections of the rat hypothalamus after incubation with vasopressin, oxytocin and neurophysin antisera, respectively. In (a), both the suprachiasmatic nucleus (left) and supraoptic nucleus (right) contain VP immunoreactive cells and fibres. The group of cell bodies in the supraoptic nucleus immediately adjacent to the optic chiasma (arrows) is localised with this antiserum. In (b), OXY immunoreactivity is not found in cells or fibres of the suprachiasmatic nucleus (arrowheads), and a different population of cells within the supraoptic nucleus contain OXY. Compare the lack of staining of the cells immediately adjacent to the optic chiasma (arrows) with the intense staining in (a). In (c), note that essentially all cells of the supraoptic nucleus are localised with neurophysin antiserum, and that many of the cells of the suprachiasmatic nucleus stain as well. OC, optic chiasma. Magnification × 120.

Recently, a prominent group of neurons in the paraventricular nucleus has been found to contain enkephalin (figure 2.2a). This group of neurons extends laterally to circumscribe the fornix. A few TRH cell bodies (figure 2.2b) also fall within this area and TRH fibres are found here as well.

An especially prominent group of neurons within the periventricular nucleus at caudal levels of the optic chiasma contain somatostatin. These neurons are located just deep to the ependyma and extend the height of the IIIrd ventricle, although they are more concentrated dorsally. Their axons project laterally and ventrally towards the mid-lateral aspect of the optic chiasma. A number of TRH cell bodies are also found within the nucleus. SP fibres are encountered here also.

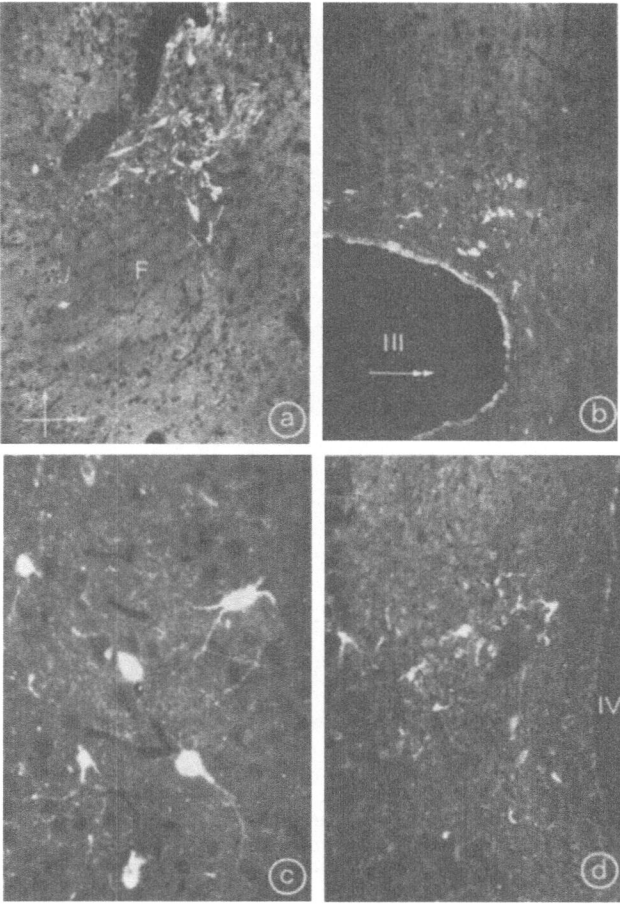

Figure 2.2 Immunofluorescence micrographs of coronal sections of the rat perifornical area (a), hypothalamic periventricular area (b), piriform cortex (c) and dorsal tegmental nucleus (d) after incubation with antiserum to methionine-enkephalin, TRH, SOM and SP, respectively. In (a), note the ENK perikarya medial to the fornix in the vicinity of a large vessel. In (b), TRH-containing cell bodies in the dorsal aspect of the periventricular area are found in small clusters. In (c), several SOM-containing cortical neurons exhibit multiple processes. In (d), note the SP-positive neuronal perikarya in the dorsal tegmental nucleus just ventral to the IVth ventricle. Arrow points medially; double arrow points dorsally. F, fornix, III, third ventricle, IV, fourth ventricle. Magnification × 120 (a, b, d) and × 300 (c).

Several peptidergic systems converge on the suprachiasmatic nucleus. A minority of these neurons contain VP and its NP (figure 2.1a, c), and a very few contain SOM. However, a striking network of VP/NP and SOM fibres terminate around neuronal soma in this nucleus (figure 2.3a, b). The VP/NP terminals are preferentially distributed to the dorsomedial aspect of this nucleus, whereas the SOM terminals are found in the ventromedial aspects. Small numbers of TRH and SP

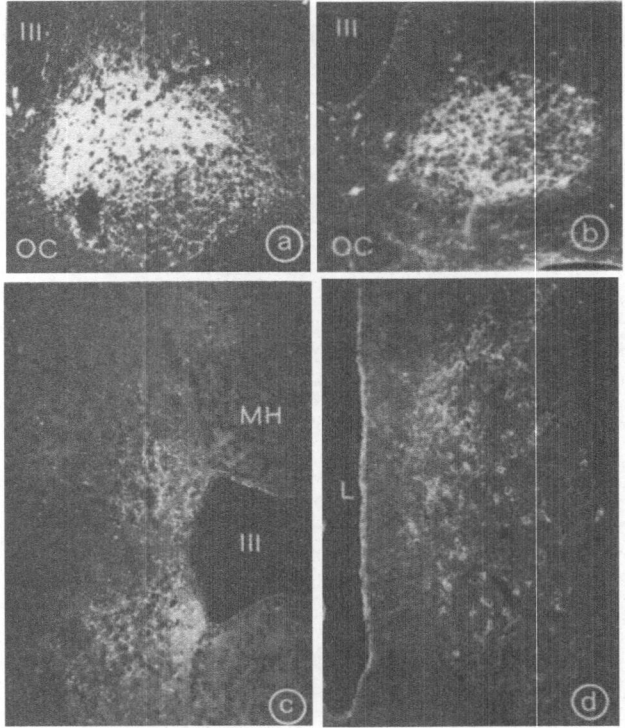

Figure 2.3 Immunofluorescence micrographs of the rat suprachiasmatic nucleus (a, b), thalamic nucleus paraventricularis rotundocellularis (c) and lateral septal nucleus (d) after incubation with antiserum to NP, SOM, NP and methionine-enkephalin, respectively. In (a) and (b), note the dense networks of fibres and terminals containing NP and SOM respectively. Note that the NP-containing elements are most prominent in the dorsal and medial aspects of this nucleus, whereas the SOM elements are found in ventral aspects as well. In (c), note the somewhat sparse distribution of NP fibres and terminals in this thalamic nucleus. Several other peptidergic systems also project to this nucleus (not shown). In (d), note the fine, basket-like terminals containing ENK surrounding some neurons of the lateral septal nucleus. III, third ventricle, OC, optic chiasm, MH, medial habenular nucleus, L, lateral ventricle. Magnification × 120.

fibres impinge on this nucleus. A moderate number of SP fibres are also found in this zone in the anterior hypothalamic nucleus. Scattered LHRH cell bodies and occasional axons are also found throughout the suprachiasmatic zone.

Tuberal zone

The retrochiasmatic component of the supraoptic nucleus contains the same peptidergic components as the main nucleus. Its fibres as well as the VP-, OXY- and NP-containing fibres from the suprachiasmatic zone of nuclei converge and proceed at the base of the hypothalamus towards the median eminence. Somatostatinergic fibres destined for the median eminence intermingle with the latter. The external layer of the median eminence is packed with peptidergic neuroendocrine terminals. Terminals containing SOM extend across the width of the median eminence, whereas LHRH fibres are more restricted and are found in the lateral aspect of the median eminence. Terminals containing TRH and ENK are found here but are not as extensive as SOM. A small component of fibres containing VP and NP terminate in the external layer close to portal capillaries, but most VP/NP as well as OXY fibres terminate in the neural lobe of the pituitary.

Directly above the median eminence, the arcuate nucleus is packed with SOM terminals. Some LHRH axons course through this nucleus and SP fibres can be seen here as well. Some investigators have localised LHRH-positive perikarya in the arcuate nucleus (see Zimmerman, 1976; Knigge, 1977). Recently, we have found a subpopulation of these cell bodies to contain enkephalin.

The ventromedial nucleus also contains ENK perikarya. Several peptidergic terminal systems can be found here, including ENK, SOM, SP (especially in the ventrolateral portion) as well as some TRH terminals.

The dorsomedial nucleus has cell bodies that contain TRH and others that contain SP. Fibres and terminals of each of these systems are in this nucleus, as are ENK-containing terminals. The rostral aspects of the perifornical area in this zone contain both ENK and TRH cell bodies.

Mamillary zone
The ventral and dorsal premamillary nuclei have both ENK and SP cell bodies. The ventral premamillary nucleus has an especially dense SOM terminal system.

Subthalamus
A system of SOM cell bodies extends from the zona incerta laterally towards the entopeduncular nucleus. A moderate number of TRH fibres and terminals are found in this zone.

Thalamus
No peptidergic perikarya have been localised in the thalamus, peptidergic projections to these nuclei are somewhat sparse. However, VP/NP (figure 2.3c), SP, SOM and ENK fibres converge on the nucleus periventricularis rotundocellularis. In addition, SP and ENK terminals are distributed densely in the intralaminar nuclei.

Epithalamus
The neuronal cell bodies of the medial habenular nucleus contain SP. Occasional scattered fibres of the other peptidergic systems are seen in both the medial and lateral habenular nuclei. The peptides TRH, LHRH (White *et al.*, 1974) have been reported in the pineal body by radioimmunoassay and SOM by immunohisto-chemistry (Pelletier *et al.*, 1975). However, we have had difficulty verifying their presence in the pineal body using immunohistochemical techniques.

Telencephalon

Deep nuclei
Corpus striatum. Cell bodies containing ENK are found throughout the caudate-putamen. Some SP cell bodies are present, mainly ventrally and caudally within the caudate. Terminals containing SOM, SP and ENK are present. Very dense ENK and SP networks are found in the globus pallidus, especially in its more ventral aspects.

Septal nuclei
Both SP and ENK perikarya are present in the septal nuclei. High concentrations
of terminals containing SP and ENK, often appearing as 'baskets' surrounding
the soma of individual septal neurons, are found in this nucleus (figure 2.3d),
as is a more modest system of TRH fibres and terminals. Weindl and Sofroniew
(1976) found the triangular nucleus of the septum to contain NP-positive cell
bodies.

Nucleus interstitialis stria terminals
This nucleus is the site of many converging peptide fibres and terminals. Thus,
rather dense networks of SOM terminals are seen in the medial aspect of this
nucleus, adjacent to the arch of the fornix. High densities of SP and some TRH
terminals can be localised within this nucleus and an enkephalin terminal field
is found in the most lateral aspect, adjacent to the internal capsule. In this lateral
area, SOM- and ENK-containing cell bodies are found. The nucleus also contains
SP cell bodies.

Nucleus accumbens
Dense islands of ENK terminals are scattered throughout this nucleus and moderately dense terminal fields containing SP, SOM and TRH have been seen.

Amygdaloid nuclei
Neuronal perikarya containing SOM are scattered throughout these nuclei, but
especially in the cortical nucleus. Cell bodies containing SP are found in the medial
and central nuclei. Terminals containing SOM, SP and ENK are found to a greater
or lesser extent in all amygdaloid nuclei, but are especially prominent in the
central nucleus, and, with regard to SP, the medial nucleus.

Olfactory bulb and tubercle
Peptides form rather minor components in these parts of the olfactory system.
Some SP fibres and terminals are present in the bulb, and SOM terminals can be
localised in the olfactory tubercle.

Cortical areas
In general, only limited peptidergic systems have been found in the cortex. However, SOM cell bodies are regularly scattered throughout the piriform and entor-
hinal cortex (figure 2.2c). A more sparse collection of SOM cell bodies is found
in the dentate gyrus of the hippocampal formation and in the neo-cortex. A rather
large number of fibres is distributed throughout these areas, but dense areas of
terminals have not been seen.

Mesencephalon
Some cell bodies in the interpeduncular nucleus contain substance P. Knigge
(1977) reports SOM-containing neurons in this area. In the lateral portion of this
nucleus SP terminals are also found in rather high concentration. The periaqueductal central grey contains cell bodies, terminals and fibres with SP and ENK.
The latter group of cell bodies is disposed ventrolaterally in the periaqueductal
grey. In parasaggital sections a few LHRH fibres extend ventral and parallel to
the aqueduct.

The dorsolateral substantia nigra has a population of ENK cell bodies. The zona compacta of the substantia nigra receives a number of ENK terminals, and ENK fibres are also closely associated with the medial lemniscus. On the other hand, SP fibres and terminals occur in very high concentrations in the zona reticulata of the substantia nigra. The motor nucleus of cranial nerve III has a component of TRH terminals.

Rhombencephalon

The cerebellum is essentially empty with respect to the peptides we have studied. However, the pons and medulla are rich in peptidergic systems. At the junction of the pons and mesencephalon, the dorsal and ventral parabrachial nuclei display cell bodies that contain ENK. In addition, several peptidergic systems send terminals to the dorsal division of the nucleus, including SP, SOM and ENK. The The dorsal tegmental nucleus contains SP cell bodies (figure 2.2d). Also at this level, the nuclei of the lateral lemniscus have a population of ENK-containing neuronal cell bodies. The motor nuclei of cranial nerves V, VII and XII contain TRH terminals. ENK terminals are also found in the latter two nuclei.

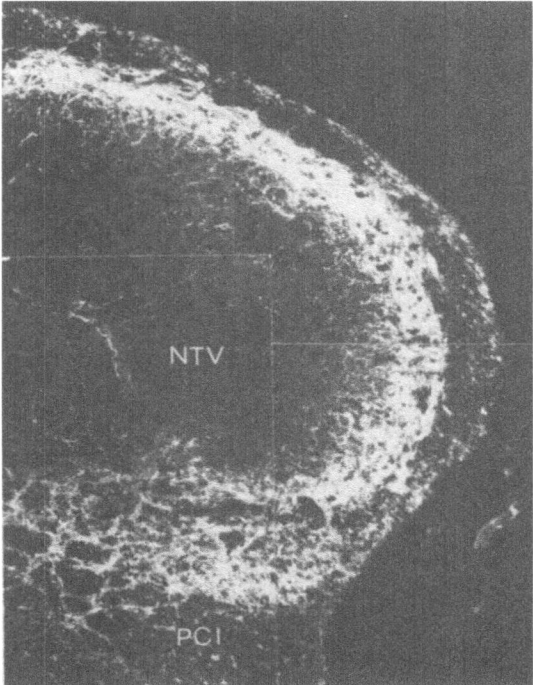

Figure 2.4 Immunofluorescence micrograph (mount) of the caudal portion of the nucleus of the spinal tract cranial nerve V after incubation with SP antiserum. Note the dense system of fibres and terminals containing SP that are found in the marginal layers (substantia gelatinosa) of this nucleus. The more central portions of the nucleus have only occasional fibres. NTV, nucleus tractus spinalis nervi trigemini, PCI, pedunculus cerebellaris inferior. Magnification × 120.

motor nuclei of cranial nerves V, VII and XII contain TRH terminals. ENK terminals are also found in the latter two nuclei.

A few cell bodies containing ENK are found in the medial vestibular nucleus and in the dorsal cochlear nucleus. Of the reticular nuclei, the ventromedial part of the nucleus reticularis gigantocellularis and the nucleus reticularis paramedianus contain ENK cell bodies. These and other nuclei of the reticular formation have a moderate number of ENK terminals as was found in the nucleus reticularis lateralis.

The medullary raphe nuclei magnus and pallidus contain both ENK and SP cell bodies. The nucleus tractus solitarii is composed of numerous ENK cell bodies, and terminals positive for ENK and SP are abundant in this region. Nucleus ambiguus has fibre and terminal input from ENK systems.

More caudally in the medulla, the nucleus commissuralis has SP and ENK positive cell bodies as well as SP and ENK terminals. The marginal zone of the caudal portion of the nucleus of the spinal tract of cranial nerve V as well as the substantia gelatinosa of that nucleus contain cell bodies with ENK immunoreactivity. This same region has both SP (figure 2.4) and SOM terminals. In much of the central grey of the pons and medulla, networks of ENK and SP were observed.

Spinal cord

In laminae II–V a number of ENK-positive cell bodies were observed. A smaller number of SP-containing cell bodies were also found in this region. Laminae I and II exhibit a high density of SP terminals and a lower density of ENK terminals. Lamina II also has a striking complement of SOM terminals. In addition, SP fibres in cross-section are seen in the tract of Lissauer.

More medial areas of the dorsal horn and the area around the central canal have SP and ENK fibres. The ventral horn has a moderate density of SP, ENK and TRH fibres. In addition, single VP, OXY and NP fibres are often disposed in a transverse plane extending from the region of the central canal to more dorso-lateral parts of the spinal grey matter.

Peripheral nervous system

The present discussion of our immunohistochemical studies in the peripheral nervous system (PNS) will be limited since these aspects were reviewed recently (Hökfelt *et al.*, 1977d). Briefly, two subpopulations of small, dorsal root ganglion neurons have been identified, one containing SP and the other SOM. Further in the periphery, SP fibres are seen in the gut, skin, glands, sympathetic ganglia and along blood vessels. SOM and ENK fibres have so far been identified only in the gut.

PEPTIDERGIC PATHWAYS

The limited data now available permit some discussion of peptidergic pathways. A thorough study of pathways requires completion of mapping studies in order to make stereotaxically placed lesions of peptidergic cell bodies and transections of their axons. In a few areas, we and others have made such studies and can identify some peptidergic pathways.

An LHRH pathway from the pre- and suprachiasmatic regions to the median eminence has been described by Barry (1976) who traced axons in serial sections and by Weiner *et al.* (1975) and Palkovits *et al.* (1976) following hypothalamic deafferentation. Lesions of the hypothalamic periventricular nucleus lead to degeneration and disappearance of the SOM terminals in the median eminence. This has been confirmed by Brownstein *et al.* (1977*a*). However, when viewing such tissue in the fluorescence microscope, it was clear that SOM terminals in the nearby arcuate and ventromedial nuclei were not affected by this lesion. Thus, we suggest the existence of a SOM pathway with cell bodies in the periventricular nucleus whose axons project to the median eminence. Further, other SOM pathways project to the arcuate and ventromedial nuclei, but the location of these cell bodies is not known.

This same periventricular lesion destroys a number of TRH perikarya in the dorsal peri- and paraventricular area. TRH terminals disappeared completely from the median eminence after this lesion. TRH fibres and perikarya in the dorsomedial nucleus are not affected by the lesion. Thus, there may also be a discrete TRH peri- and paraventricular–median eminence pathway.

Lesions that include the paraventricular nucleus cause most of the VP and NP terminals to disappear from the external layer of the median eminence. Lesions of the suprachiasmatic nucleus cause no change in VP and NP staining in the external layer terminal system. We regard these findings as evidence for a VP and NP paraventricular–median eminence (external layer) pathway and thereby confirm the findings of Antunes *et al.* (1977).

Lesions of the paraventricular nucleus also destroy some ENK perikarya. Some of the ENK-containing terminals in the median eminence were destroyed after this lesion. Thus, there may also be an ENK paraventricular/perifornical–median eminence pathway.

Some SP pathways have also been described. The existence of an SP habenulo-interpeduncular tract has been established by radioimmunoassay (Hong *et al.*, 1976) and immunohistochemistry (Hökfelt *et al.*, 1977*f*). In addition, a striato-nigral SP pathway has also been described (Brownstein *et al.*, 1977*b*; Hong *et al.*, 1977; Kanazawa *et al.*, 1977).

Several peptidergic pathways have been established in the spinal cord and in primary afferent neurons. After transection of dorsal roots, the SOM and a proportion of the SP terminals in the dorsal horn disappear, indicating a SOM and SP primary afferent pathway. Further evidence of this comes from the differential distribution of SOM and SP in neuronal cell bodies of the dorsal root ganglion.

Transection of the thoracic spinal cord leads to disappearance in the lumbar cord of ventral horn SP and TRH as well as the sparse VP, OXY and NP fibres. These peptides must, therefore, be descending supraspinal systems. The cell bodies of origin for these pathways have not been identified, but it is interesting to note that autoradiographic studies have described pathways descending from the paraventricular nucleus to the spinal cord (Conrad and Pfaff, 1976; Saper *et al.*, 1976).

ENK terminals in the spinal cord grey matter do not disappear after either dorsal root section or proximal transection of the cord. Since ENK cell bodies have now been demonstrated at all levels of the cord, it is likely that these represent a system of ENK interneurons.

DISCUSSION

The preceding description of the regional distribution of several neuropeptides is based on immunohistochemistry and represents only a brief survey of these systems. In fact, a detailed description of each peptidergic system now seems to be so complex that the patterns of several peptidergic systems cannot be simultaneously illustrated in a schematic fashion. The distribution of each peptide as revealed by immunohistochemistry must now be mapped photomicrographically as has been done, for example, for the enzyme tyrosine hydroxylase (Hökfelt *et al.*, 1976) and schematically as for SP (Hökfelt, Ljungdahl and Nilsson, in preparation).

We have chosen to discuss areas where peptide systems are prominent and where several peptide projections seem to converge, bearing in mind that many interesting aspects have been excluded.

Immunohistochemical techniques can give details of the nature of structures containing peptides, although this information can be used only in a semiquantitative fashion. The high sensitivity of radioimmunoassays combined with microdissection techniques offer an excellent tool for quantitative studies (see Brownstein, chapter 3). Taken together, radioimmunoassay and immunohistochemical findings offer a more adequate understanding of the peptide systems.

One of the primary goals of our immunohistochemical investigation is to localise peptidergic neuronal perikarya. In some cases, such as VP, OXY and NP, this can be achieved in brains from normal rats and extensive descriptions of these cell bodies have been reported (Livett *et al.*, 1971; Burlet *et al.*, 1973; Elde, 1973; Zimmerman *et al.*, 1973; Silverman, 1975; Vandesande and Dierickx, 1975; Watkins, 1975). Barry (1976) and Silverman (1976) have identified LHRH neuronal perikarya in the pre- and suprachiasmatic zones of the hypothalamus. Other investigators have reported LHRH neurons in the arcuate nucleus (Naik, 1976; Zimmerman, 1976). Recently, Knigge (1977) has shown that these discrepancies may be due to different combining sites among several antisera used for LHRH immunohistochemistry.

Neuronal perikarya containing TRH, ENK and SP have been unusually difficult to localise with immunohistochemical techniques on normal brains. One reason often proposed to explain this problem is that there may be too little of the peptide present in perikarya in normal circumstances. Inhibitors of axonal transport such as colchicine have been used for a number of years (Dahlström, 1970). Barry *et al.* (1973) had also used this alkaloid in early experiments to enhance the staining of LHRH perikarya. Therefore, we have administered this agent intracerebrally to animals before they were killed. Tissues viewed after this treatment were found to contain abundant neuronal perikarya containing TRH, ENK and SP as well as the other more easily localised peptides. In normal animals and in colchicine-treated animals, all staining could be abolished with specifically blocked antiserum, indicating that colchicine administration is a valid means of increasing staining of peptidergic perikarya.

A preliminary account of the distribution of peptide-containing cell bodies is summarised in table 2.2. Since mapping studies are still in progress for many of these peptides, it is likely that additional groups of cell bodies may yet be encountered.

The activity of peptidergic neurons is likely to be controlled largely by afferent pathways to their cell bodies and dendrites. Therefore, it will become increasingly

Table 2.2 Nuclei containing specific peptidergic cell bodies

Vasopressin	Oxytocin	Neurophysins	TRH
Supraopticus	Supraopticus	Supraopticus	Periventricularis (hypothalami)
Suprachiasmaticus	Paraventricularis	Suprachiasmaticus	Paraventricularis
Paraventricularis		Paraventricularis	Periformicalis
		Triangularis septi*	Dorsomedialis (hypothalami)
			Lateralis (hypothalami)

Somatostatin	Substance P	Enkephalins	
Caudatus	Interstitialis striæ terminalis	Caudatus	Substantia nigra
Amygdaloidei	Amygdaloidei	Septi lateralis	Lemniscus lateralis
Zona incerta	Medialis habenulae	Interstitialis striæ terminalis	Parabrachialis dorsalis
Periventricularis (hypothalami)	Hypothalamici	Preopticus periventricularis	Parabrachialis ventralis
Cortex piriformis	Interpeduncularis	Preopticus medialis	Vestibularis medialis
Cortex entorhinalis	Tegmenti dorsalis	Paraventricularis	Magnus raphes
Gyrus dentatus		Periformicalis	Pallidus raphes
Neocortex		Arcuatus	Cochlearis dorsalis
		Ventromedialis (hypothalami)	Reticularis gigantocellularis
		Premamillaris dorsalis	Reticularis gigantocellularis
		Premamillaris ventralis	Tractus solitarii
		Substantia grisea centralis	Reticularis paramedianus
		Colliculus superior	Tractus spinalis nervi trigemini
			Substantia grisea medullae spinalis

*Weindl and Sofroniew (1976)

important to discover the nature of the neuronal pathways projecting to these peptidergic cell groups. Possible interactions between these and other neurotransmitter systems have been discussed (Hökfelt *et al.*, 1977*b, d*).

It is interesting to observe that of the peptidergic cell bodies so far discussed, the majority of them are confined to the brainstem and subcortical areas. However, SOM cell bodies occur in large numbers in limbic-cortical and neocortical areas. In addition, vasoactive intestinal polypeptide (VIP) has recently been localised in cortical and hypothalamic neurons (Fuxe *et al.*, 1976, 1977; Larsson *et al.*, 1976) as has a peptide that cross-reacts with an antibody to gastrin (see Hökfelt *et al.*, 1977*b, d*).

As mentioned in the introduction, numerous immunohistochemical reports describe the localisation of neuroendocrine peptides in terminals of the hypothalamus-pituitary neuroendocrine axis. Our recent findings in this area are discussed more thoroughly elsewhere (Elde and Hökfelt, 1978) and only mentioned briefly here. It is interesting to note that the ENK pathway described to the median eminence may represent a newly discovered neuroendocrine control system. This system may act to regulate prolactin release in accordance with the findings of Lien *et al.* (1976) who reported an increase in prolactin release after administration of exogenous enkephalin.

The localisation of ENK fibre and terminal systems (Elde *et al.*, 1976*b*) as discussed here, has recently been confirmed by Simantov *et al.* (1977; see also, Snyder, this volume, chapter 6). The regional distribution of these terminals coincides rather well with the distribution of opiate receptors as determined by autoradiography (Pert *et al.*, 1976) or biochemical assays (Hiller *et al.*, 1973; Kuhar *et al.*, 1973; Pert and Snyder, 1973; see also Snyder, chapter 6). It is also interesting to note that SP terminals are often found in many ENK-rich terminal areas. This overlap is especially striking in areas suspected of mediating mechanisms of pain and analgesia (Hökfelt *et al.*, 1977*g*).

The widespread occurrence of peptidergic systems, especially SP and ENK, suggests that they participate in many modalities of central and peripheral nervous system function. To mention one example, ENK cell bodies and terminals, as well as SP and SOM terminals are found in the dorsal portion of the nucleus parabrachialis. This nucleus appears to be involved in the control of respiration (von Euler *et al.*, 1976). Respiratory depression is of course a prominent aspect of opiate activity (Flórez *et al.*, 1968) and it is tempting to suggest that ENK may have an inhibitory role in controlling respiratory drive. Both SOM and SP afferents may serve to control what may be ENK interneurons.

CONCLUSIONS

Immunohistochemical studies strongly suggest that certain biologically active peptides are contained within discrete populations of neuronal cell bodies, fibres and terminals. The peptides ENK and SP are found in most regions of the nervous system. Neurons containing SOM and TRH are less extensively distributed; VP, OXY and NP systems are mainly confined to the hypothalamus and are found outside of this region in rather small numbers. Neurons containing LHRH are almost exclusively found within the hypothalamus. From the widespread distribution of some of these peptides, it is suggested that they may be extracellular messengers participating in diverse modalities of nervous system function.

ACKNOWLEDGEMENTS

R.E. was supported by a National Research Service Award (NS 05047-01) from the National Institute of Neurological, Communicative Disorders and Stroke. This work was supported by grants from the Swedish Medical Research Council (04X-2887, 04X-2886, 04X-3521, 04X-04495, 04X-715, 25X-5065, 19X-00034), Magnus Bergwalls Stiftelse, Knut and Alice Wallenbergs Stiftelse, by a grant (1R01 HL 18994-01, 555) from the U.S. Department of Health, Education and Welfare, through the National Heart and Lung Institute, by U.S. Public Health Service Grants MH-02717 and MH-25504-3, and by National Science Foundation Grant CB-8465. The technical assistance of W. Hjort, A. Nygårds and L. Persson is gratefully acknowledged.

REFERENCES

Amoss, M., Burgus, R., Blackwell, R., Vale, W., Fellows, R. and Guillemin, R. (1971). *Biochem. biophys. Res. Commun.*, 44, 205-10
Antunes, J. L., Carmel, P. W. and Zimmerman, E. A. (1977). *Brain. Res.*, (in press)
Barry, J., Dubois, M. P. and Poulain, P. (1973). *Z. Zellforsch.*, 146, 351-66
Barry, J. (1976). In *Hypothalamus and Endocrine Functions* (ed. F. Labrie, J. Meites and G. Pelletier), Plenum Press, New York, pp. 451-474.
Brazeau, P., Vale, W., Burgus, R., Ling, N., Butcher, M., Rivier, J. and Guillemin, R. (1973). *Science*, 179, 77-79
Brownfield, M. S. and Kozlowski, G. P. (1977). *Cell Tiss. Res.*, 178, 111-27
Brownstein, M., Arimura, A., Sato, H., Schally, A. V. and Kizer, J. S. (1975). *Endocrinology*, 96, 1456-61
Brownstein, M. J., Palkovits, M., Saavedra, J. M. and Kizer, J. S. (1976). In *Frontiers in Neuroendocrinology*, Vol. 4: *Distribution of hypothalamic hormones and neurotransmitters within the diencephalon* (ed. L. Martini and W. F. Ganong), Raven Press, New York, pp. 1-23
Brownstein, M. J., Arimura, A., Fernandez-Durango, R., Schally, A. V., Palkovits, M. and Kizer, J. S. (1977a). *Endocrinology*, 100, 246-49
Brownstein, M. J., Mroz, E. A., Tappaz, M. L. and Leeman, S. E. (1977b). *Brain Res.*, (in press)
Burgus, R., Dunn, T. F., Desiderion, D., Ward, D. N., Vale, W. and Guillemin, R. (1970). *Nature*, 226, 321-25
Burlet, A., Marchetti, J. and Duhielle, J. (1973). *C.r. Séanc Soc. Biol. (D) (Paris)*, 167, 924-28
Chang, M. M., Leeman, S. E. and Niall, H. D. (1971). *Nature new Biol.*, 232, 86-87
Conrad, L. C. A. and Pfaff, D. W. (1976). *J. comp. Neurol.*, 169, 221-61
Coons, A. H. (1958). In *General Cytochemical Methods* (ed. J. F. Danielli), Academic Press, New York, pp. 399-422
Dubois, M. P. (1976). *Ann. Biol. anim. Biochem. Biophys.*, 16, 177-94
Du Vigneaud, V. (1956). In *The Harvey Lectures 1954-55*. Academic Press, New York, pp. 1-26
Elde, R. P. (1973). *Anat. Rec.*, 175, 255
Elde, R. P. (1974). Ph.D. thesis, University of Minnesota
Elde, R. P., Efendić, S., Hökfelt, T., Johansson, O., Luft, R., Parsons, J. A., Roovete, A. and Sorenson, R. L. (submitted)
Elde, R. and Hökfelt, T. (1978). In *Frontiers in Neuroendocrinology*, Vol. 5 (ed. W. F. Ganong and L. Martini), Raven Press, New York, pp. 1-33
Elde, R. P., Hökfelt, T., Johansson, O., Efendić, S. and Luft, R. (1976a). *Neurosci. Abstr.*, 11, 759
Elde, R., Hökfelt, T., Johansson, O. and Terenius, L. (1976b). *Neuroscience*, 1, 349-51
Elde, R. P. and Parsons, J. A. (1975). *Am. J. Anat.*, 144, 541-48
Euler, C. von, Marttila, I., Remmers, J. E. and Trippenbach, T. (1976). *Acta physiol. scand.*, 96, 324-37
Flórez, J., McCarthy, L. E. and Borison, H. L. (1968). *J. Pharmac. exp. Ther.*, 163, 448-55

Fuxe, K., Hökfelt, T., Johansson, O., Ganten, D., Goldstein, M., Perez de la Mora, M., Possani, L., Tapia, R., Teran, L., Palacios, R., Said, S. and Mutt, V. (1976). In *Colloque de Synthese des Actions Thématiques 22 et 35. Neuromédiateurs et Polypeptides Hypothalamiques a Action Relàchante ou Inhibitrice* (ed. R. Mornex and J. Barry) Institut National de la Santé et de la Recherche Médicale, Paris, in press

Fuxe, K., Hökfelt, T., Said, S. and Mutt, V. (1977). *Neurosci. Lett.*, in press

Guillemin, R. (1977). In *ARNMD: The Hypothalamus* (ed. S. Reichlin), Raven Press, New York, (in press)

Hartman, B. K. (1973). *J. Histochem. Cytochem.*, **21**, 312–32

Hiller, J. M., Pearson, J. and Simon, E. J. (1973). *Res. Commun. chem. Path. Pharmac.*, **6**, 1052–62

Hökfelt, T., Efendić, S., Johansson, O., Luft, R. and Arimura, A. (1974). *Brain Res.*, **80**, 165–69

Hökfelt, T., Efendić, S., Hellerström, C., Johansson, O., Luft, R. and Arimura, A. (1975a). *Acta Endocr. (Kbh.)*, Suppl. 200, 5–41

Hökfelt, T., Elde, R. P., Johansson, O., Luft, R. and Arimura, A. (1975b). *Neurosci. Lett.*, **1**, 231–35

Hökfelt T., Fuxe, K. and Goldstein, M. (1975c). *Ann. N. Y. Acad. Sci.*, **254**, 407–32

Hökfelt, T., Fuxe, K., Goldstein, M., Johansson, O., Fraser, H. and Jeffcoate, S. (1975d). In *Anatomical Neuroendocrinology* (ed. W. E. Stumpf and L. D. Grant), Karger, Basel, pp. 381–92

Hökfelt, T., Fuxe, K., Johansson, O., Jeffcoate, S. L. and White, N. (1975e). *Eur. J. Pharmac.*, **34**, 389–92

Hökfelt, T., Fuxe, K., Johansson, O., Jeffcoate, S. and White, N. (1975f). *Neurosci. Lett.*, **1**, 133–39

Hökfelt, T., Johansson, O., Efendić, S., Luft, R. and Arimura, A. (1975g). *Experientia*, **31**, 852–854

Hökfelt, T., Johansson, O., Fuxe, K., Löfström, A., Goldstein, M., Park, D., Ebstein, R., Fraser, H., Jeffcoate, S., Efendić, S., Luft, R. and Arimura, A. (1975h). In *CNS and Behavioural Pharmacology. Proc. Sixth Int. Congr. Pharmac.*, Vol. 3 (ed. J. Tuomisto and M. K. Paasonen), Forssan Kirjapaino Oy, pp. 93–110

Hökfelt, T., Kellerth, J–O., Nilsson, G., and Pernow, B. (1975i). *Brain Res.*, **100**, 235–52

Hökfelt, T., Kellerth, J–O., Nilsson, G. and Pernow, B. (1975j). *Science*, **190**, 889–890

Hökfelt, T., Elde, R., Johansson, O., Luft, R., Nilsson, G. and Arimura, A. (1976a). *Neuroscience*, **1**, 131–36

Hökfelt, T., Johansson, O., Fuxe, K., Goldstein, M. and Park, D. (1976b). *Med. Biol.*, **54**, 427–53

Hökfelt, T., Kellerth, J–O., Ljungdahl, Å., Nilsson, G., Nygårds, A. and Pernow, B. (1977a). In *Neuroregulators and Hypotheses of Psychiatric Disorders* (ed. J. Barchas, E. Costa and E. Usdin), Oxford University Press, New York, pp. 299–311

Hökfelt, T., Meyerson, B., Nilsson, G., Pernow, B. and Sachs, Ch. (1976c). *Brain Res.*, **104**, 181–86

Hökfelt, T., Elde, R., Fuxe, K., Johansson, O., Ljungdahl, Å., Goldstein, M., Luft, R., Nilsson, G., Said, S., Fraser, H., Jeffcoate, S. L., White, N., Ganten, D. and Rehfeld, J. (1977b). In *ARNMD: The Hypothalamus* (ed. S. Reichlin), Raven Press, New York, (in press)

Hökfelt, T., Elde, R. P., Johansson, O., Kellerth, J–O., Ljungdahl, A., Nilsson, G., Pernow, B. and Terenius, L. (1977c). In *Proc. Collegium Internationale Neuropharmacologium* (ed. Radouco-Thomas), (in press)

Hökfelt, T., Elde, R. P., Johansson, O., Ljungdahl, A., Schultzberg, M., Fuxe, K., Goldstein, M., Nilsson, G., Pernow, B., Terenius, L., Ganten, D., Jeffcoate, S. L., Rehfeld, J. and Said, S. (1977d). In *American College of Neuropharmacology*, Raven Press, New York, (in press)

Hökfelt, T., Elde, R., Johansson, O., Terenius, L. and Stein, L. (1977e). *Neurosci. Lett.*, **5**, 25–31

Hökfelt, T., Johansson, O., Kellerth, J–O., Ljungdahl, A., Nilsson, G., Nygårds, A. and Pernow, B. (1977f). In *Substance P Nobel Symposium*, Vol. 37 (ed. U. S. von Euler and B. Pernow), Raven Press, New York, (in press)

Hökfelt, T., Ljungdahl, Å., Terenius, L., Elde, R. and Nilsson, G. (1977g). *Proc. nath. Acad. Sci. U.S.A.*, **74**, 3081–85

Hong, J. S., Costa, E. and Yang, H-Y. T. (1976). *Brain Res.*, **118**, 523–25

Hong, J. S., Yang, H-Y. T., Racagni, G. and Costa, E. (1977). *Brain Res.,* **122,** 541–44
Hughes, I., Smith, T. W., Kosterlitz, H. W., Fothergill, L. H., Morgan, B. A. and Morris, H. R. (1975). *Nature,* **258,** 577–79
Jeffcoate, S. L., Fraser, H. M., Gunn, A. and White, N. (1973). *J. Endocr.,* **59,** 191–92
Jeffcoate, S. L., Hollan, D. T., Fraser, H. M. and Gunn, A. (1974). *Immunochemistry,* **11,** 75–77
Kanazawa, I., Emson, P. C. and Cuello, A. C. (1977). *Brain Res.,* **119,** 447–53
Knigge, K. M. (1977). In *ARNMD: The Hypothalamus* (ed. S. Reichlin), Raven Press, New York, (in press)
Kuhar, M. J., Pert, C. B. and Snyder, S. H. (1973). *Nature,* **245,** 447–50
Larsson, L-I., Fahrenkrug, J., Schaffalitzky de Muckadell, O., Sundler, F., Håkanson, R. and Rehfeld, J. F. (1976). *Proc. natn. Acad. Sci. U.S.A.,* **73,** 319–200
Lembeck, F. and Zetler, G. (1962). *Int. Rev. Neurobiol.,* **4,** 159–215
Lien, E. L., Fenichel, R. L., Garsky, V., Sarantakis, D. and Grant, N. H. (1976). *Life Sci.,* **19,** 837–40
Livett, B. G., Uttenthal, L. O. and Hope, D. B. (1971). *Phil. Tans. R. Soc. B.,* **261,** 371–378
McCann, S. M., Krulich, L., Quijada, M., Wheaton, J. and Moss, R. L. (1975). In *Anatomical Neuroendocrinology* (ed. W. E. Stumpf and L. D. Grant), Karger, Basel, pp. 192–199
Mroz, E. A., Brownstein, M. J. and Lemman, S. E. (1976). *Brain Res.,* **113,** 597–599
Naik, D. V. (1976). *Cell Tiss. Res.,* **173,** 143–66
Nair, R. M. G., Barrett, J. F., Bowers, C. Y. and Schally, A. V. (1970). *Biochemistry,* **9,** 1103–1106
Palkovits, M. (1975). In *Anatomical Neuroendocrinology* (ed. W. E. Stumpf and L. D. Grant), Karger, Basel, pp. 72–80
Palkovits, M., Brownstein, M. and Kizer, S. J. (1976). *International Symposium on Cellular and Molecular Bases of Neuroendocrine Processes* (ed. E. Endroczy), Akadémiai Kiadó, Budapest, pp. 575–99
Parsons, J., Erlandsen, S., Hegre, O., McEvoy, R. and Elde, R. P. (1976). *J. Histochem. Cytochem.,* **24,** 872–82
Pease, D. C. (1962). *Anat. Rec.,* **142,** 342
Pelletier, G., Leclerc, R., Dubé, D., Labrie, F., Puviani, R., Arimura, A. and Schally, A. V. (1975): *Am. J. Anat.,* **142,** 397–01
Pelletier, G., Leclerc, R. and Dubé, D. (1976). *J. Histochem. Cytochem.,* **24,** 864–71
Pernow, B. (1953). *Acta physiol. scand.,* **29,** suppl. 105, 1–90
Pert, C. B. and Snyder, S. H. (1973). *Science,* **179,** 1011–14
Pert, C. B., Kuhar, M. J. and Snyder, S. H. (1976). *Life Sci.,* **16,** 1849–54
Rodbard, D. (1974). *Clin. Chem.,* **20,** 1255–73
Saper, C. B., Loewy, A. L., Swanson, L. W. and Cowan, W. M. (1976). *Brain Res.,* **117,** 305–12
Schally, A. V., Arimura, A., Baba, Y., Nair, R. M. G., Matsuo, J., Redding, T. W., Debeljuk, L. and White, W. F. (1971). *Biochem. biophys. Res. Comman.,* **43,** 393–99
Silverman, A. J. (1975). *Am. J. Anat.,* **144,** 433–44
Silverman, A. J. (1976). *Endocrinology,* **99,** 30–41
Simantov, R., Kuhar, M. J., Uhl, G. R. and Snyder, S. H. (1977). *Proc. natn. Acad. Sci. U.S.A.,* **74,** 2167–71
Sternberger, L. A. (1974). *Immunocytochemistry.* Prentice-Hall, Englewood Cliffs, N. J
Vale, W., Brazeau, P., Rivier, C., Brown, M., Boss, B., Rivier, J., Burgus, R., Ling, N. and Guillemin, R. (1975). *Recent Prog. Horm. Res.,* **31,** 365
Vandesande, F. and Dierickx, K. (1975). *Cell Tiss. Res.,* **164,** 153–62
Walter, R., Schlesing, D. H., Schwartz, I. L. and Capra, J. D. (1971). *Biochem. Biophys. Res. Commun.,* **44,** 293–98
Watkins, W. B. (1975). *Int. Rev. Cytol.,* **41,** 241–284
Weindl, A. and Sofroniew, M. V. (1976). *Pharmakopsych.,* **9,** 226–34
Weiner, R. I., Pattou, E., Kerdelhue, B. and Kordon, C. (1975). *Endocrinology,* **97,** 1597–600
White, W. F., Hedlund, M. T., Weber, G. F., Rippel, R. H., Johnson, E. S. and Wilber, J. F. (1974). *Endocrinology,* **94,** 1422–90
Zimmerman, E. A., Hsu, K. G., Robinson, A. G., Carmel, P. W., Frank, A. G. and Tannenbaum, M. (1973). *Endocrinology,* **92,** 931
Zimmerman, E. A. (1976). In *Frontiers in Neuroendocrinology* (ed. L. Martini and W. R. Ganong), Raven Press, New York, pp. 25–62

3

Are hypothalamic hormones central neurotransmitters?

Michael J. Brownstein (Laboratory of Clinical Science,
National Institute of Mental Health, Bethesda, Maryland 20014, U.S.A.)

INTRODUCTION

Chemical neutotransmitters share a number of properties. They are found in, synthesised by, and released from neurons. When they are released from nerves, they have specific, drug sensitive effects on target cells; and these effects are also seen when they are applied to target cells exogenously. Finally, neurotransmitters are rapidly metabolised or removed from the synaptic cleft so that their actions are brief. 'Hypothalamic hormones' (releasing hormones and release inhibiting hormones) have not traditionally been thought of as neurotransmitters. Although it is true that these hormones are released from nerves into the portal vessels to act on cells of the anterior and intermediate lobes of the pituitary, it is unlikely that their only sites of action are in the pituitary. Data suggesting that the hypothalamic hormones function as chemical neurotransmitters are reviewed below.

THE HYPOTHALAMIC HORMONES

Only three peptides which regulate the secretion of tropic hormones by the hypophysis have been isolated and chemically characterised: luteinising hormone releasing hormone (LHRH), thyrotropin releasing hormone (TRH), and growth hormone release inhibiting hormone (somatostatin). Hypothalamic extracts also appear to contain factors which stimulate the release of corticotropin (that is, corticotropin releasing factor—CRF), growth hormone (GH-RF), prolactin (PRF) and melanocyte stimulating hormone (MSH-RF) and which inhibit the release of prolactin (PIF) and melanocyte stimulating hormone (MSH-IF).

Luteinising hormone releasing hormone
In 1960 McCann, Taleisnik and Friedman showed that the hypothalamus has luteotropin releasing activity. A decade later LHRH was isolated and its structure shown to be pyroGlu-His-Tryp-Ser-Tyr-Gly-Leu-Arg-Pro-Gly-NH$_2$ (Baba *et al.*,

37

1971; Matsuo *et al.*, 1971). This peptide stimulates the release of follicle stimulating hormone (FSH) as well as LH and, at present, there are dwindling numbers of investigators who believe in the existence of an FSH-RH separate from LHRH.

Thyrotropin releasing hormone
Compelling evidence for the existence of a factor that released TSH was first provided by Shibusawa *et al.* (1956) and by Schreiber *et al.* (1961). This factor (TRH) has been identified as pyroGlu–His–Pro–NH$_2$ (Bøler *et al.*, 1969; Burgus *et al.*, 1969). In addition to releasing thyrotropin, TRH also causes prolactin release (Vale *et al.*, 1973).

Growth hormone release inhibiting hormone and growth hormone releasing factor
Somatostatin has been obtained in pure form and its amino acid composition is H–Ala–Gly–Cys–Lys–Asn–Phe–Phe–Trp–Lys–Thr–Phe–Thr–Ser–Cys–OH (Brazeau *et al.*, 1973; Burgus *et al.*, 1973; Ling *et al.*, 1973). The synthetic peptide inhibits the release of growth hormone and, under certain circumstances, of thyrotropin and prolactin from cells of the anterior pituitary (Vale *et al.*, 1974*a*). In addition to being present in the brain, somatostatin is found in the stomach, the intestine and the pancreas (Hökfelt *et al.*, 1975*a*). It appears to inhibit the secretion of glucagon, insulin, gastrin and gastric acid (Bloom *et al.*, 1974; Mortimer *et al.*, 1974).

Destruction of the pituitary stalk blocks the secretion of growth hormone induced by insulin, 2-deoxyglucose, or arginine; and electrical stimulation of the ventromedial hypothalamic nuclei, or administration of hypothalamic extracts, causes the pituitary rapidly to liberate growth hormone. It seems clear that the brain must make one or more growth hormone releasing factors. Discrepancies between the bioassay and radioimmunoassay for growth hormone and the presence of somatostatin in hypothalamic extracts have hampered efforts to purify GH-RF.

Corticotropin releasing factor
Hume and Wittenstein (1950) showed that there was a factor in the hypothalamus which produced eosinophilia when injected into animals. Subsequently, Guillemin and Rosenberg (1955) and Saffran, Schally and Benfey (1955) demonstrated the presence of CRF in hypothalamic extracts. Porter and Jones (1956) then found CRF in portal blood. In spite of the fact that CRF was the first releasing factor to be discovered, its identity has not yet been established.

Prolactin releasing and release inhibiting factors
In mammals the predominant effect of the hypothalamus on prolactin secretion is an inhibitory one (cf. the net stimulatory effect in birds). Section of the hypophysial stalk results in an outpouring of prolactin from the pituitary. Similarly, when pituitaries are removed from the sella turcica and cultured they secrete large amounts of prolactin into the culture medium. The release of other pituitary hormones decreases in these circumstances.

The hypothalamic factor which inhibits the release of prolactin appears to be a relatively small molecule. Whether this factor is a catecholamine (specifically, dopamine) (Takahara, Arimura and Schally, 1974), or a peptide (Dular *et al.*,

1974), or whether there are several PIFs remains to be determined. As mentioned earlier, TRH has prolactin releasing activity, but the physiological importance of this is unknown.

Melanocyte-stimulating hormone releasing factor and release inhibiting factor
MSH release from mammalian and amphibian pituitary glands is inhibited tonically by the hypothalamus. Both MSH releasing, and release inhibiting, activity is present in hypothalamic extracts, however. The C-terminal portion of oxytocin, Pro-Leu-Gly-NH$_2$, reacts in some, but not all, assays for MSH-IF (Celis, Taleisnik and Walter, 1971; Nair, Kastin and Schally, 1971). So does tocinoic acid, Cys-Tyr-Ileu-Gln-Asn-Cys, the ring structure of oxytocin (Hruby *et al.*, 1972). Pro-Leu-Gly-NH$_2$ and tocinoic acid may both originate from the action of enzymes in hypothalamic tissue on oxytocin (Celis *et al.*, 1971). It has been suggested that a third peptide found in hypothalamus, H-Pro-His-Phe-Arg-Gly-NH$_2$, is an MSH-IF (Nair, Kastin and Schally, 1972).

RADIOIMMUNOASSAYS AND BIOASSAYS: A WORD OF CAUTION

Many of the studies that are cited in the following pages were based on the use of either radioimmunoassays or bioassays. Both methods have their pitfalls. Radio-immunoassays are sensitive, simple, fast and fairly reliable. They can be extraordinarily specific, but need not necessarily be. Radioimmunoassays for peptides can fail to detect amino acid replacements, deletions of part of the peptide molecule, or additions to it.

Some bioassays are reasonably easy to perform; others are difficult. On the whole, bioassays are somewhat less sensitive than radioimmunoassays and are somewhat less reliable. A bioassay system can respond simultaneously to more than one releasing, or release inhibiting, factor in an extract.

It is important to verify results obtained by means of one assay method by using others. To be confident that the 'TRH-like' or 'somatostatin-like' material that is outside of the hypothalamus is indeed TRH or somatostatin, as many of the properties of these molecules as possible should be established. In the meantime, there is room for skepticism.

EVIDENCE THAT 'HYPOTHALAMIC HORMONES' MAY FUNCTION AS NEUROTRANSMITTERS

Distribution within the central nervous system

Luteinising hormone releasing hormone
The highest concentrations of LHRH are found in the medial basal nuclei of the posterior two-thirds of the hypothalamus and in the median eminence (McCann, 1962; Schneider *et al.*, 1959; Crighton *et al.*, 1970; Palkovits *et al.*, 1974; Wheaton, *et al.*, 1975). Smaller amounts are present over the optic chiasma in the anterior hypothalamus and preoptic area. The majority of the neurons which synthesise LHRH and whose processes terminate in the median eminence may not be in the posterior hypothalamus. Knife cuts or lesions just behind the chiasma, thus separating the anterior and posterior parts of the hypothalamus, cause marked

decreases in the level of LHRH and in LH releasing activity in the latter (Martini *et al.*, 1968; Schneider *et al.*, 1969; Brownstein *et al.*, 1975*b*, 1976; Weiner *et al.*, 1975). The LHRH in the anterior hypothalamus and preoptic area most of which is in the supraoptic crest, (Kizer *et al.*, 1976) does not fall postoperatively. Thus, the LHRH that is rostral to the posterior border of the chiasma seems to be in cells that are also rostral to this level. It is possible that the LHRH in the posterior hypothalamus may be in cells which are under the trophic influence of neurons whose axons travel backwards through the anterior hypothalamus, but a more parsimonious suggestion is that there are relatively few LHRH-containing cells in the posterior hypothalamus. Consistent with this hypothesis are findings that destruction of cell bodies in the arcuate nucleus by treatment with monosodium glutamate does not change the level of LHRH in the median eminence (Kizer, personal communication); that nerve terminals with immunoreactive LHRH have been visualised in the median eminence (Barry *et al.*, 1973; Baker *et al.*, 1974; King *et al.*, 1974; Kordon *et al.*, Kozlowski and Zimmerman, 1974; Pelletier *et al.*, 1974; Hökfelt *et al.*, 1975*c*; Naik, 1975; Sétaló *et al.*, 1975); and that LHRH-containing perikarya have been seen in the preoptic area, septum and parolfactory region (Barry *et al.*, 1973, 1974; Naik, 1975).

Thyrotropin releasing hormone
Not unexpectedly, a high level of radioimmunoassayable TRH and TSH releasing activity was shown to be present in the median eminence (Brownstein *et al.*, 1974; Krulich *et al.*, 1974). The medial part of the ventromedial nucleus, the periventricular nucleus, the arcuate nucleus and the dorsomedial nucleus have moderately high concentrations of TRH as well. TRH is not confined to the medial basal hypothalamic nuclei, however. It is found in measurable amounts in all of the hypothalamic nuclei. Furthermore, TRH and TSH releasing activity are present throughout the brain and in the spinal cord (Jackson and Reichlin, 1974; Oliver *et al.*, 1974; Winokur and Utiger, 1974). There are especially high amounts in the preoptic area and septum, and lower concentrations in the brainstem, mesencephalon, basal ganglia and cerebral cortex.

Seventy-five per cent of the TRH which was normally in the medial basal hypothalamus disappeared after this region was surgically isolated from the rest of the brain, but the level of TRH did not change elsewhere (Brownstein *et al.*, 1975*c*). Therefore, the TRH which is outside of the medial basal hypothalamus is not in processes of hypothalamic cells. On the other hand, it seems likely that a large part of the TRH in the medial basal hypothalamus is in axons provided by other regions of the brain. Alternatively, the TRH may be in hypothalamic cells which respond to denervation by making or storing less TRH.

Knife cuts behind the optic chiasma result in substantial decreases in hypothalamic TRH just as total medial basal hypothalamic deafferentation does. Consequently, the neurons which synthesise TRH or regulate its synthesis must either be rostral to the posterior hypothalamus or send their processes into this area from a rostral direction.

TRH has been visualised in axons and terminals in the spinal cord, brainstem, mesencephalon and hypothalamus by means of immunocytochemistry (Hökfelt *et al.*, 1975*b*), and recently in specific populations of neuronal perikarya.

Growth hormone release inhibiting hormone

Radioimmunoassayable somatostatin and growth hormone release inhibiting activity are found throughout the brain (Brownstein *et al.*, 1975*a*; Vale *et al.*, 1974*b*). In fact, only about 25 per cent of the somatostatin in the brain is in the hypothalamus. The median eminence is very rich in the peptide; the arcuate, periventricular, ventral pre-mamillary and ventromedial nuclei have relatively high levels of radioimmunoassayable somatostatin. Although the median eminence, arcuate nucleus, periventricular nucleus, and ventral premamillary nucleus have high growth hormone release inhibiting activities, the ventromedial nucleus does not. Perhaps this is because the ventromedial nucleus is rich in both somatostatin and a growth hormone releasing factor; the two might cancel one another in a bioassay.

Complete deafferentation of the medial basal hypothalamus results in a profound decrease in somatostatin there (Brownstein *et al.*, 1977). This finding is perfectly compatible with the idea that neurons in the anterior periventricular nucleus provide the posterior hypothalamus with its somatostatin. This idea is based on recent immunocytochemical demonstrations of somatostatin-containing neuronal perikarya in the periventricular nucleus (Elde and Parsons, 1975; Hökfelt *et al.*, 1975*a*; Alpert *et al.*, 1976).

Isolation of the hypothalamus did not cause significant decreases in somatostatin in extrahypothalamic regions studied. The origin of somatostatin in extrahypothalamic structures remains to be determined.

Corticotropin releasing factor

Corticotropin releasing factor was bioassayed by measuring the release of ACTH from dispersed pituitary cells and was found to be present in the hypothalamus and in extrahypothalamic areas of the brain of normal and Brattleboro (diabetes insipidus) rats (Krieger *et al.*, 1977*a*). The highest concentrations were in the median eminence, arcuate nucleus, dorsomedial nucleus, ventromedial nucleus and periventricular nucleus. The thalamus had one-fifth and the cerebral cortex one-tenth the level of CRF measured in the hypothalamus as a whole.

CRF increased in the median eminence and medial basal hypothalamus after complete deafferentation of the posterior hypothalamus; thus, CRF is probably synthesised by cells in the posterior hypothalamus.

Prolactin releasing factor

Using a bioassay, Vale and his colleagues have shown prolactin releasing activity in extracts of hypothalamus and in extracts of extrahypothalamic nervous tissue (Vale *et al.*, 1974*b*). Whether the factor(s) responsible for the releasing activity is TRH or some other peptide(s) will not be known until it is obtained in a pure form.

BIOSYNTHESIS OF 'HYPOTHALAMIC HORMONES.

Vasopressin

Relatively little is known about the biosynthesis of mammalian centrally acting peptides with the exception of vasopressin. On the basis of a number of studies, Sachs and his co-workers have concluded that vasopressin and its neurophysin are derived from a common, ribosomally synthesised precursor protein. The evidence that supports this conclusion is as follows:

(1) Vasopressin synthesis takes place only in neuronal perikarya in the hypo-thalamus, not in axons and terminals in the neurohypophysis (Sachs, 1960; Sachs *et al.*, 1971).

(2) Incorporation of [^{35}S] cysteine into vasopressin *in vivo* or *in vitro* can only be demonstrated 1.5 hours or more after administration of the isotope. The synthesis of labelled vasopressin can be blocked by puromycin if the drug is given during the 'pulse' period, but not if it is given after the 'pulse' period (before the appearance of labelled vasopressin). Apparently, puromycin inhibits the synthesis of a precursor. Once the precursor is formed, however, puromycin does not affect its maturation (Sachs and Takabatake, 1964; Takabatake and Sachs, 1964; Sachs *et al.*, 1969).

(3) Vasopressin and neurophysin are synthesised with similar kinetics (Sachs *et al.*, 1971), and inhibition of neurophysin synthesis by analogues of amino acids found in neurophysin but not in vasopressin inhibits synthesis of vasopressin (Sachs *et al.*, 1969). Furthermore, Brattleboro rats, which are congenitally defi-cient in vasopressin, lack the vasopressin-associated neurophysin (Valtin *et al.*, 1974; Pickering *et al.*, 1975).

Taken together, the above data suggest that neurophysin and vasopressin are syn-thesised as parts of the same molecule.

Recently, Gainer, Sarne and Brownstein have extended Sachs' findings (Gainer *et al.*, 1977*a*, *b*). We have presented direct evidence for the existence of protein precursors for the neurophysins. In normal animals there appear to be two proteins of molecular weight 20 000 synthesised by cells of the supraoptic nucleus shortly after a pulse of [^{35}S] cysteine. One of these, the isoelectric point (pI) of which is 6.1, is synthesised only by normal rats; the other (pI 5.4) is synthesised by both normal and Brattleboro rats (Brownstein and Gainer, 1977). One hour after the cysteine is given, two additional proteins are found in the supraoptic nucleus of normal animals. These proteins have molecular weights of about 17 000 and pI 5.1 and 5.6. Only the pI 5.1 species is present in Brattleboro rates (Brown-stein and Gainer, in preparation). Two hours after the injection of cysteine, labelled products begin to arrive in the posterior pituitary. These products are relatively small in size. Instead of the large proteins synthesised in the cell bodies, two proteins of molecular weights around 12 000 are delivered to the neuro-hypophysis of normal rats (pI 4.6 and 4.8). Only the pI 4.6 protein is found in the posterior lobe of the Brattleboro rat. The precursors, intermediates and neuro-physins themselves bind to anti-rat neurophysin antibodies (Brownstein, Robinson, and Gainer, 1977).

We have concluded that the two neurophysins are synthesised from 20 000 molecular weight precursors via 17 000 molecular weight intermediates. The oxy-tocin-associated neurophysin (which is present in the Brattleboro animal) has pI 4.6; its precursor and intermediate have pI 5.4 and 5.1, respectively. The vaso-pressin-associated neurophysin (pI 4.8) seems to stem from a precursor and inter-mediate with pI 6.1 and 5.6, respectively. Whether the two precursors (and inter-mediates) give rise to vasopressin and oxytocin along with the neurophysins is still unknown.

Our studies of vasopressin and neurophysin synthesis, and those of Sachs, were facilitated by several factors. The anatomy of the hypothalamo–neurohypophysial system is well characterised. Therefore, we could inject labelled precursor adjacent to the cell bodies of the neurons selected for study and sample cell body-, axon- and terminal-rich areas at selected times afterwards. Since the protein species which are transported from the cell bodies to the posterior lobe of the pituitary are the biologically relevant ones, injecting near the cell bodies and sampling axon- and terminal-containing regions made the job of chromatographic separation easier. Because vasopressin, oxytocin, and the neurophysins are rich in cysteine, and because [^{35}S]-labelled cysteine of very high specific activity is available, we were able to generate enough labelled product to study without much difficulty. We were aided in doing this by earlier investigations of vasopressin and neurophysin secretion. When animals are given 2 per cent saline to drink, the incorporation of cysteine into precursors, intermediates and neurophysins increases four- or fivefold.

Thyrotropin releasing hormone

The case of TRH is completely different from that of vasopressin. The locations of neuronal perikarya which synthesise TRH are unknown; the rate of formation of TRH appears to be slow; the factors which stimulate TRH synthesis and secretion are poorly understood and the labelled amino acids available for studying TRH biosynthesis did not until recently have exceptionally high specific activities. These problems vex those interested in the biosynthesis of TRH and other 'hypothalamic hormones'. The small amount of product formed from labelled precursor and the large number of other peptides labelled when the amino acid is infused into the lateral ventricle makes the task of purifying any given radioactive product a very difficult one. While immunoaffinity chromatographic methods and immuno-precipitation promise to be useful, these methods must be used with extreme care when they are applied to small peptides. The methods are no better than the antibodies used, and precautions (such as purifying the antibodies) must be taken to avoid non-specific binding. The bound products must still be rigorously identified.

Among the early attempts to study the biosynthesis of TRH were those of Mitnick and Reichlin (1971) whose work has been criticised because their chromatographic procedures did not separate TRH from radioactive contaminants. More recently, McKelvy (1974) and Grimm–Jørgensen (McKelvy and Grimm-Jørgensen, 1975) have used a lengthy separation procedure, including derivatisation, to show that guinea pig hypothalamic cultures and whole newt brain incorporate [^{3}H] proline into TRH. Whether TRH is synthesised by means of an RNA template has yet to be proven.

Little is known about the synthesis of biologically active peptides by the parvicellular neurons of the hypothalamus, and synthesis of 'hypothalamic hormones' by extrahypothalamic neurons has not yet been demonstrated in mammals.

RELEASE OF 'HYPOTHALAMIC HORMONES'

TRH and LHRH are found in synaptosomes prepared from hypothalamic tissue. TRH and LHRH can be released from hypothalamic synaptosomes in a calcium-dependent manner by potassium and dopamine (Bennett *et al.*, 1974; Schaeffer *et al.*, submitted for publication). (Dopamine is effective in releasing LHRH and TRH from synaptosomes prepared from median eminence tissue, but not synaptosomes prepared from the remainder of the medial basal hypothalmus.) The dopamine induced release is mimicked by apomorphine and blocked by chlorpromazine. Radioimmunoassayable TRH in septal synaptosomes can also be released by potassium, but not by dopamine. Therefore, presynaptic dopamine receptors may be involved in regulating TRH release at some synapses but not at others.

Release of hypothalamic hormones at discrete extrahypothalamic synapses has not been shown to occur, though there is little doubt that they can be released from nerve endings which they occupy.

INACTIVATION OF 'HYPOTHALAMIC HORMONES'

The most important mechanism for inactivating hypothalamic peptides seems to be cleavage by peptidases. There have been numerous reports of the existence of peptidase activity in mammalian hypothalamus as well as other regions of the brain (see Marks, 1976). TRH, LHRH and somatostatin are rapidly degraded by tissue peptidases. In fact, their rapid degradation has hampered studies of synthesis and release. To some degree, the problem of degradation can be overcome by use of inhibitors of the peptidases such as bacitracin (McKelvy *et al.*, 1976). Alternatively, studies can be conducted using tissues which are relatively poor in enzymes that degrade the peptide to be examined.

Uptake of peptides by the cells that release them or by cells adjacent to the synaptic cleft has not been shown to play any part in terminating the actions of these agents.

'HYPOPHYSIAL HORMONES'

While a discussion of hypophysial hormones is only peripherally related to my stated task, it seems worthwhile to mention two hypophysial hormones in passing, melanocyte stimulating hormone (MSH) and adrenocorticotropic hormone (ACTH). In the hypothalami of hogs (Guillemin *et al.*, 1962) and dogs (Schally *et al.*, 1962) there are materials which react in bioassays for corticotropin and melanotropin. These substances are chromatographically similar to $ACTH_{1-39}$ and α-MSH and β-MSH. Recently, we have found immunoreactive (and bioreactive) ACTH-like materials in the rat hypothalamus (Krieger *et al.*, 1977b). These materials are not confined to the hypothalamus, however. They are present in relatively large amounts in several limbic structures, even after the animals are hypophysectomised. In light of the behavioural effects of ACTH and its analogues (de Wied, 1974 and chapter 16 of this volume) the roles of 'hypophysial hormones' in the nervous system may be broader than originally suspected.

CONCLUSIONS

It is premature to state unequivocally that 'hypothalamic (and hypophysial) hormones' comprise a class of chemical neurotransmitters. Their presence in neurons is not disputed, but the location of the perikarya of these neurons is only beginning to be known. Studies of the biosynthesis of most of the peptides are still in their infancy and studies of the release of the peptides to date have been rather indirect, except those involving release into the portal system. Enzymic processes for terminating the action of the hormones have been demonstrated but not identified.

Although the hypothalamic hormones have not been proven to function as central neurotransmitters, the evidence to date is at least consistent with this point of view.

REFERENCES

Alpert, L. C., Brawer, J. R., Patel, Y. C. and Reichlin, S. (1976). *Endocrinology*, 98, 255–58
Baba, Y., Matsuo, H. and Schally, A. V. (1971). *Biochem. biophys. Res. Commun.*, 44, 459–63
Baker, B. L., Dermody, W. C. and Reel, J. R. (1974). *Am. J. Anat.*, 139, 129–34
Barry, J., Dubois, M. P. and Carette, B. (1974). *Endocrinology*, 95, 1416–23
Barry, J., Dubois, M. P. and Poulain, P. (1973). *Z. Zellforsch*, 146, 351–66
Bloom, S. R., Mortimer, L. H., Thorner, M . O., Besser, G. M., Hall, R., Gomez-Pan, A., Roy, V. M., Russell, R. C. G., Coy, D. H., Kastin, A. J. and Schally, A. V. (1974). *Lancet*, ii, 1106–09
Bøler, J., Enzmann, F., Folkers, K., Bowers, C. Y. and Schally, A. V. (1969). *Biochem. biophys. Res. Commun.*, 37, 705–10
Bennett, G. W., Edwardson, J. A., Holland, D., Jeffcoate, S. L. and White, N. (1975). *Nature*, 257, 323–25
Brazeau, P., Vale, W., Burgus, R., Ling, N., Butcher, M., Rivier, J. and Guillemin, R. (1973). *Science*, 179, 77–79
Brownstein, M., Arimura, A., Sato, H., Schally, A. V. and Kizer, J. S. (1975a). *Endocrinology*, 96, 1456–61
Brownstein, M., Arimura, A., Schally, A. V., Palkovits, M. and Kizer, J. S. (1976). *Endocrinology*, 98, 662–65
Brownstein, M. J., Arimura, A., Fernandez-Durango, R., Schally, A. V., Palkovits, M. and Kizer, J. S. (1977). *Endocrinology*, 100, 246–49
Brownstein, M. J. and Gainer, H. (1977). *Proc. natn. Acad. Sci. U.S.A.*, 74, 4046–49
Brownstein, M. J., Paklovits, M. and Kizer, J. S. (1975b). *Life Sci.*, 17, 679–82
Brownstein, M. J., Palkovits, M., Saavedra, J., Bassiri, R. and Utiger, R. D. (1974). *Science*, 185, 267–69
Brownstein, M. J., Robinson, A. and Gainer, H. (1977). *Nature*, 269, 259–61
Brownstein, M., Utiger, R., Palkovits, M. and Kizer, J. S. (1975c). *Proc. natn. Acad. Sci. U.S.A.*, 72, 4177–79
Burgus, R., Dunn, T. F., Desiderio, D. and Guillemin, R. (1969). *C. r. hebd. Séanc Acad. Sci., Paris (D)*, 269, 1870–73
Burgus, R., Ling, N., Butcher, M. and Guillemin, R. (1973). *Proc. natn. Acad. Sci. U.S.A.*, 70, 684–88
Celis, M. E., Taleisnik, S. and Walter, R. (1971). *Proc. natn. Acad. Sci. U.S.A.*, 68, 1428–33
Crighton, D. B., Schneider, H. P. G. and McCann, S. M. (1970). *Endocrinology*, 87, 323–329
de Wied, D. (1974). In *The Neurosciences Third Study Program* (ed. F. O. Schmitt and F. G. Worden) M.I.T. Press, Cambridge, Mass., pp. 653–66
Dular, R., LaBella, F., Vivian, S. and Eddie, L. (1974). *Endocrinology*, 94, 563–67
Elde, R. P. and Parsons, J. A. (1975). *Am. J. Anat.*, 144, 541–48
Gainer, H., Sarne, Y. and Brownstein, M. J. (1977a). *Science*, 195, 1354–56.
Gainer, H., Sarne, Y. and Brownstein, M. J. (1977b). *J. Cell. Biol.*, 73, 366–81
Guillemin, R. and Rosenberg, B. (1955). *Endocrinology*, 57, 599–607
Guillemin, R., Schally, A. V., Lipscomb, H. S., Andersen, R. N. and Long, J. M. (1962). *Endocrinology*, 70, 471–77
Hökfelt, T., Efendic, S., Hellerstrom, C., Johansson, O., Luft, R. and Arimura, A. (1975a). *Acta Endocr.*, 80, suppl. 200
Hökfelt, T., Fuxe, K., Johannsson, O., Jeffcoate, S. and White, N. (1975b). *Neurosci. Lett.*, 1, 133–39
Hökfelt, T., Fuxe, K., Goldstein, M., Johansson, O., Park, D., Fraser, H. and Jeffcoate, S. L. (1975c). In *Anatomical Neuroendocrinology* (ed. W. E. Stumpf and L. D. Grant), Karger, Basel, pp. 381–92
Hruby, V. J., Smith, C. W., Bower, S. A. and Hadley, M. E. (1972). *Science*, 176, 1331–32
Hume, D. M. and Wittenstein, G. J. (1950). In *Proceedings of the First Clinical ACTH Conference* (ed. J. R. Mote), Blakiston, Philadelphia, pp. 134–46

46   *Centrally Acting Peptides*

Jackson, I. M. D. and Reichlin, S. (1974). *Endocrinology*, 95, 854–62
King, J. C., Parsons, J. A., Erlandsen, S. L. and Williams, T. H. (1974). *Cell Tiss. Res.*, 153, 211–17
Kizer, J. S., Palkovits, M. and Brownstein, M. (1976). *Endocrinology*, 98, 309–15
Kordon, C., Kerdelhue, B., Pattou, E. and Jutisz, M. (1974). *Proc. Soc. exp. Biol. Med.*, 147, 122–27
Kozlowski, G. P. and Zimmerman, E. A. (1974). *Anat. Rec.*, 178, 396
Krieger, D. T., Liotta, A. and Brownstein, M. J. (1977a). *Endocrinology*, 100, 227–37
Krieger, D. T., Liotta, A. and Brownstein, M. J. (1977b). *Proc. natn. Acad. Sci. U.S.A.*, 74, 648–52
Kurlich, L., Quijada, M., Hefco, E. and Sundberg, D. K. (1974). *Endocrinology*, 95, 9–17
Ling, N., Burgus, R., Rivier, J., Vale, W. and Brazeau, P. (1973). *Biochem. biophys. Res. Commun.*, 52, 786–791
McCann, S. M. (1962). *Am. J. Physiol.*, 202, 395–400
McCann, S. M., Taleisnik, S. and Friedman, H. M. (1960). *Proc. Soc. exp. Biol. Med.*, 104, 432–34
McKelvy, J. F. (1974). *Brain Res.*, 65, 489–502
McKelvy, J. F. and Grimm-Jørgensen, Y. (1975). In *Hypothalamic Hormones: Chemistry, Physiology, Pharmacology and Clinical Uses* (ed. M. Motta, P. T. Crosignani and L. Martini), Academic Press, New York, pp. 13–26
McKelvy, J. G., LeBlanc, P., Landes, C., Perrie, C., Grimm-Jørgensen, Y. and Kordon, C. (1976). *Biochem. biophys. Res. Commun.*, 73, 507–15
Marks, N. (1976). In *Subcellular Mechanisms in Reproductive Neuroendocrinology* (ed. F. Naftolin, K. J. Ryan and I. J. Davies), Elsevier, New York, pp. 129–47
Martini, L., Fraschini, F. and Motta, M. (1968). *Recent Prog. Horm. Res.*, 24, 439–85
Matsuo, H., Baba, Y., Nair, R. M. G., Arimura, A. and Schally, A. V. (1971). *Biochem. biophys. Res. Commun.*, 43, 1334–39
Mitnick, M. A. and Reichlin, S. (1971). *Science*, 172, 1241–43
Mortimer, H., Tunbridge, W. M. G., Carr, D., Yeomans, L., Lind, T., Coy, D. H., Bloom, S. R., Kastin, A., Mallinson, C. N., Besser, G. M., Schally, A. V. and Hall, R. (1974). *Lancet*, i, 679–01
Naik, D. V. (1975). *Cell Tiss. Res.*, 157, 437–55
Nair, R. M. G., Kastin, A. J. and Schally, A. V. (1971). *Biochem. biophys. Res. Commun.*, 43, 1376–81
Nair, R. M. G., Kastin, A. J. and Schally, A. V. (1972). *Biochem. biophys. Res. Commun.*, 43, 1420–25
Oliver, C., Eskay, R. L., Ben-Jonathan, N. and Porter, J. C. (1974). *Endocrinology*, 95, 540–53
Palkovits, M., Arimura, A., Brownstein, M. J., Schally, A. V. and Saavedra, J. M. (1974). *Endocrinology*, 96, 554–58
Pelletier, G., Labrie, F., Puviani, R., Arimura, A. and Schally, A. V. (1974). *Endocrinology*, 95, 314–15
Pickering, B. T., Jones, C. W., Burford, G. D., McPherson, M., Swann, R. W., Heap, P. F. and Morris, J. F. (1975). *Ann. N.Y. Acad. Sci.*, 248, 15–35
Porter, J. C. and Jones, S. C. (1956). *Endocrinology*, 58, 62–67
Sachs, H. (1960). *J. Neurochem.*, 5, 297–303
Sachs, H. and Takabatake, Y. (1964). *Endocrinology*, 75, 943–48
Sachs, H., Saito, S. and Sunde, D. (1971). In *Subcellular Organization and Function in Endocrine Tissues* (ed. H. Heller and K. Lederis), Cambridge University Press, New York, pp. 325–36
Sachs, H., Fawcett, P., Takabatake, Y. and Portanova, R. (1969). *Recent Prog. Horm. Res.*, 25, 447–91
Saffran, M., Schally, A. V. and Benfey, B. G. (1955). *Endocrinology*, 57, 439–44
Schally, A. V., Lipscomb, H. S., Long, J. H., Dear, W. E. and Guillemin, R. (1962). *Endocrinology*, 70, 478–80
Schneider, H. P. B., Crighton, D. B. and McCann, S. M. (1969). *Neuroendocrinology*, 5, 271–80
Schreiber, V., Eckertova, A., Franz, Z., Koci, J., Rybau, M. and Kmentova, V. (1961). *Experientia*, 17, 264–65
Sétaló, G., Vigh, S., Schally, A. V., Arimura, A. and Flerko, B. (1975). *Endocrinology*, 96, 135–42

Shibusawa, K., Saito, S., Nishi, K., Yamamoto, T., Tomizawa, K. and Abe, C. (1956). *Endocr. jap.*, 3, 116–124

Takabatake, Y. and Sachs, H. (1964). *Endocrinology*, 75, 934–42

Takahara, J., Arimura, A. and Schally, A. (1974). *Endocrinology*, 95, 462–70

Vale, W., Blackwell, R., Grant, G. and Guillemin, R. (1973). *Endocrinology*, 93, 26–33

Vale, W., Rivier, C., Brazeau, P. and Guillemin, R. (1974a). *Endocrinology*, 95, 968–77

Vale, W., Rivier, C., Palkovits, M., Saavedra, J. M. and Brownstein, M. J. (1974b). *Endocrinology*, 94, A128

Valtin, H., Stewart, J. and Sokol, H. W. (1974). *Handbk Physiol.*, 7, 131–71

Weiner, R. I., Pattou, E., Kerdelhue, B. and Kordon, C. (1975). *Endocrinology*, 97, 1597–1600

Wheaton, J. E., Krulich, L. and McCann, S. M. (1975). *Endocrinology*, 97, 30–38

Winokur, A. and Utiger, R. D. (1974). *Science*, 185, 265–267

4
Diffuse neuroendocrine system: peptides common to brain and intestine and their relationship to the APUD concept

A. G. E. Pearse (Royal Postgraduate Medical School, London W12 0HS, U.K.)

INTRODUCTION

Although 'peptides common to brain and intestine' makes an attractively concise title it requires amplification before one can begin to discuss the common peptides, and the possible implications of their existence. The phrase must therefore be expanded to read:

> 'peptides common to the central and peripheral divisions of the nervous system on the one hand, and to the gastroenteropancreatic and related endocrine systems on the other'.

Whenever, in the cause of brevity, reference is made to 'brain and intestine' this amplification is hereafter to be understood.

A peptide produced by an endocrine cell is likely, though not necessarily accurately in the physiological sense, to be called a hormone. Peptides synthesised by neurons and transported via their axons are likely, but possibly inaccurately, to be classified as neurotransmitters. In both cases the requisite experimental evidence for either function may well be lacking and it is therefore expedient, as well as correct, to regard all the substances we have to consider in the context of this paper as peptides, unless their claim to one or other of the alternative titles has been shown to be respectable.

It is seldom that one has to write on a subject which has little or no directly antecedent history but the fact remains that only in 1975 was the first of a now rapidly growing series properly established, on the basis of experimental investigation, as a peptide common to brain and intestine. Before 1975 there was nothing but speculation, and little enough of that. Now, with hindsight, one can see that if

49

the detection of Substance P by von Euler and Gaddum, in 1931, had been followed up in a manner appropriate to the importance of the problem, instead of nearly half a century a tenth of the time might have sufficed for its elucidation.

Table 4.1   Peptides common to brain and intestine

Peptide	'Original' location(s)	'New' location(s)
Substance P	Substantia nigra Habenula Dorsal root ganglia, etc. Posterior columns	Gastrointestinal EC_1 cells
Somatostatin	Cerebral cortex Nucleus periventricularis Neurohypophysis Autonomic nervous system Pineal gland	Islet D cells Gastrointestinal D cells
Neurotensin	Hypothalamus	Intestinal 'E' cells
Enkephalin	Telencephalon Diencephalon	Antral and duodenal G cells Myenteric plexus neurons
Gastrin	Antral and duodenal G cells	Cerebral cortex (grey matter)
CCK	Intestinal I cells	Cerebral cortex (grey matter) Hypothalamus
VIP	Intestinal and pancreatic D_1 cells	Cerebral cortex Hypothalamus Autonomic nervous system

Such is the current rate of discovery in the field of the peptides common to brain and intestine that any tabulation will probably be out of date before it appears in print. Nevertheless, in table 4.1, a list of common peptides is presented. The localisation accepted as 'primary' for the first four peptides in the list is the central nervous system (CNS). For the remaining peptides their primary localisation has usually been thought of as intestine, with the pancreas playing only a minor role.

BRAIN PEPTIDES IN GUT AND PANCREAS

Substance P

This, the doyen of the common peptides, was isolated, purified and sequenced by Chang and Leeman (1970), and by Chang *et al.* (1971), almost 40 years after its original detection in alcoholic extracts of equine brain and intestine by von Euler and Gaddum (1931). Using radioimmunoassay, it has been demonstrated in the brain, with the highest concentrations in substantia nigra, habenula and hypothalamus (Kanazawa and Jessel, 1976), and in spinal cord sensory ganglia, dorsal roots and substantia gelatinosa (Takahashi and Otsuka, 1975; Cuello *et al.*, 1976).

Immunocytochemical techniques have been used to demonstrate the endecapeptide, or rather Substance P-like immunoreactivity (SPLI), not only in the CNS and primary sensory neurons (Hökfelt *et al.*, 1975c, 1976) but also in one type of enterochromaffin cell (EC_1) found in the gastrointestinal tract and bile ducts (Pearse and Polak, 1975; Heitz *et al.*, 1976, 1977). It seems likely that Substance P is widely distributed in primary sensory neurons and their processes, in the EC_1 cells of the gastrointestinal tract and, probably, in enterochromaffin cells having a much wider pattern of distribution (for example, in lung, thymus and genitourinary tract) than is indicated by the evidence available at present.

Somatostatin

Named for its growth hormone release inhibiting properties, somatostatin was isolated from ovine hypothalamus and characterised as a tetradecapeptide by Brazeau *et al.* (1973). It was later found to release a much larger number of endocrine peptides including insulin, glucagon and gastrin but not parathyrin. Somatostatin, or somatostatin-like immunoreactivity (SLI) has been demonstrated principally in the median eminence, the arcuate nucleus, the ventromedial nucleus of the hypothalamus, and the nucleus periventricularis, both by radioimmunoassay and by immunocytochemistry (Hökfelt *et al.*, 1974; Pelletier *et al.*, 1974; Brownstein *et al.*, 1975; Sétaló *et al.*, 1975; Elde and Parsons, 1975; Palkovits *et al.*, 1976; Baker and Yu, 1976). According to Elde and Parsons (1975) the only cell bodies containing somatostatin were those of the nucleus periventricularis but Alpert *et al.* (1976) found positively reacting cell bodies also in the ventromedial nucleus and, in one instance, in the putamen. The arcuate nucleus contains positively reacting fibres only.

Following the demonstration by Arimura *et al.* (1975), using radioimmunoassay, that somatostatin was present in stomach and pancreas in concentrations equal to those in the hypothalamus, several groups of workers were able to localise the peptide in endocrine or 'endocrine-like cells' in both regions. Dubois (1975) and Dubois *et al.* (1975) described its presence in rat pancreas and in human foetal pancreas, in cells different from those containing insulin and glucagon. Goldsmith *et al.* (1975) localised the peptide, also in rat pancreas, in cells belonging to the D group or a sub-group thereof and Hökfelt *et al.* (1975a) indicated that the D cells alone were implicated. Polak *et al.* (1975) demonstrated the peptide in the D cells of the pancreas and of the upper intestine in dog, pig and man. Hökfelt *et al.* (1975b) found in rat intestine a new class of autonomic nerve fibres that contained immunoreactive somatostatin. They reported, at the same time, the presence of SLI in some unidentified cells in the lamina propria. These may well be the neuroendocrine cells described in that location by Matsuo and Seki (1976). Consideration of the physical associations of the D cells, with cells producing insulin, glucagon and gastrin, leads to the supposition that the actions of the peripherally produced variety of somatostatin are paracrine (Feyrter, 1938), that is to say directed towards neighbouring cells or tissues, rather than truly endocrine.

Neurotensin

This is a biologically active tridecapeptide that was isolated from acid/acetone extracts of bovine hypothalamus, and purified and sequenced by Carraway and Leeman (1973, 1975). It has now been shown, by these same authors (1976) to be present in rat hypothalamus in amounts (3–4 pmol per region) insufficient to produce the peripheral effects noted on its administration by intravenous injection. Kitabgi *et al.* (1976), however, found large amounts of a tridecapeptide in bovine intestine which they were able partially to characterise as neurotensin. Immunocytochemical studies by Orci *et al.* (1976) demonstrated the peptide in some unidentified endocrine cells in the ileum of the dog. The cells were few in number and widely distributed; their scarcity was in agreement with the low levels of the peptide found in the intestine by radioimmunoassay.

Enkephalins

The last of the 'brain' peptides in table 4.1 are the two enkephalins (Hughes *et al.*, 1975; Simantov and Snyder, 1976). These are closely related pentapeptides (H–Tyr–Gly–Gly–Phe–Leu–OH and H–Tyr–Gly–Gly–Phe–Met–OH) and the sequence of amino acids in the second, methionine-enkephalin, is identical with residues 61–65 of the pituitary peptide β-lipotropin. In spite of this, and with contrary results to those achieved with anti-endorphins, met-enkephalin has not been identified immunocytochemically in the pituitary gland. Enkephalin-like immunoreactivity has, however, been detected in the antral gastrin cells, and also in the myenteric plexus and in nerve fibres in the gut wall of three mammalian species (Polak *et al.*, 1977).

GUT PEPTIDES IN BRAIN

Gastrin

This hormone, or more accurately gastrin-like immunoreactivity, was found in human and animal brain (chiefly cerebral cortex) by Vanderhaegen *et al.* (1975). The concentration in cerebral grey matter, was as high as 200 ng per g dry weight. Since the peptide responsible for immunoreactivity was found to elute later than gastrin on gel-filtration, it must be considered to be closely related to, rather than identical with, gastrin.

Cholecystokinin (CCK)

CCK-like immunoreactivity was identified by Dockray (1976) in whole brain extracts from rats and in extracts of cerebral cortex from dog and pig. Fractionated on Sephadex G-25, more than 70 per cent of the total immunoreactivity present in the extracts eluted in a similar position to the C-terminal octapeptide of CCK. In terms of CCK-8 this main component was present in amounts of 50–150 pmol/g in the brains of the three species tested. CCK-like immunoreactivity has been demonstrated immunocytochemically in nerve fibres and cells in the hypothalamus and elsewhere in the nervous system.

Vasoactive intestinal peptide (VIP)

This 28-residue peptide was extracted from pig intestine by Said and Mutt (1972) and later sequenced by them (Mutt and Said, 1974). It has been identified by radioimmunoassay throughout the whole length of the gut, from oesophagus to rectum (Bloom and Bryant, 1973) and it was shown to be present in an otherwise unidentified gut endocrine cell by Polak *et al.* (1974). Buffa *et al.* (1977) later demonstrated VIP-immunoreactive cells in the pancreatic islets and also in the gastrointestinal mucosa of dog, guinea pig and man. They were tentatively identified with the D_1 cells of the Wiesbaden classification (Solcia *et al.*, 1973).

Three recent papers have recorded the presence of VIP in neural tissues. Said and Rosenberg (1976) found VIP-immunoreactivity in three neuroblastoma cell lines in culture and in acid alcohol extracts of the CNS and sympathetic ganglia and Bryant *et al.* (1976) found high VIP levels in the cerebral cortex and hypothalamus of both man and pig. Parallel immunocytochemical studies showed that VIP was present not only in endocrine cells in the mucosa but in fine fibres in the lamina propria and muscularis layers and in cell bodies in the myenteric plexus. Larsson *et al.* (1976), on the other hand, localised VIP in gastrointestinal nerves and in the ventromedial hypothalamus, but they found no reactivity in any endocrine cell in the mucosa.

ORIGINS OF THE APUD CONCEPT

The two sides of the equation, set forth at some length above, are united by an essentially simple concept. This was first formulated more than 10 years ago to explain the possession, by neurons and by endocrine cells producing polypeptide hormones, of a set of common cytochemical and ultrastructural characteristics. As originally expressed, the concept contained the postulate that all the endocrine cells were derived from a common neuroectodermal ancestor and it was suggested (Pearse, 1966*a*, *b*) that 'the amine storage mechanism and the presence of cholin-esterase', the two most salient features of the series of endocrine polypeptide-producing cells which shortly after become the APUD series (Pearse, 1968, 1969), together pointed 'towards a cell of neural origin, perhaps coming from the neural crest'.

DERIVATION OF THE APUD CELLS

During the years immediately following the formulation of the APUD concept much effort was devoted to elucidation of the precise relationship of the neural crest to the widely distributed endocrine cells of the APUD series, many of whose members had by now been recognised as belonging to the diffuse endocrine system of clear cells (Helle Zellen) described by Feyrter (1938, 1953).

Using the biological marker system and the allograph technique to produce chick–quail chimaeras, Le Douarin (1969) and her co-workers (Le Douarin and Le Lièvre, 1970; Teillet and Le Douarin, 1970; Le Douarin and Teillet, 1971; Le Douarin *et al.*, 1972) showed that not only the melanocytes, adrenomedullary and other chromaffin cells, and the neurons of the sympathetic ganglia, came from the neural crest but also the type 1 cells of the carotid body and the C (calcitonin) cells of the ultimobranchial gland. These findings were confirmed, in respect of the last two cells, by Pearse *et al.* (1973) and by Pearse and Polak (1971*a*).

It was also postulated (Pearse and Polak, 1971*b*) that the APUD cells of the gut and pancreas arose from the neural crest, allowing that there was still a gap between the earliest demonstration of APUD cells in the foregut and its derivatives and the APUD cells coming from the definitive neural crest. This interpretation of the facts has been rendered untenable by subsequent experimental embryological investigations, carried out by Andrew (1974) and by Pictet *et al.* (1976), which have demonstrated that the crest proper cannot contribute *endocrine* cells to the foregut or its early derivatives such as the pancreas. Andrew (1975) on the other hand has shown that APUD cells (indicated by their amine-handling quality) are present in the developing foregut of the chick at an earlier stage (16 somites) than was previously recognised, and that these cells are almost certainly precursors of the APUD cells of the dorsal pancreatic anlage in which, at 32 and 42 somites, respectively, Beaupain and Dieterlen-Lièvre (1974) and Dieterlen-Lièvre and Beaupain (1974) demonstrated glucagon and insulin by immunocytochemical techniques.

Because of these, and many other findings, the 40 endocrine cells that at present constitute the APUD series are now divided, in terms of their origin, into three distinct groups:

(a) cells derived from neuroectoderm or placodes
(b) cells derived from the definitive neural crest
(c) cells derived from neuroendocrine-programmed epi- or ectoblast

In the first group are the endocrine cells of the pineal, hypothalamus, pituitary and parathyroid glands and in the second, the C cells of the thyroid and ultimobranchial glands, the small intensely fluorescent (SIF) cells of Eränkö and Härkönen (1963), the carotid body type 1 cells, the two adrenomedullary cells and the melanoblasts and melanocytes. The origin of all these cell types can be considered completely proven except for the parathyroid chief cell which is still regarded by some authorities as having an endodermal origin. Two possibilities exist for the origin of the 18 endocrine cells of the gastroenteropancreatic system. Either they arise, together with all the other cells of the fore (and hind) gut, in mammals from the yolk-sac endoderm, or in birds as derivatives of the invaginating layer of the epiblast of Hensen's node region, or else they are secondary invaders 'dropping out' from the ectoblast. In either case it is necessary to postulate that they are initially programmed for neuroendocrine function and that they await the microenvironmental influences which subsequently determine their precise mode of activity and their precise product(s).

It has thus become necessary to redefine the APUD concept, in the light of experimental findings of the past few years, in order to provide the proper structure for further experimental inquiry.

PRESENT STATUS OF THE APUD CONCEPT

The APUD concept is now expressed in the following terms:

The cells of the APUD series, producing peptides active as hormones or as neurotransmitters, are all derived from neuroendocrine-programmed cells originating from the ectoblast. They constitute a third (endocrine or neuroendocrine) division of the nervous system whose cells act as third-line effectors to support, modulate or amplify the actions of neurons in the somatic and autonomic divisions, and possibly as tropins to both neuronal and non-neuronal cells.

The implications of the concept must affect the majority of disciplines in the biomedical sciences. In the context of the present symposium its chief importance lies, perhaps, in explaining the nature of the relationship between the many peptides now found to be common to brain and intestine and in emphasising the pressing need for elucidation of their physiological roles.

Are the various peptides to be regarded on the one hand as hormones and on the other as neurotransmitters, or is there an overlap between the two roles? Alternatively, is there a spectrum of activities? In amplification of this last suggestion we can observe that the functions of the APUD cells are divisible into six categories as shown in table 4.2. Prime examples of each of these functions are (1) the SIF and carotid body cells (neurocrine), (2) the neurosecretory cells of the hypothalmus (neuroendocrine), (3) the cells of the anterior hypophysis (endocrine), (4) the D, or somatostatin, cells of the pancreas and gastric antrum (paracrine), (5) the melanocytes (epicrine) and (6) the endocrine cells of anuran cutaneous glands (exocrine).

Table 4.2   Functions of APUD cells

Neurocrine	Into neurons
Neuroendocrine	Via axons
Endocrine	Into the bloodstream
Paracrine	Into intercellular space
Epicrine	Into somatic cells
Exocrine	To the externum

'The ancestral neurone' is a 'functionally versatile entity endowed with the means for long distance as well as more or less localised chemical signalling' (Scharrer, 1976). Descendants of the ancestral neuron, the APUD cells have mostly lost their ability to carry out long range signalling by way of their (vanished) axons. At the same time they have retained all the other features of primitive neurosecretory neurons, whether they are still directly innervated through synaptic terminals or whether, as is the common mode, they are only indirectly subject to nervous control.

With the exception of steroid and iodothyronine production, the whole field of endocrinology is now neuroendocrinology, and the six modes of APUD-cell expression provide a basis for its total integration into the nervous system.

REFERENCES

Alpert, L. C., Brawer, J. R., Patel, Y. C. and Reichlin, S. (1976). *Endocrinology*, 98, 255–58
Andrew, A. (1974). *J.Embryol.exp.Morphol.*, 31, 589–98
Andrew, A. (1975). *Gen.comp.Endocr.*, 26, 485–95
Arimura, A., Sato, H., Dupont, A., Nishi, N. and Schally, A.V. (1975). *Science*, 189, 1007–09
Baker, B. L. and Yu, Y.-Y. *Anat. Rec.*, 186, 343–56
Beaupain, D. and Dieterlen-Lièvre, F. (1974). *Gen.comp.Endocr.*, 23, 421–31
Bloom, S. R. and Bryant, M. G. (1973). *Gut*, 14, 823
Brazeau, P., Vale, W., Burgus, R., Ling, N., Butcher, M., Rivier, J. and Guillemin, R. (1973). *Science*, 179, 77–79
Brownstein, M., Arimura, A., Sato, H., Schally, A. V. and Kizer, J. S. (1975). *Endocrinology*, 96, 1456–61
Bryant, M. G., Polak, J. M., Modlin, I., Bloom, S. R., Alburquerque, R. H. and Pearse, A. G. E. (1976). *Lancet*, i, 991–93
Buffa, R., Capella, C., Solcia, E., Frigerio, B. and Said, S. I. (1977). *Histochemistry*, 50, 217–27

56   *Centrally Acting Peptides*

Carraway, R. and Leeman, Susan E. (1973). *J. biol. Chem.*, **248**, 6854–61
Carraway, R. and Leeman, Susan E. (1975). *J.biol.Chem.*, **250**, 1907–11
Carraway, R. and Leeman, Susan E. (1976). *J.biol.Chem.*, **251**, 7035–44
Chang, M. M. and Leeman, Susan, E. (1970). *J.biol.Chem.*, **245**, 4784–90
Chang, M. M., Leeman, Susan E., and Niall, H. D. (1971). *Nature new Biol.*, **232**, 86
Cuello, A. C., Polak, J. M. and Pearse, A. G. E. (1976). *Lancet*, ii, 1054–56
Dieterlen-Lièvre, F. and Beaupain, D. (1974). *Gen.comp.Endocr.*, **22**, 62–69
Dockray, G. J. (1976). *Nature*, **264**, 568–70
Dubois, M.-P. (1975). *Proc.natn.Acad.Sci. U.S.A.*, **72**, 1340–1343
Dubois, P. M., Paulin, C., Assan, R., and Dubois, M.-P. (1975). *Nature*, **256**, 731–32
Elde, R. P. and Parsons, J. A. (1975). *Am.J.Anat.*, **144**, 541–48
Eränkö, O. and Härkönen, M. (1963). *Acta physiol.scand.*, **58**, 285–86
Euler, U. S. von and Gaddum, J. H. (1931). *J.Physiol.*, *Lond.*, **72**, 74–87
Feyrter, F. (1938). *Uber diffuse endokrine epitheliale Organe*. Barth, Leipzig
Feyrter, F. (1952). *Acta neuroreg.*, **4**, 409–24
Feyrter, F. (1953). *Uber die peripheren endokrinen (parakrinen)* Drüsen des Menschen. Maudrich, Wien-Dusseldorf
Goldsmith, P. C., Rose, J. C., Arimura, A. and Ganong, W. F. (1975). *Endocrinology*, **97**, 1061–64
Heitz, Ph., Polak, J. M., Kasper, M., Timson, C. M. and Pearse, A. G. E. (1977). *Histochemistry*, **50**, 319–25
Heitz, Ph., Poalk, J. M., Timson, C. M. and Pearse, A. G. E. (1976). *Histochemistry*, **49**, 343–47
Hökfelt, T., Efendic, S., Johansson, O., Luft, R. and Arimura, A. (1974). *Brain Res.*, **80**, 165–69
Hökfelt, T., Efendic, S., Hellerström, C., Johansson, O., Luft, R. and Arimura, A. (1975a). *Acta endocr.*, **80**, suppl. 200, 1–41
Hökfelt, T., Johansson, O., Efendic, S., Luft, R., and Arimura, A. (1975b). *Experientia*, **31**, 852–54
Hökfelt, T., Kellerth, J. O., Nilsson, G. and Pernow, B. (1975c). *Science*, **190**, 889–90
Hökfelt, T., Meyerson, B., Nilsson, G., Pernow, B. and Sachs, C. (1976). *Brain Res.*, **104**, 181–86
Hughes, J., Smith, T. W., Kosterlitz, H. W., Fothergill, L. A., Morgan, B. A. and Morris, H. R. (1975). *Nature*, **258**, 577–79
Kanazawa, I. and Jessell, T. (1976). *Brain Res.*, **117**, 362–67
Kitabgi, P., Carraway, R. and Leeman, S. E. (1976). *J.biol.Chem.*, **251**, 7053–58
Larsson, L.-I., Fahrenkrug, J., Schaffalitzky de Muckadell, O., Sundler, F., Häkanson, R. (1976). *Proc.natn.Acad.Sci. U.S.A.*, **73**, 3197–200
Le Douarin, N. (1969). *Bull.biol.Fr.Belg.*, **103**, 435–52
Le Douarin, M. and Le Lièvre, C. (1970). *C.r.hebd.Séanc.Acad.Sci.*, *Paris*, *D*, **270**, 2857–60
Le Douarin, N., Le Lièvre, C. and Fontaine, J. (1972). *C.r.hebd.Séanc.Acad.Sci.*, *Paris*, *D*, **275**, 583–86
Le Douarin, N. and Teillet, M. A. (1971). *C.r.hebd.Séanc.Acad.Sci.*, *Paris*, *D*, **272**, 481–84
Matsuo, Y. and Seki, A. (1976). *Asian med. J.*, **19**, 589–630
Mutt, V. and Said, S. I. (1974). *Eur. J. Biochem.*, **42**, 581–89
Orci, L., Baetens, O., Rufener, C., Brown, M., Vale, W. and Guillemin, R. (1976). *Life Sci.*, **19**, 559–62
Palkovits, M., Brownstein, M. J., Arimura, A., Sato, H., Schally, A. V. and Kizer, J. S. (1976). *Brain Res.*, **109**, 430–34
Pearse, A. G. E. (1966a). *Nature*, **211**, 598–600
Pearse, A. G. E. (1966b). *Vet.Rec.*, **79**, 587–90
Pearse, A. G. E. (1968). *Proc. R. Soc. B.*, **170**, 71–80
Pearse, A. G. E. (1969). *J.Histochem.Cytochem.*, **17**, 303–13
Pearse, A. G. E. and Polak, J. M. (1971a). *Histochemie*, **27**, 96–102
Pearse, A. G. E. and Polak, J. M. (1971b). *Gut*, **12**, 783–88
Pearse, A. G. E. and Polak, J. M. (1975). *Histochemistry*, **41**, 373–75
Pearse, A. G. E., Polak, J. M., Rost, F. W. D., Fontaine, J., Le Lièvre, C., and Le Douarin, N. (1973). *Histochemie*, **34**, 191–203
Pelletier, G., Labrie, F., Arimura, A. and Schally, A. V. (1974). *Am. J. Anat.*, **140**, 445–50

Pictet, R. L., Rall, L. B., Phelps, P. and Rutter, W. J. (1976). *Science, 191,* 191–92

Polak, J. M., Pearse, A. G. E., Garaud, J.-C. and Bloom, S. R. (1974). *Gut, 15,* 720–24

Polak, J. M., Pearse, A. G. E., Grimelius, L., Bloom, S. R., and Arimura, A. (1975). *Lancet,* i, 1220–21

Polak, J. M., Sullivan, S. N., Bloom, S. R., Facer, P. and Pearse, A. G. E. (1977). *Lancet,* i, in press

Said, S. I. and Mutt, V. (1972). *Eur.J.Biochem.,* 28, 199–204

Said, S. I. and Rosenberg, R. N. (1976). *Science,* 192, 907–08

Scharrer, B. (1976). *Perspectives in Brain Research, Progress in Brain Research 45* (ed. M. A. Corner and D. F. Swaab), Elsevier, Amsterdam, pp. 125–137

Sétaló, G., Vigh, S., Schally, A. V., Arimura, A. and Flerkó, B. (1975). *Brain Res.,* 90, 352–56

Simantov, R. and Snyder, S. H. (1976). *Life Sci.,* 18, 781–88

Solcia, E., Pearse, A. G. E., Grube, D., Kobayashi, S., Bussolati, G., Creutzfeldt, W. and Gepts, W. (1973). *Rc.Gastroenterol.,* 5, 13–16

Takahashi, T. and Otsuka, M. (1975). *Brain Res.,* 87, 1–11

Teillet, M.-A. and Le Douarin, N. (1970). *C.r.hebd.Séanc.Acad.Sci., Paris, D,* 270, 3095–98

Vanderhaegen, J. J., Signeau, J. C. and Gepts, W. (1975). *Nature,* 257, 604–605

5

Electrophysiological analysis of peptide actions in neural tissue

Leo P. Renaud and Ante Padjen (Division of Neurology,
Montreal General Hospital and Department of Pharmacology
and Experimental Therapeutics, McGill University, Montreal, Quebec, Canada)

INTRODUCTION

The functional characterisation of endogenous neural substances demands the concerted efforts of investigators in numerous disciplines, a point that is adequately illustrated in the proceedings of this symposium. Over the past 20 years, electrophysiological techniques have provided a most valuable insight into the function of diverse excitable tissues, especially the nervous system. Because of its excellent time resolution, relatively simple technology and often unequivocal interpretation, electrophysiology has continued to be a most useful method for real-time monitoring of neuronal activity especially at the cell and/or membrane level. Electrophysiological techniques have been used to elucidate basic mechanisms of synaptic transmission, and continue to be of major importance in the sometimes frustrating analysis of complex neural organisation and behaviour. A large portion of our knowledge of central nervous system (CNS) pharmacology is derived from the analysis of drug effects on the electrical activity of the nervous system.

The presence of specific peptides in neural tissue has been known for several decades. However, recent progress in their isolation, structural characterisation and synthesis has allowed both their neural distribution to be verified quantitatively by radioimmunoassay and immunohistochemistry, and their effects to be tested on neural tissue. The latter is the task of the electrophysiologist, whose first approach is to examine whether peptides modify neuronal behaviour, and then to attempt to characterise the nature of the observed interactions. These items form the subject matter of this chapter.

ELECTROPHYSIOLOGICAL APPROACH TO PEPTIDE ACTION

There are many ways to introduce agents into the CNS—cf. Myers (1974)—but microiontophoresis is the major technique for application of agents into the extracellular space around single neurons, or directly into the cell. This technique has been described in several reviews (Curtis, 1964; Krnjević, 1971; Bloom, 1974; Kelly, 1975; Kelly *et al.*, 1975). Since iontophoresis is usually combined with extra- or intracellular recording, all of the electrophysiological advantages of neuronal identification (afferent and/or efferent connections) are preserved. However, the technique has a drawback in terms of calculating the quantity of drug administered. In the case of peptides, because of the difficulty of their release from micropipettes, one must question the value of a negative result when using the microiontophoretic technique.

Alternative approaches to drug administration have been developed using the technique of perfusion of isolated neuronal tissue. This offers several advantages: (1) the drug concentration is known; (2) the neuronal environment can be modified, for example ion substitutes; (3) one can exclude indirect responses by using magnesium and low levels of calcium (to block transmitter release) or tetrodotoxin (TTX) (to block the spread of activity). Another obvious advantage is that this approach obviates anaesthesia. Table 5.1 outlines other considerations in the choice of a system for use *in vivo* or *in vitro*. Although the level of organisation is obviously optimal *in vivo*, it is often well preserved in many systems used *in vitro* (for example, organ or tissue culture, and tissue slices). Depending on the complexity of the system, there are corresponding differences in the ability to identify the neuronal structures; *in vivo*, it is often possible to ascertain that the recorded potentials originate from a particular structure (axon, soma, dendrite), although visual observation *in vitro* offers the ideal situation.

Table 5.1 Some considerations involved in the choice of system

	In vivo	*In vitro*
Functional organisation	Optimal	Well preserved (organ culture, tissue slices)
Identification of structure	Possible but difficult	Usually simple
Drug delivery and recording		
iontophoresis (quantification difficult)	Localisation possible	Increased accessibility and resolution
perfusion (known concentration)	Diffuse	Modified environment
anaesthesia	Seldom avoidable	Not necessary
Proximity to membrane events	Monitoring difficult (intracellular)	Greatly facilitated
	Often indirect (extracellular)	Decreased indirect responses (Mg^{2+}, TTX)
Relevance for behaviour	Direct	Indirect

In electrophysiology, membrane related events (changes in potential or conductance) are the actual detectors of drug–receptor interaction and these can best be recorded with an intracellular electrode. With preparations *in vivo*, this is difficult to maintain over a long time. Consequently, extracellular recordings are most often used, although these are rather indirect witnesses of membrane changes since the only measurable parameter (frequency of action potentials) is a consequence of events at the membrane level.

Behavioural correlations are likely to have more significance when the drug is tested *in vivo*, especially in the chronically implanted animal where drug application can be compared with the behaviourally synchronised neural activity (see, for example, Foote *et al.*, 1975). The interpretation and significance of the data obviously must be considered in perspective with the type of tissue and experimental conditions that yield the results. For this reason, some of the studies *in vitro* may not be directly relevant to peptide actions in intact organisms where natural event-related and peptide-mediated processes may occur. Some invertebrate neurons (see Barker, 1976, 1977; Barker and Smith, 1977) offer distinct advantages for the analysis of peptide actions in neural tissue, and provide evidence of their possible mode of action. Some generalisations can be derived from these interesting studies but the relevance of these findings to vertebrate CNS physiology requires further investigation.

The account that follows is an attempt to highlight selective results of peptide actions obtained in several laboratories using two basic methodologies: the isolated frog or rat spinal cord, and microiontophoresis with extracellular or intracellular recordings. The methods for the isolated cord preparation are described in detail elsewhere (Konishi and Otsuka, 1974*a*; Otsuka and Konishi, 1975; Barker *et al.*, 1975).

PEPTIDE ACTION IN NEURAL TISSUE: SUBSTANCE P

The following sections illustrate the use of different electrophysiological approaches to determine the role of Substance P (SP) in the CNS. Most of these studies are from investigations in the spinal cord, although SP is known to be present in many other areas of the CNS.

In the spinal cord, SP is found in highest concentration in the substantia gelatinosa. When the dorsal root is ligated, spinal cord SP (but not γ-aminobutyric acid (GABA) or glutamate) content is decreased in the dorsal horn, whereas levels of SP (but not of glutamate) increase six-fold proximal to the ligation, suggesting that SP is transported centrally (Takahashi and Otsuka, 1975). The presence of SP-containing primary afferent fibres is strongly supported by the findings of Hökfelt and his colleagues (see Elde, chapter 2) using immunohistochemical techniques. In the rat, approximately 25 per cent of cell bodies of neurons in the dorsal root ganglion, and many fine nerve fibres in the substantia gelatinosa, demonstrate SP immunoreactivity; as might be expected, a significant reduction in SP-positive fibres follows dorsal root ligation (Hökfelt *et al.*, 1976).

Isolated frog spinal cord

Otsuka and his colleagues have done much to promote the concept of a peptide as a primary sensory neuro-transmitter. In 1972, using a logical but infrequently followed approach, they reported that extracts of bovine dorsal roots exerted a potent depolarising action on the spinal motoneurons of the isolated hemisected and perfused frog spinal cord (Otsuka *et al.*, 1972*a*). Subsequent studies indicated that this agent is identical to the undecapeptide SP isolated by Leeman and colleagues (Takahashi *et al.*, 1974) and that SP and related peptides with common C-terminal sequences all had a more potent excitation on isolated frog spinal motoneurons than L-glutamate (Konishi and Otsuka, 1974*a*). On an equimolar basis, SP is about 200 times, physalaemin 1500 times and eledoisin 2000 times more potent as a depolarising agent. These peptides have a direct depolarising action on frog moto-neurons, since their effects persist after blockade of synaptic transmission by calcium deficient media or tetrodotoxin.

Nicoll (1976) has confirmed and extended some pharmacological observations of Otsuka and colleagues. SP is approximately 5 times as potent as glutamate and exerts a very prolonged direct depolarising action on frog motoneurons. It has very little effect on either glutamate- or GABA-evoked responses, suggesting that it does not interact or modulate the action of these putative amino acid neurotransmitters. Intracellular recordings show that SP-induced depolarisation is associated with a reduction in membrane resistance (although this is less marked than the resistance change evoked by L-glutamate) suggesting an increase in ionic conductance, pre-sumably for sodium ions. Since isolated ventral roots are not affected by SP and glutamate, one can conclude that receptors for these substances are located on the soma–dendrite membrane of motoneurons.

Isolated rat spinal cord
Provided that appropriate precautions are taken, a similar experimental technique can be successfully applied to the isolated hemisected spinal cord of the newborn rat (Otsuka and Konishi, 1974). In this preparation, SP again is a more potent

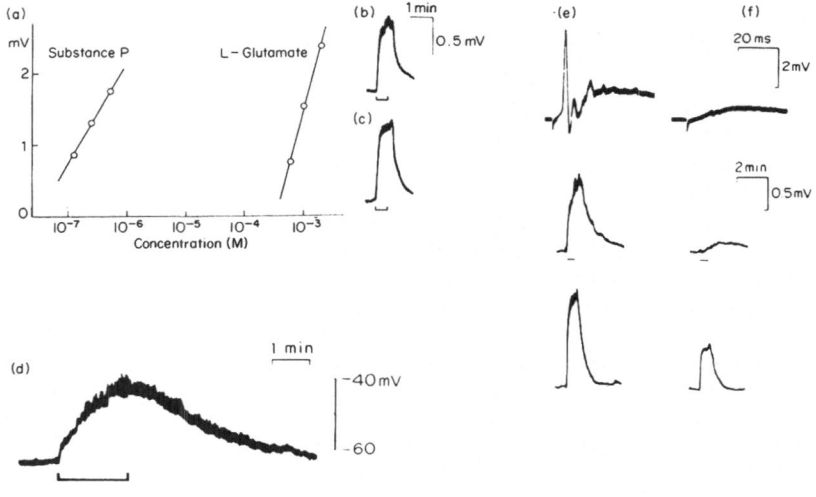

Figure 5.1   Effect of Substance P (SP) on isolated spinal cord of the newborn rat. (a) Dose-response curves compare the potency of SP and L-glutamate-evoked depolarisation of L4 ventral root. (b, c) Actual recorded ventral root depolarising potentials durmg application (horizontal line) of SP (2×10^{-7} M) and L-glutamate (7×10^{-7} M), respectively. (d) Intra-cellular recording from a motoneuron illustrating depolarisation evoked by SP (5×10^{-5} M) applied during the time indicated by the horizontal line. (e) From the top downwards, the records indicate ventral root responses evoked by stimulation of the dorsal root, and perfusion with SP (2×10^{-7} M) and L-glutamate (10^{-3} M). (f) Same responses as in (e), recorded after addition of Lioresal (5×10^{-6} M). Note that the dorsal root evoked response is absent, the response to SP is almost abolished and the response to L-glutamate is markedly reduced. (Otsuka and Konishi, 1975, © 1975 Cold Spring Harbor Laboratory)

depolarising agent on spinal motoneurons than L-glutamate, and other amino acids including L-aspartate, glycine and GABA. As in the frog, the depolarising action of SP persists after blockade of synaptic transmission (figure 5.1a), providing evidence for a direct effect on motoneurons (Konishi and Otsuka, 1974b). Intracellular recordings show that SP evokes a prolonged depolarisation associated with repetitive firing of motoneurons (figure 5.1b).

The isolated perfused frog or rat spinal cord is very suitable for comparing and testing the actions of SP analogues. In the rat preparation the deca- nona- and octapeptide analogues of SP exert a motoneuron-depolarising action equally as effective as the undecapeptide SP, whereas the penta- tetra- and tripeptide analogues are virtually inactive (Otsuka and Konishi, 1974). Since the hepta- and hexapeptide analogues were 8–10 times more potent than the undecapeptide, it is possible that the undecapeptide is converted to the biologically more active shorter peptides at central receptor sites. Little more can be added at present since the metabolism and/or degradation of SP in neural tissue is yet to be determined.

Microiontophoresis: extracellular observations
SP has been applied by microiontophoresis to neurons in several sites in the CNS: the cuneate nucleus (Krnjević and Morris, 1974), spinal interneurons and motoneurons (Henry *et al.*, 1974, 1975; Henry, 1976), substantia nigra (Davies and Dray, 1976) mesencephalic reticular formation (Walker *et al.*, 1976), cerebral cortex (Phillis and Limacher, 1974a; Henry *et al.*, 1975; Phillis, 1976) and amygdala (Ben Ari, personal communication).

Excitant actions
In general, spontaneously active central neurons are more responsive than quiescent neurons to the application of SP. Using iontophoretic currents of 25–300 nA, an increase in neuronal excitability is the most common response pattern. This excitation differs from the brisk on–off excitatory pattern associated with microiontophoresis of L-glutamate, or after stimulation of cutaneous afferent fibres. With SP there is a characteristic delayed onset of 10–30 s, and a slow build-up of cell discharge frequency during which there is some decrease in the sensitivity of neurons to L-glutamate (figure 5.2a). The response persists for several tens of seconds following the application of SP. This pattern of SP action closely resembles the findings in the frog and rat spinal cord.

Several other polypeptides related to SP have been tested on central neurons. In the cerebral cortex SP is the most potent excitatory polypeptide tested, followed by physalaemin, bradykinin, eledoisin and bombesin in that order (Phillis and Limacher, 1974a, b). The more complex polypeptide insulin has only a weak excitatory effect on cortical neurons (Phillis, 1977).

Depressant actions
A small percentage of neurons in the cuneate nucleus, spinal cord and substantia nigra (Henry *et al.*, 1975; Davies and Dray, 1976) are depressed by SP. This is manifested in different ways. In cells that are otherwise unresponsive to SP, the excitatory action of glutamate may be reduced by application of SP (figure 5.2). In Renshaw cells and certain other central neurons, SP evokes a reduction in

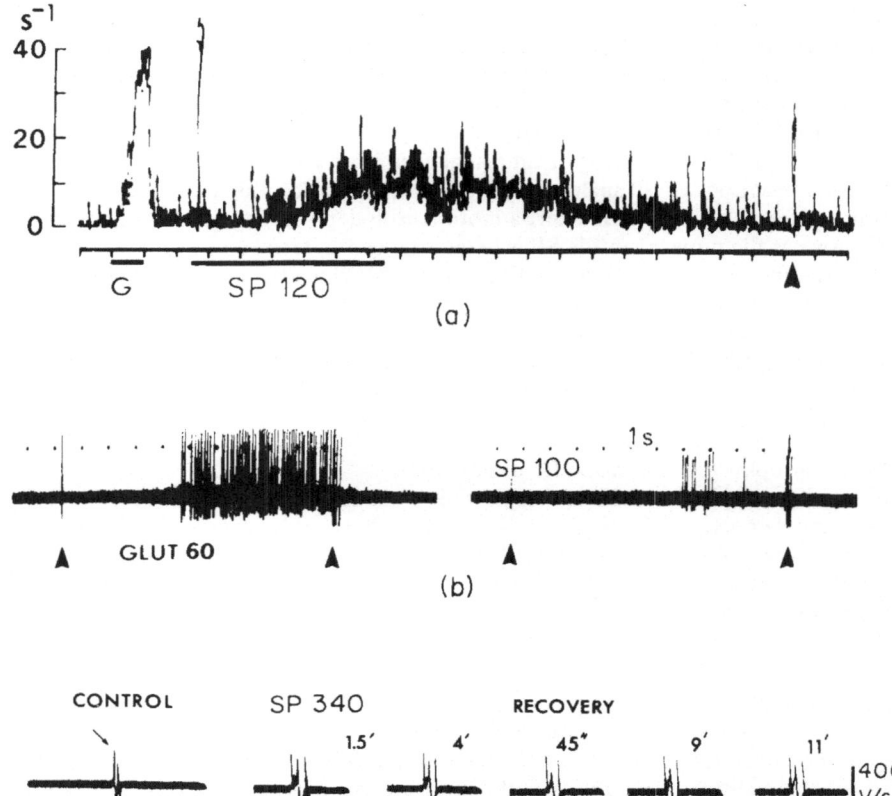

Figure 5.2   Influence of SP iontophoresis on central neurons. (a) Ratemeter record from a
spinal cord dorsal horn neuron of the Dial-anaesthetised cat. This cell was activated by microiontophoretic application of L-glutamate (G, 60 nA) and substance P (SP, 120 nA), and by an
air jet applied to the peripheral receptor field (arrowhead). Note the different time courses of
action in each instance (Henry, Krnjević and Morris, 1975). (b) On the left, action potentials
recorded from a dorsal horn neuron activated by microiontophoresis of L-glutamate (60 nA)
applied between the arrowheads. On the right, during simultaneous application of SP (100 nA)
note the decrease in excitability, associated with a depression in spike amplitude (Henry,
Krnjević and Morris, 1975). (c) Intracellular records from a cat lumbar motoneuron during
extracellular SP iontophoresis (340 nA). From the bottom upwards, the traces illustrate the
antidromic spike at low gain (D.C. recording, calibration on the left) and at higher gain, and the
first derivative of the spike (peak marked by the arrow). Note the depolarisation (indicated by
the upward shift in the lower traces), spike depression and prolongation during SP application
(second and third columns) and subsequent partial recovery after 11 min (Krnjević, 1977 in
Substance P (ed. U.S. von Euler and B. Pernow) Raven Press, New York, ©1977 Raven Press

spontaneous activity. A common feature of all SP-evoked depressant responses is a tendency for a reversible reduction in spike amplitude of the affected neuron, suggestive of a depolarising rather than a hyper-polarising type of block (Krnjević, 1977).

Substance P and acetylcholine

In certain regions (cortex, substantia nigra) many of the neurons that are excited by SP also demonstrate a muscarinic pattern of excitation to acetylcholine. This observation suggests possible relationship between these two agents. For example, SP might be considered to promote the release of acetylcholine through an action on presynaptic cholinergic fibres, or interact with postsynaptic acetylcholine receptors. However, in most instances where acetylcholine causes a muscarinic excitation of central neurons, this is no clear correlation between acetylcholine sensitivity and SP sensitivity. In the cortex, for example, SP is a more effective excitant with a more prolonged time course of action compared with that of acetylcholine; cells are responsive to SP that do not respond to acetylcholine. The receptor sites for acetylcholine and SP appear independent since iontophoretic application of atropine reduces or abolishes acetylcholine-induced excitation without affecting SP evoked responses (Phillis and Limacher, 1974*a*).

A different relationship seems to exist between acetylcholine and SP in the spinal cord. By itself, SP exerts a slight depressant effect on the spontaneous or glutamate evoked discharges of Renshaw cells (Krnjević, 1977), but there is a striking and seemingly selective depression of acetylcholine evoked excitation of Renshaw cells (Ryall *et al.*, personal communication; Lekić and Krnjević, personal communication). Thus there may be some interaction between SP and acetylcholine at nicotinic cholinergic receptor sites.

Microiontophoresis: intracellular observations

Intracellular records during the extracellular iontophoresis of SP have been obtained from lumbosacral neurons of Dial anaesthetised cats (Krnjević, 1977; Krnjević, Puil and van Meter, unpublished observations). These findings are in general agreement with the extracellular results reported above, and indicate a postsynaptic site of action. Application of SP with currents of 100–340 nA generally causes a membrane depolarisation with a slow onset associated with an increase in membrane resistance (decrease in membrane conductance) over a prolonged time course. Low ejection currents enhance the amplitude of spontaneous excitatory postsynaptic potentials (EPSPs); larger applications are associated with a marked depolarisation leading to the generation of spikes from subthreshold EPSPs. The latter response is usually accompanied by a decrease in spike amplitude, a decrease in the rate of the rise time and falling phase of the action potential, and an increase in post-spike hyperpolarisation (figure 5.2). The reversal level of these excitant effects is more negative than the resting membrane potential, consistent with an interference with potassium or chloride conductance. The results therefore indicate that a SP-induced depolarisation of the postsynaptic membrane may arise through a reduction in membrane potassium permeability. In this respect, the results on the frog spinal cord differ since Nicoll (1976) has found an increase in membrane conductance.

Microiontophoresis: primary afferent terminals
Preliminary data obtained from studies on primary afferent terminal excitability in the cuneate nucleus support the notion that SP has no significant role on primary afferent terminal depolarisation (Krnjević, Morris and Fox, unpublished observations; Krnjević, 1977). When applied by microiontophoresis into the cuneate nucleus with currents of 200 nA, SP facilitates input–output transmission through the nucleus (Krnjević and Morris, 1976). This is associated with a mild depressant effect on the terminals, whereas applications with currents of 600 nA depressed both transmission and terminal excitability.

Mauthner fibre–giant fibre synapse
The Mauthner fibre–giant fibre synapse of the hatchetfish is a unique vertebrate cholinergic motor relay synapse where both presynaptic and postsynaptic events can be simultaneously recorded close to the synaptic areas (Auerbach and Bennett, 1961; Spira *et al.*, 1970; Highstein and Bennett, 1975). In this preparation, both presynaptic and postsynaptic actions of SP have been reported. SP does not change the resting potential or input resistance of either the presynaptic or postsynaptic fibres, but measurements of the dynamic aspects of this synapse revealed two distinct effects of SP. At *low* concentration (10^{-6} M) the effects are mostly postsynaptic; there is a rapid reduction in the amplitude of miniature postsynaptic potentials (MPSPs), and EPSPs, and a slight decrease in the frequency of the MPSPs. At *higher* concentrations (10^{-4} M) definite presynaptic effects are observed, that is, a reduction in the MPSP frequency and quantal EPSP content, (Steinacker and Highstein, 1976). These observations contrast with those reported for SP in other systems, since the sum effect of SP at this synapse is to decrease synaptic efficiency. This action, however, could explain some depressant effects of SP reported during microiontophoretic studies in mammals. However, the data obtained in this preparation may not be entirely relevant to the effects of SP in other areas of the vertebrate CNS.

Substance P antagonist–experience with baclofen
Any investigation of neurotransmitters is simplified if specific pharmacological antagonists to the action of the putative transmitter agents are available (cf. Werman, 1966). Unfortunately, great difficulty has been encountered in a search for specific antagonists for amino acids and peptides in particular. In 1975, Saito *et al.* announced that Lioresal exhibited a specific antagonism of SP. Baclofen (Lioresal) is the β-chlorophenyl derivative of GABA. It is known as a centrally active muscle relaxant that relieves certain forms of spasticity in man (Knuttsson *et al.*, 1973; McLellan, 1973). In the cat spinal cord, baclofen was reported to depress selectively mono- and polysynaptic reflexes without altering inhibitory postsynaptic potentials or other electrical properties of spinal motoneurons (Pierau and Zimmermann, 1973). This suggested an effect mainly on primary afferent transmission rather than on postsynaptic neurons (but CF. Davidoff and Sears, 1974; Krnjević, 1977). In the isolated rat spinal cord, baclofen blocks mono- and polysynaptic reflexes, and selectively decreases the SP-evoked depolarisation of motoneurons (figure 5.2c). However, baclofen has a non-selective postsynaptic depressant effect on central neuronal excitability, and it affects primary afferent terminal excitability. Therefore it is unlikely that baclofen is a *specific* SP antagon-

ist (Curtis *et al.*, 1974; Davidoff and Sears, 1974; Davies and Watkins, 1974; Davies and Dray, 1976; Fotherby *et al.*, 1976; Henry and Ben-Ari, 1976; Phillis, 1976; Krnjević, 1977).

Substance P and primary afferent transmission

Beside the distribution and immunohistochemical localisation of SP in the spinal cord, the evidence that has accumulated to date only indirectly suggests a primary afferent neurotransmitter role for SP. Records from dorsal horn neurons and interneurons of the spine indicate that cutaneous or proprioceptive stimuli usually evoke a brisk response from these cells, similar to the response of these cells to L-glutamate applied by microiontophoresis. This analogy has led to the suggestion that glutamate may be the primary afferent neurotransmitter agent. SP and related peptides also evoke excitation of spinal neurons, but with a pattern of excitation that is relatively slow in onset and prolonged in action, unlike that evoked by natural stimuli. On the other hand, it is important to note that nociceptive spinal neurons do respond to natural stimuli with a slow build-up of activity followed by a prolonged after discharge (Calvillo *et al.*, 1974; Handwerker *et al.*, 1975; Henry, 1976). Furthermore, these neurons are consistently sensitive to SP (Henry, 1976, 1977). This identity of action favours the involvement of SP either as a mediator or modulator of transmission in spinal nociceptive pathways. This would agree with the proposal advanced by Iversen (chapter 9) to explain the observed interaction between opioid peptides and SP. It is far from certain whether SP can be secreted alone, or in conjunction with a more conventional neurotransmitter agent. The recent findings of Pickel *et al.* (1977) indicate that SP storage sites within nerve terminals are independent of the numerous small vesicles present in the axon terminals. Thus, nerve terminals may contain and possible release more than one neurotransmitter (Burnstock, 1976). The available data do not allow one to conclude with any certainty whether SP initiates synaptic events or modulates the activity of some other neurotransmitter.

SOMATOSTATIN

The isolation and structural characterisation of somatostatin (Brazeau *et al.*, 1973) and development of specific radioimmunoassays has led to the discovery that this peptide is distributed not only in hypothalamus but also in different parts of the nervous system as well as in non-neural tissues (see chapters 2 and 3 by Elde and Brownstein, respectively). In the nervous system, the high content and localisation of somatostatin in the dorsal root entry zone of the spinal cord and in the hypothalamus suggests at least two areas in which attention should be focused for investigating the role of somatostatin. Two different electrophysiological methods have been applied to the investigation of somatostatin action on neural tissue (a) studies on the isolated frog spinal cord *in vitro* and (b) microiontophoretic studies in rat, cat and rabbit *in vivo*. The following is an account of the available results.

Isolated frog spinal cord

Somatostatin is present in considerably larger quantities in the amphibian nervous system than in the mammalian nervous system (Vale *et al.*, 1976). In the frog spinal cord, somatostatin levels in the dorsal quadrant (334 ± 33 fmol/mg; $n = 6$) are 60

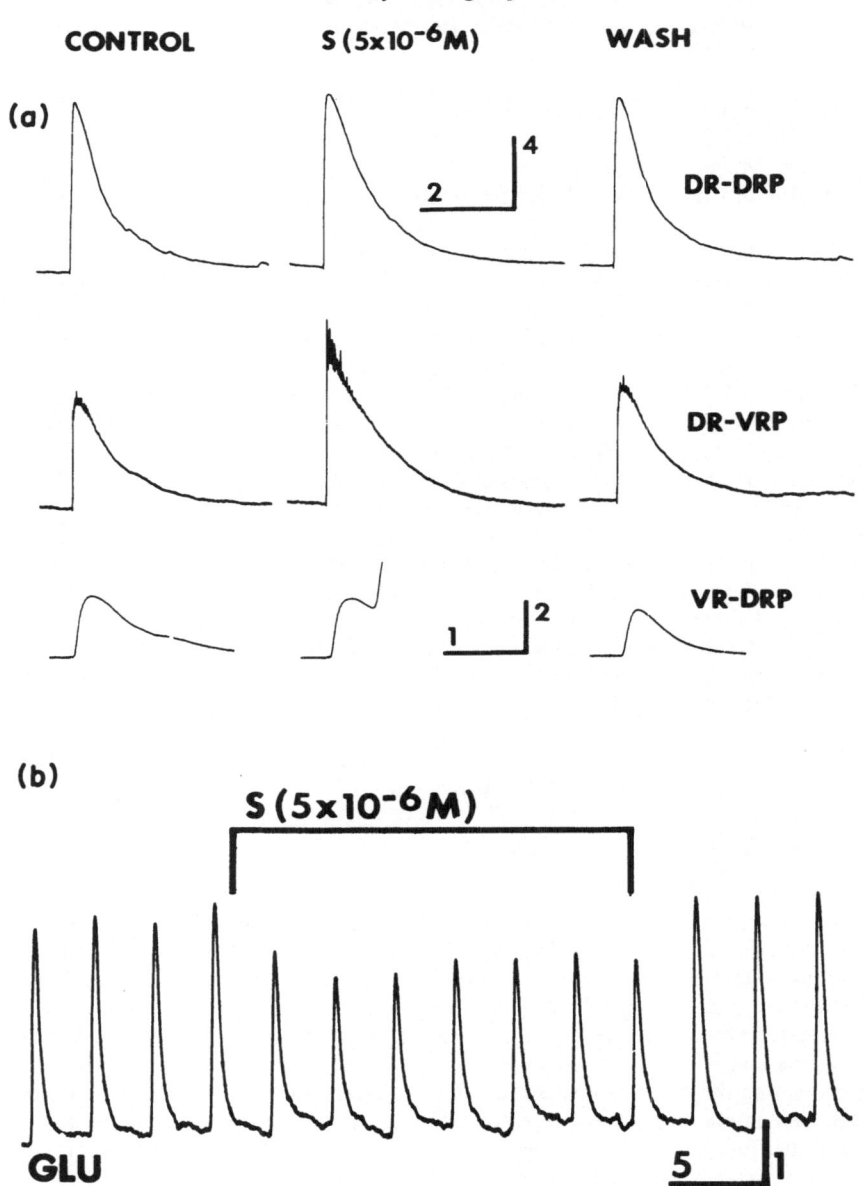

Figure 5.3 Effect of somatostatin on frog spinal cord. (a) Effect of somatostatin on synaptic potentials recorded simultaneously in the same preparation using sucrose gap technique (for methodology see Barker *et al.*, 1975). Vertical columns: left, control responses; middle, 30 min after beginning of perfusion with somatostatin (s, 5×10^{-6} M); right, 1 hour after removal of somatostatin. Calibration mark in mV and s. (b) Effect of depolarising glutamate responses on ventral roots recorded by sucrose gap in the presence of 10 mM magnesium and 0.5 mM calcium. Calibration: 5 min, 1 mV (Padjen, unpublished observations).

per cent higher than in the ventral quadrant (208 ± 16 fmol/mg) (Epelbaum, Padjen and Renaud, unpublished observations). This finding is in general agreement with the immunohistochemical findings in mammalian spinal cord (cf. Elde, chapter 2).

Application of somatostatin to the hemisected frog spinal cord, while recording root potentials with the sucrose gap technique (for methodology see Barker *et al.*, 1975), produces several different effects (Padjen, unpublished observations). The most striking observation is a dose-dependent increase in the ventral root potential obtained by dorsal root stimulation (DR-VRP; see figure 5.3). This effect is observed with threshold concentrations of 2×10^{-6} M (figure 5.3a) and increases three- to four-fold at concentrations of 10^{-5} M. At the same time there is no change in the baseline, and dorsal root potentials are either marginally affected (DRP, evoked by dorsal root stimulus, DR-DRP, figure 5.3) or not at all (ventral root evoked dorsal root potential, VR-DRP, figure 5.3). This increase in DR-VRP becomes apparent 10–20 min after the onset of perfusion with somatostatin, and outlasts the application by 30–60 min. Furthermore, this response shows no desensitisation or tachyphylaxis. Preliminary studies with somatostatin analogues have shown a direct correlation between neural actions and those involved in growth hormone suppression in monolayer pituitary culture. Thus the des-carboxy-des-amino somatostatin analogue is as active as cyclic somatostatin in increasing DR-VRP, whereas the retro-somatostatin inactive analogue has no effect in concentrations up to 10^{-5} M.

To characterise further the mechanism of action of somatostatin, experiments were performed in the presence of high magnesium and low calcium, or tetrodotoxin. In such conditions, somatostatin produced a small and variable but immediate hyperpolarisation which shows desensitisation and tachyphylaxis. In parallel with this direct action of somatostatin, responses to glutamate, but not to GABA, were selectively depressed in both ventral and dorsal roots.

At present, it is difficult to reconcile this direct depressant action of somatostatin on motoneurons and primary afferents, with its facilitatory action on synaptic transmission although the former effect has partially desensitised by the time the DR-VRP potential begins to display augmentation. Since there is no evidence of an increase in postsynaptic excitability of motoneurons, primary afferent terminals, or interneurons, a presynaptic site of action of somatostatin may be involved. It is of interest that calcium uptake into synaptosomes is facilitated by somatostatin (Tan *et al.*, 1977). Such an effect would have to be rather site selective to the DR-VRP pathway. In fact somatostatin *is* localised in primary afferent fibres in mammals (cf. Elde, chapter 2).

Microiontophoresis of somatostatin
Release in vitro
Somatostatin will migrate as a cation when dissolved in solutions at pH 4.0–6.5. In an effort to obtain quantitative data, the release of somatostatin from micropipettes containing 5–10 mM solutions in distilled water or 0.2 N acetic acid (to prevent degradation) has been examined *in vitro*, using both a bioassay (monolayer pituitary culture) and a radioimmunoassay (sensitive to 4–10 pg/ml). In common with similar studies on SP (Krnjević and Morris; 1974) on LHRH (Kelly and Moss, 1976) and on TRH (Wilbur *et al.*, 1976) this has proven to be a rather frustrating experience. Using micropipettes prepared according to the same dimensions as

those used in the experiments *in vivo*, some electrodes released somatostatin in an approximately linear fashion at the rate of 0.1-1.0 fmol/s. The test conditions *in vitro* were more drastic than the short (10-30 s) applications used during experiments *in vivo*; many electrodes demonstrated a tendency to block after very short periods of current application. This tendency to block possibly explains some of the negative results obtained with microiontophoresis, and indicates that negative results *in vivo* may be meaningless. It also explains the difficulties encountered with reproduction of similar results during consecutive applications to any one neuron.

Studies in vivo
Somatostatin has been applied by microiontophoresis to cells at several levels of the neural axis. In rats anaesthetised with pentobarbitone or urethane we have observed a predominantly depressant effect on 50 per cent of the neurons tested in the cerebral cortex, and up to 80 per cent of tested neurons in the cerebellar cortex and hypothalamus (Renaud *et al.* 1975, 1976). The response of somatostatin-sensitive neurons is usually brisk in onset, readily reversible and proportional to the amount of ejecting current. The decrease in excitability, whether on spontaneously active cells or those whose activity is maintained by administration of L-glutamate, is often associated with an increase in the size of the action potential (figure 5.4a). This may indicate an underlying hyperpolarisation. Similar depressant responses during somatostatin microiontophoresis have been reported from cat spinal cord nociceptive neurons (Miletić *et al.*, 1977; Randić and Miletić, 1977). Somatostatin analogues that have similar or reduced biological activity also are active in the CNS (figure 5.5a), whereas somatostatin analogues that are inactive in pituitary cell cultures (for example, retro-somatostatin) also appear to be inactive in microiontophoresis experiments (Renaud, unpublished observations).

In the cerebral cortex of the restrained unanaesthetised rabbit adapted for chronic recordings, Havlicek *et al.*, (1977) have reported opposite results during somatostatin microiontophoresis. In this preparation, somatostatin evokes only an increase in neuronal excitability, similar to that observed with L-glutamate (figure 5.4b). Possibly the absence of anaesthetics and/or species differences are important in any explanation of this striking contrast in experimental observations.

THYROTROPIN RELEASING HORMONE

The peptide (pyroGlu-His-Pro-NH_2) is a phylogenetically ancient molecule. Its presence in both hypothalamic and extrahypothalamic brain regions, and in the brain of species lacking a thyroid gland indicates a role for brain TRH unrelated to its action on TSH secretion (Jackson and Reichlin, 1974). Thus it is quite possible that TRH is involved in neurotransmission. Recent evidence supports this opinion. TRH is found in neuronal structures (Hökfelt *et al.*, 1975*a, b*) and after subcellular fractionation TRH is found in synaptosomes and synaptic vesicles (Winokur *et al.*, 1977) from which TRH can be released by appropriate stimulation (Bennett *et al.*, 1975). The brain contains stereo-selective high affinity binding sites for TRH (Burt and Snyder, 1975) and TRH can provoke a variety of behavioural phenomena, considered to result from a direct action of TRH on neural tissue (Prange *et al.*, 1977).

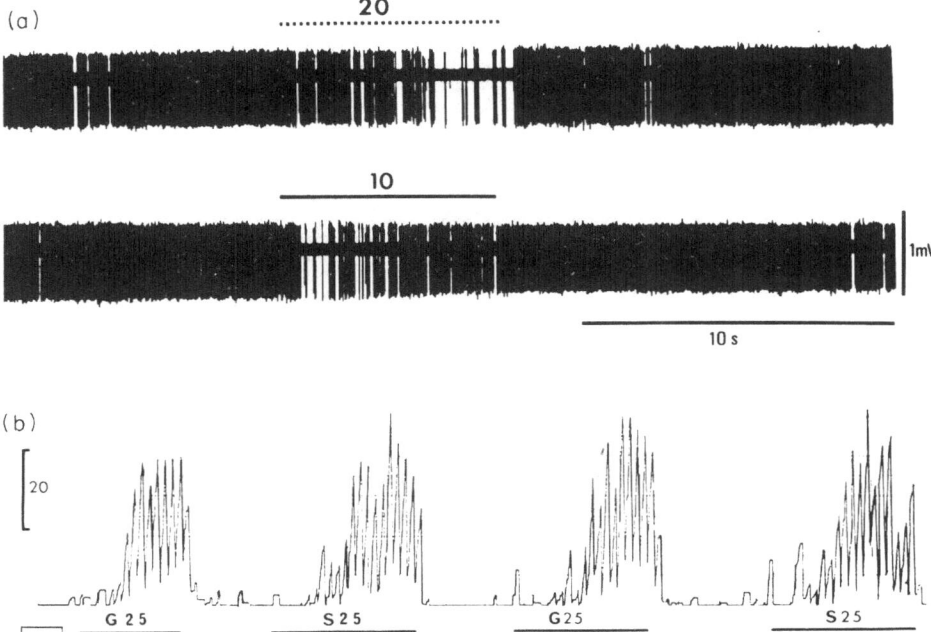

Figure 5.4 Effect of somatostatin on neuronal excitability. (a) The upper and lower continuous oscilloscope traces illustrate spike discharges from a spontaneously active Purkinje neuron in the cerebellar cortex of a pentobarbital-anaesthetised rat. The two traces illustrate the decrease in action potential frequency associated with microiontophoretic application (horizontal bars, numbers refer to nA) of cyclic somatostatin and an active somatostatin analogue AY 25,111 respectively (Renaud, unpublished observations). (b) Ratemeter records obtained from the cerebral cortex of an unanaesthetised rabbit, illustrating the increase in firing frequency of a single neuron during microiontophoresis of L-glutamate (G, 25 nA) and somatostatin (S, 25 nA). The vertical bar indicates firing frequency of 20 Hz (Ioffe, Havilicek, Friesen and Chernick, unpublished observations).

Electrophysiological analyses indicate that TRH and certain TRH analogues are biologically active in neural tissue. Two approaches, isolated frog spinal cord sucrose gap recording and microiontophoresis have been used in an attempt to characterise these actions.

Isolated frog spinal cord
Sucrose gap recording reveals that TRH has a small depolarising action on frog motoneurons (Nicoll, 1977). This depolarising action of TRH differs from that of glutamate in that it does not surpass threshold levels for induction of spontaneous firing. It exerts a background facilitory action on which subliminal excitatory potentials induce action potentials. The depolarising action is of long duration (figure 5.5a) and often shows tachyphylaxis. Blockade of synaptic transmission with magnesium or tetrodotoxin reduces the size of the TRH evoked response by 10–40 per cent, suggesting that there are both direct and indirect components to

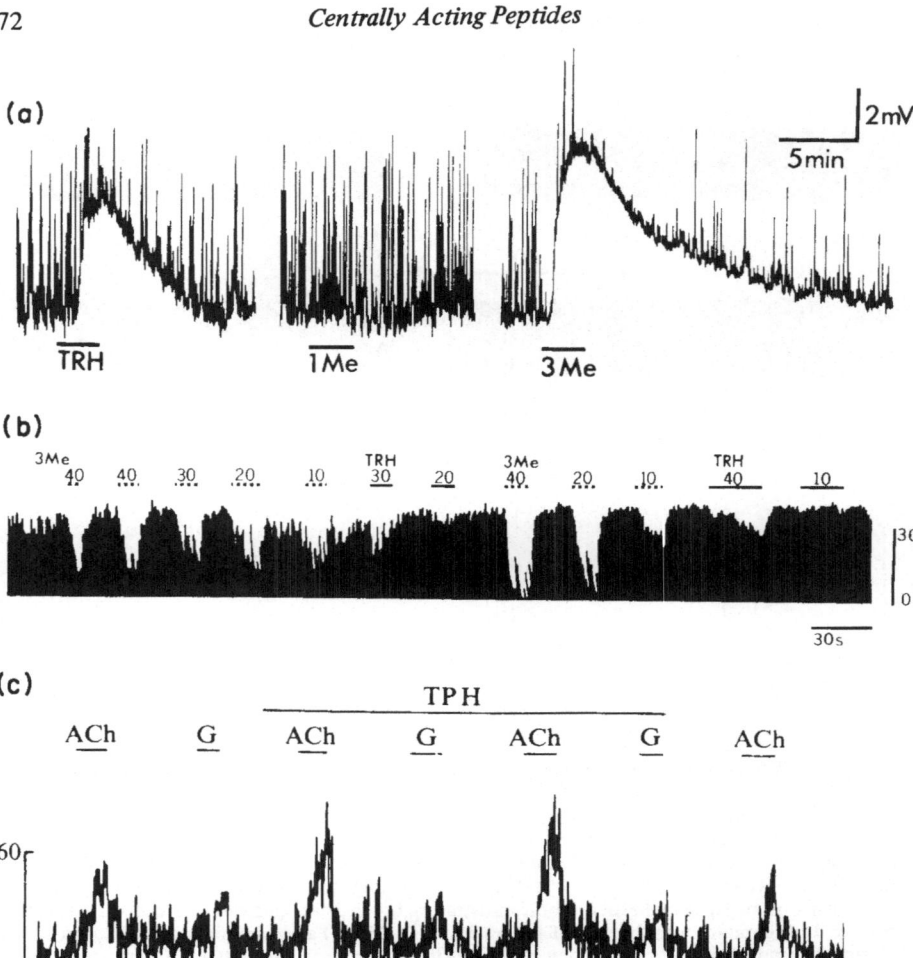

Figure 5.5 TRH influence on neural tissue. (a) Ventral root responses in isolated frog spinal cord recorded with sucrose gap during perfusion with solutions containing TRH (5×10^{-4} M), 1-methyl-histidine-TRH (1Me 5×10^{-4} M) and 3-methyl-histidine-TRH (3Me, 5×10^{-5} M) in normal Ringers solution. Approximately 45 min elapses between each of the responses in order to minimise desensitisation (Nicoll, unpublished observations). (b) Ratemeter records from a spontaneously active cerebellar cortical neuron in a pentobarbital-anaesthetised rat illustrates the depressant effects of TRH, and its more potent analogue 3-methyl-histidine-TRH (3Me) when applied by microiontophoresis during the time intervals indicated by the horizontal solid and interrupted bars. The vertical bar on the right indicates the height of an individual column containing 30 spikes (lower panel of Figure 5 in Renaud, Martin and Brazeau, 1976). (c) Ratemeter record of the activity of a cortical neuron in a pentobarbital-anaesthetised rat during microiontophoresis (horizontal bars) illustrating the excitatory action of acetylcholine, (ACh, 20 nA) and glutamate (G, 20 nA) and selective enhancement of acetylcholine but not glutamate excitation during simultaneous microiontophoresis of TRH, at a current of 25 nA (Yarbrough, 1976).

the depolarisation. Intracellular recordings from motoneurons confirm the depolarising action of TRH and reveal a small increase in membrane conductance, possibly to sodium ions.

Further studies with TRH analogues indicate that the [1-methyl-His]-TRH analogue is devoid of activity, while the [3-methyl-His]-TRH analogue is more potent than native TRH in inducing these effects on spinal cord motoneurons (figure 5.5a) in agreement with the microiontophoretic data.

Microiontophoretic studies

Microiontophoretic application of TRH from solutions of 5–10 mM alters the excitability of a certain percentage of neurons in the cuneate nucleus, cerebral and cerebellar cortices, hypothalamus and preoptic area (Dyer and Dyball, 1974; Renaud and Martin, 1975a, b; Steiner, 1975; Moss, 1977; Moss et al., 1977). In general, TRH-sensitive neurons are more frequently encountered in the hypothalamus than elsewhere. Characteristic of the action of TRH on the majority of these responsive neurons is a relatively brief onset of action and ready reversibility. Where depressions in firing frequency are observed (figure 5.5b) these occur with relatively low iontophoretic currents (5–20 nA) on cells that are either spontaneously active or whose activity is maintained by application of L-glutamate. Repeated applications of the peptide did not appear to be associated with desensitisation. Simultaneous application of the amino acid antagonists bicuculline and strychnine failed to alter the TRH induced responses (Renaud and Martin, 1975b) although these same agents abolished the response to iontophoretic application of GABA and glycine, respectively (Kelly and Renaud, 1973). TRH enchances the excitability of a small percentage of central neurons (Dyer and Dyball, 1974; Moss, 1977) with an action that also occurs over short time periods and seldom outlasts the period of application.

TRH antagonises the barbiturate-induced sleeping time in experimental animals (Breese et al., 1975) and TRH can potentiate acetylcholine (figure 5.5c) and carbachol, but not glutamate-evoked excitation of cerebral cortical neurons (Yarbrough, 1976). These actions can be antagonised by atropine, suggesting that TRH may interact with cholinergic mechanisms in a manner yet to be defined.

Preliminary microiontophoretic studies conducted with TRH analogues indicate that the inactive DDD-TRH analogue has no influence on neuronal excitability, whereas the more potent [3-methyl-His]-TRH analogue is more effective than native TRH in its ability to decrease central neuronal excitability (figure 5.5b; also Renaud et al., 1976).

Inhibitory tripeptide in spinal cord

It has recently been reported (Lote et al., 1976) that the cat spinal cord contains an endogenous peptide His–Gly–Lys. When applied by microiontophoresis, this peptide evokes a powerful and selective depression in spontaneous or glutamate-evoked activity of approximately 40 per cent of dorsal horn spinal neurons, and 50 per cent of medial brainstem neurons. In each instance the effects occur with low microiontophoretic currents, are brisk in onset and are rapidly reversed when the application ends, that is, similar to the depressant responses observed with TRH, LHRH and somatostatin (cf. Renaud et al., 1975, 1976). The role of this new peptide remains undefined. Nevertheless, there are already clear indications for regional specificity, since this peptide is relatively inactive when applied to most cuneate neurons (affecting less than 10 per cent of cells) whereas more than 50 per cent of medial brainstem neurons are depressed.

LUTEINISING HORMONE RELEASING HORMONE

The distribution of LHRH in the CNS is quite restricted compared with that of TRH and somatostatin, and confined mainly to the mediobasal hypothalamus preoptic areas (Brownstein, chapter 3). In addition to its role in the regulation of gonadotropin release, LHRH is capable of inducing reproductive behaviour, apparently through a direct action on the CNS (Moss and McCann, 1973; Pfaff, 1973). Several laboratories have attempted to ascertain how LHRH might influence neural tissue, using electrophysiological methods.

Microiontophoretic studies

Despite its restricted distribution, LHRH can enhance or depress the activity of a certain population of cells not only within the hypothalamus but also in the preoptic area, septum, cerebral and cerebellar cortices (Dyer and Dyball, 1974; Kawakami and Sakuma, 1974; Renaud *et al.*, 1975, 1976; Moss, 1976, 1977). Most of the peptide-sensitive neurons tend to be located in the hypothalamus. Excitation is observed in particular from cells in the arcuate nucleus that do not project to the region of the median eminence, that is, are non-tuberoinfundibular neurons (Moss, 1977). However, a significant number of cells in the ventromedial nucleus exhibit decreased activity during LHRH microiontophoresis (Renaud *et al.*, 1975, 1976). All the responses tend to be rapid in onset, proportional to the iontophoretic current and are rapidly reversed. LHRH analogues are also effective when tested on LHRH-sensitive neurons (Renaud *et al.*, 1976; Moss, 1977; Moss *et al.*, 1977) although it appears that the structure–activity potency ratios established for LH release from the pituitary are not necessarily valid when applied to neural tissue (cf. Renaud *et al.*, 1976).

ANGIOTENSIN II

The circulating octapeptide angiotensin II has both peripheral and central actions. In the periphery, this peptide constricts smooth muscle and promotes aldosterone secretion. In the CNS, angiotensin II is thought to participate in the regulation of thirst and cardiovascular homeostasis (Severs and Daniels–Severs, 1973; Fitzsimons, 1975; Severs and Summy-Long, 1975). The CNS contains specific angiotensin II binding sites with the highest binding in the hypothalamus, midbrain and thalamic regions (McLean *et al.*, 1975). Many but not all of these receptor sites are present in and around circumventricular structures, for example, the area postrema and subfornical organ. The significance of angiotensin II receptors has been in doubt because there is little evidence that circulating angiotensin II penetrates the blood-brain barrier (Osborne *et al.*, 1971; Bolicer and Loew, 1971). It has generally been assumed that altered blood–brain barrier permeability in the vicinity of the circumventricular structures permits circulating angiotensin II to exert its influence on neural tissues. The subfornical organ, for example, is considered to participate in drinking behaviour (Simpson and Routtenberg, 1973) and electrophysiological analysis has provided useful supporting information regarding the role of the subfornical organ. Angiotensin II, applied either to the IIIrd venticle wall or directly microiontophoresed on to neurons in the subfornical organ, elicits an increase in firing frequency of these cells (Felix and Akert, 1974; Phillips and Felix, 1975; Buranarugsa and Hubbard, 1976).

It now appears that the brain contains its own renin–angiotensin system (Ganten *et al.*, 1976) and angiotensin II peptidergic neural pathways (Fuxe *et al.*, 1976). The brain contains renin and angiotensin I and II (Ganten *et al.*, 1971a, b; Fischer-Ferraro *et al.*, 1971; Ganten *et al.*, 1972), angiotensinogen (Ganten *et al.*, 1971a,

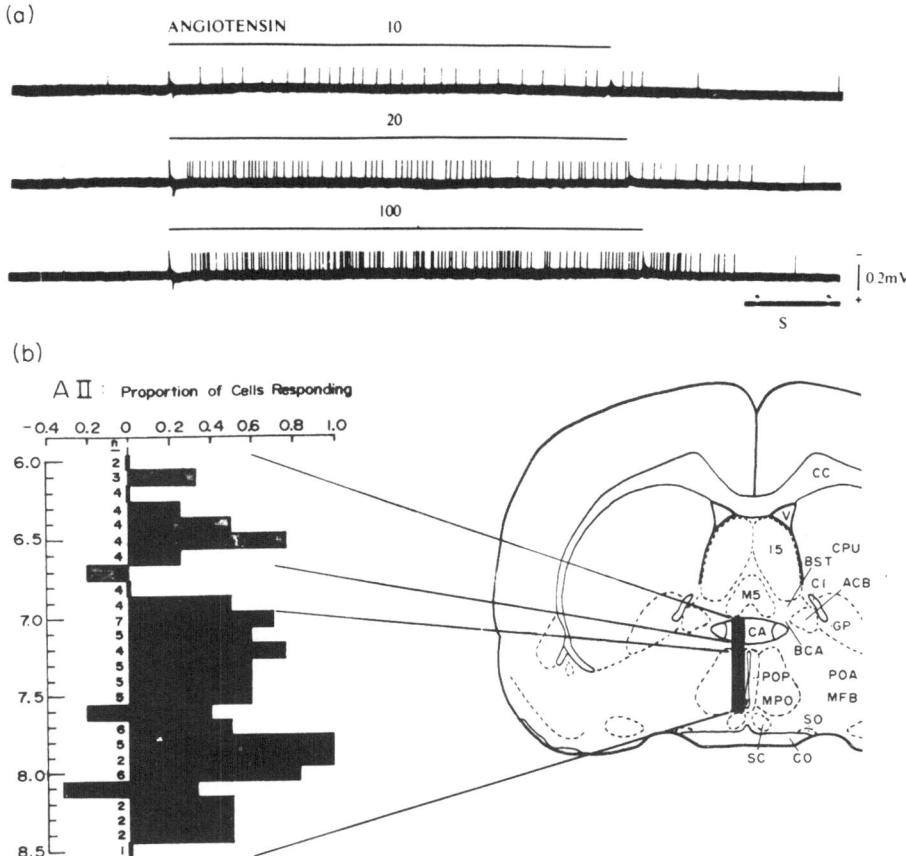

Figure 5.6 Angiotensin II excitation of central neurons. (a) Three continuous oscilloscope traces from a neuron in the lateral hypothalamus of a urethane–chloralose-anaesthetised rat, illustrating the excitant action of angiotensin II applied with three different iontophoretic currents (numbers refer to nA) during the period indicated by the horizontal lines (Figure 8 in Wayner, Ono and Nolley, 1973). (b) Influence of angiotensin II on neurons in the preoptic area of urethane-anaesthetised rats. The black bar in the coronal brain section on the right illustrates the approximate location of the neurons tested. The graph on the left illustrates the proportion of angiotensin II-sensitive neurons at the various depths below the cortical surface. The number of tested cells is contained along the vertical axis (York, unpublished observations).

1972; Reid and Ramsay, 1975), angiotensin-converting enzyme (Roth *et al.*, 1969; Yang and Neff, 1972: Poth *et al.*, 1975), and angiotensinase (Abrash *et al.*, 1971; Goldstein *et al.*, 1972). The brain renin–angiotensin system appears to operate independently of the renal system, and is not affected by nephrectomy. Exogenous renin injected into the CNS is rapidly converted to angiotensin II (Epstein *et al.*, 1973; Reid and Ramsay, 1975). Since renin is present in synaptosomes (Minnich *et al.*, 1972) but not in the cerebrospinal fluid (Reid and Ramsay, 1975) it is likely that angiotensin II formation does occur in brain but at some as yet undefined site, probably intraneuronally.

Recent immunohistochemical studies have demonstrated the presence of angiotensin II or angiotensin II-like immunoreactivity, that persists after nephrectomy in nerve terminals and axons in the brain and spinal cord (Fuxe *et al.*, 1976). Particularly dense accumulations of angiotensin II-containing terminals are found in the median eminence of the hypothalamus, locus coeruleus, spinal intermediolateral column of autonomic motoneurons and substantia gelatinosa of the spinal cord and medulla. These accumulations point to specific sites where angiotensin II may be involved in neuronal function and synaptic transmission, and should be sites for further study in future electrophysiological experiments.

Electrophysiology has now provided additional reasons to re-examine the role of circumventricular structures in the mediation of angiotensin II-induced changes in blood pressure and water intake. Many central neurons located outside the blood–brain barrier are sensitive to angiotensin II applied by microiontophoresis. Neurosecretory neurons in the supraoptic nucleus have been examined both with extracellular recordings *in vivo* (Nicoll and Barker, 1971*a*) and with intracellular recordings in organ culture (Sakai *et al.*, 1974), neurons in the rat hypothalamus (figure 5.6a) zona incerta, thalamus (Wayner *et al.*, 1973) medial preoptic area (Gronan and York, 1976; see figure 5.6b) and cerebral cortex (Phillis and Limacher, 1974*a*, *b*; Phillis, 1977). Moreover sucrose gap recordings of frog spinal cord neurons show that these are also sensitive to angiotensin II; both the dorsal and ventral roots are depolarised by angiotensin, presumably through an indirect effect on interneurons since high magnesium and tetrodotoxin abolish this response (Otsuka *et al.*, 1972*b*; Nicoll, 1976). These observations indicate that there are functional angiotensin II receptors on a variety of central neurons. These receptors may have physiologically significant roles; for example, the increase in firing of supraoptic neurons initiated by the action angiotensin II may promote vasopressin release. The action of angiotensin on other central neurons could be related to drinking behaviour. Both actions would be complimentary in the conservation of water balance during osmotic stress.

PEPTIDES: NEUROTRANSMITTERS OR NEUROMODULATORS?

Immunohistochemistry has demonstrated conclusively that specific peptides are localised in neurons, but clarification of their role in brain requires precise answers to several questions (cf. Werman, 1966). We must demonstrate the *release of peptides by neuronal activation*. Although there is evidence for the release of certain peptides such as Substance P (Otsuka and Konishi, 1976), such experiments are difficult to perform because of the complexity of the nervous system and numerous technical problems, for example inability to activate selectively a particular peptidergic pathway, assay specificity and sensitivity, etc. We also need to know the *influence of peptides on neuronal activity.* This area belongs to the electrophysiologist.

Our present knowledge supports the concept that peptides may be neurotransmitters (Nicoll and Barker 1971*b*; Dyer and Dyball, 1974; Renaud *et al.*, 1975, 1976; see also Bloom, 1972; Renaud, 1977*b*). Alternatively, these peptides may have other functions that could be descriptively categorised as neuromodulators or neurohormones (Krnjević and Morris, 1974; Barker and Gainer, 1975*a*, *b*; Nicoll, 1976; Yarbrough, 1976; see also Barker, 1977; Krnjević, 1977). This latter concept has been deduced rather intuitively to describe '. . . any compound of cellular . . . origin that affects the excitability of nerve cells . . .' (see Florey, 1967, for full explanation). Similar roles have been considered in the case of certain transmitter systems, for example central muscarinic cholinergic (Krnjević *et al.*, 1971) and catecholaminergic pathways (Freedman *et al.*, 1976). At the cellular level, their action has been identified with an unusual decrease in membrane permeability (Krnjević *et al.*, 1971), similar to observations on sympathetic ganglia (cf. Weight and Padjen, 1973; see also Weight, 1974; Schulman and Weight, 1976). Recent intracellular observations during SP iontophoresis suggests a similar mechanism of action (Krnjević, 1977; but see Nicoll, 1976). This definition allows the incorporation of such phenomena of co-transmitters, as might be implied by the recent studies on SP (Pickel *et al.*, 1976; see also Burnstock, 1976). Some of these modulatory functions could be exerted through a presynaptic site of action (Steinacker and Highstein, 1976; Bloom, 1975).

A most important point is the *identity of action between the exogenously applied peptide and a 'natural' physiological event.* Although classical synaptic mechanisms in the CNS are well described (Eccles, 1964; Krnjević, 1974), in the case of possible modulatory interactions among neurons, we have no indication of their existence using 'dry' electrophysiological techniques. This again emphasises the need to identify and examine peptidergic pathways. The *availability of specific antagonists* would greatly facilitate this investigation. At the moment this seems to be applicable only in the case of the opiate peptides.

ELECTROPHYSIOLOGY OF PEPTIDERGIC PATHWAYS

Most of our current information on the dynamics of peptide secretion is derived from the studies on the hypothalamic magnocellular system where specialised neurosecretory neurons are known to secrete neurohypophyseal peptides into the circulation. Two functionally distinct peptide-containing pathways have been identified in the hypothalamus (figure 5.7a): (1) the neurohypophyseal system, originating from the magnocellular neurons of the supraoptic and paraventricular nuclei, for the secretion of oxytocin and vasopressin from nerve terminals in the neurohypophysis; (2) the tuberoinfundibular system, originating from parvicellular neurons in the mediobasal hypothalamus and medial preoptic region, for the release of hypophysiotropic peptides into the pituitary portal circulation (Harris, 1955; Szentágothai *et al.*, 1968; Green, 1969; Halász, 1969). Owing to their peptide content and specialised role, electrophysiological analyses of both systems have been a topic of current interest.

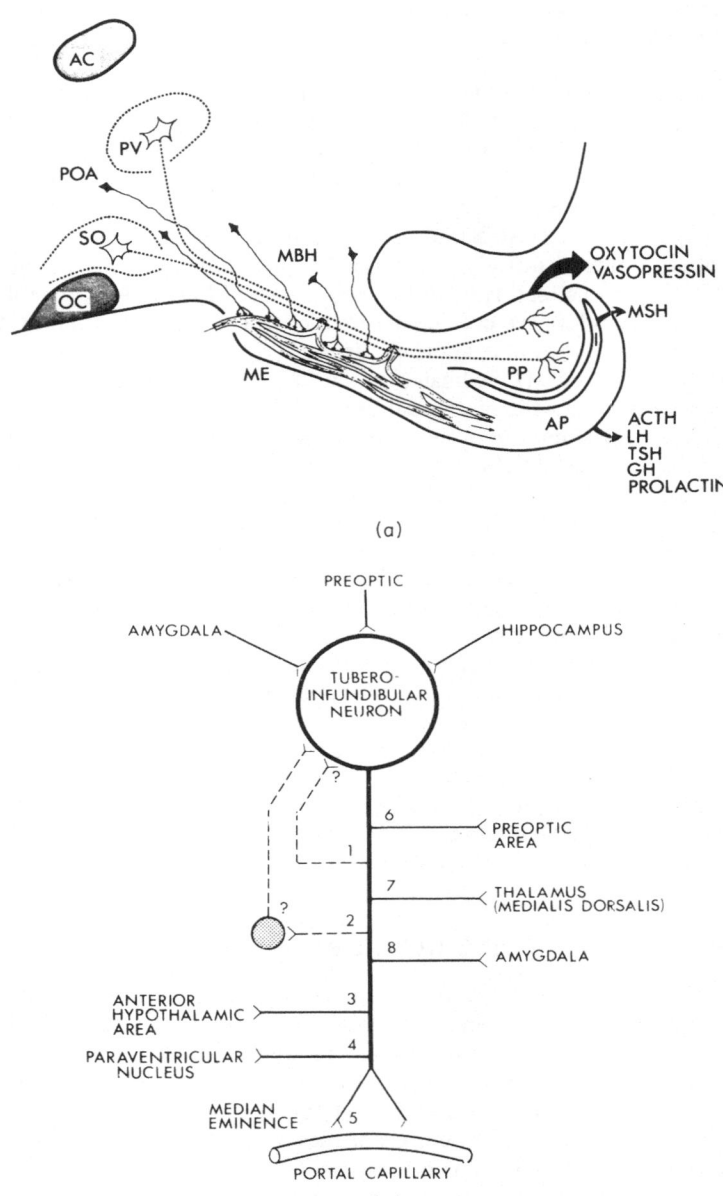

(a)

(b)

Figure 5.7　Peptidergic pathways in the hypothalamus. (a) A simplified sketch of the hypothalamus in saggital section illustrates the two classical neurosecretory pathways related to the
pituitary: the *neurohypophyseal* tract (dotted lines) originating from the magnocellular paraventricular (PV) and supraoptic (SO) neurons to the posterior pituitary (PP), responsible for
liberation of oxytocin and vasopressin; the parvicellular *tuberoinfundibular* pathway originating
from cells in the mediobasal hypothalamus (MBH) and preoptic area (POA) and terminating on

Neurohypophyseal pathways (oxytocin and vasopressin)

Cross and Green (1959) pioneered the electrophysiology of the magnocellular neurosecretory system when they demonstrated cellular spike discharges identical to those recorded from other central neurons. Several investigators have since characterised the activity patterns of magnocellular neurosecretory neurons in several species using the technique of antidromic invasion of axon terminals from the neurohypophysis as a means of cellular identification (Kandel, 1964; Barker *et al.*, 1971; Dreifuss and Kelly, 1972; Koizumi and Yamashita, 1972; Hayward and Jennings, 1973; Arnauld *et al.*, 1975; Cross *et al.*, 1975). These same neurosecretory neurons have provided one of the first clues for peptide involvement at central synapses, since these cells display recurrent postsynaptic inhibition after antidromic stimulation (Kandel, 1964; Barker *et al.*, 1971; Koizumi and Yamashita, 1972; Dreifuss and Kelly, 1972; Negoro *et al.*, 1973). This observation implies functional axon collaterals within this peptidergic network of the hypothalamus, Dale's postulate (Dale, 1935) may also apply to these neurons in that the same substance that is liberated at neurohypophyseal axon terminals may be also liberated at these central axon collaterals during activity in the neurohypophyseal system. If oxytocin or vasopressin are liberated from these central axon collateral terminals to mediate recurrent inhibition then certain hypothalamic neurons might be expected to be sensitive to these peptides. Vasopressin does in fact depress the activity of many cells in the supraoptic nucleus, whereas it tends to promote an increase in activity of cells located outside of this structure (Nicoll and Barker, 1971*b*). Since recurrent inhibition of supraoptic neurons is generally resistant to known antagonists of conventional neurotransmitter agents, vasopressin has been suggested as a possible transmitter in the recurrent inhibitory pathways (Nicoll and Barker, 1971*b*). However, the validity of this assumption has been questioned because inhibition still occurs in the vasopressin deficient Brattleboro rat (Dyball, 1974). The duration of the recurrent inhibitory period is, decreased, but not abolished, by GABA antagonists (Nicoll and Barker, 1971*b*), and this could indicate that recurrent inhibition is mediated by an amino acid neurotransmitter (for example GABA) rather than a peptide. Microiontophoretic application of oxytocin tends to enhance the activity of cells in the supraoptic and paraventricular nuclei (Moss, *et al.*, 1972). Thus, oxytocin is also an unlikely neurotransmitter candidate for a direct recurrent inhibitory pathway. Nevertheless, the data from both single unit (Nicoll and Barker, 1971; Moss *et al.*, 1972) and behavioural studies (de Weid, chapter 16) imply some role for these peptides in neurobiological function. In this context it would help to know whether cells in the neurohypophyseal system do have axon

---

portal capillaries in the median eminence (ME), responsible for elaborating hypothalamic relea-leasing factors that regulate anterior pituitary (AP) secretion. AC, anterior commissure; I, inter-mediate lobe; OC, optic chiasma. (b) A schematic summary of the connections of mediobasal hypothalamic tuberoinfundibular neurons based on the available experimental data. The principal axon of the tuberoinfundibular neuron is illustrated as the heavy vertical line terminating on median eminence portal capillaries with some terminal axon branches (No. 5). Other *intrahypothalamic* axon collaterals mediate recurrent inhibition either directly (1) or indirectly through an inhibitory interneuron (2), and may also reach the anterior hypothalamic area (3) or paraventricular nucleus (4). *Extrahypothalamic* axon collaterals innervate the medial preoptic area (6), the midline thalamic nuclei (7) or amygdala (8). These cells receive afferents from the: amygdala, preoptic area and hippocampus (Renaud, 1977).

collaterals to sites elsewhere in the brain. The presence of a network of peptidergic axon collaterals certainly would add significance to the microiontophoretic studies that demonstrate vasopressin or oxytocin sensitive neurons in extra hypothalamic areas.

Tuberoinfundibular pathways

Recent electrophysiological investigations have led to a reasonable definition of the distribution and central connections of another putative peptidergic neural network, that is, the tuberoinfundibular system (Renaud *et al.*, 1978). Based on antidromic invasion techniques after stimulation of the median eminence, parvicellular tubero-infundibular neurons have been located in mediobasal hypothalamic and medial preoptic areas (Makara *et al.*, 1972; Sawaki and Yagi, 1973; Harris and Sanghera, 1974; Moss *et al.*, 1975; Renaud, 1976*a*, *b*, 1977*a*, *b*). In the hypothalamus, there is often a close correspondence between the distribution of these electrically identified neurons, and the peptidergic neurons identified with immunohistochemical techniques (cf. Elde, chapter 2). This would suggest that the electrophysiologist is in fact investigating peptidergic neurons and pathways, and providing that this basic assumption is correct, some functional aspects of these neurons can be considered. Tuberoinfundibular neurons do receive afferent connections from several extrahypothalamic structures (Renaud, 1976*c*, 1977*a*) and there is evidence for axon collaterals (Harris and Sanghera, 1974; Renaud, 1975, 1976*a*, *b*, 1977*b*; Sawaki and Yagi, 1976; Yagi and Sawaki, 1977). However, these collaterals are not only involved in *local* recurrent inhibitory and facilitatory circuits, but also extend to several *extrahypothalamic* structures (figure 5.7b). Once again Dale's postulate suggests that the same substance(s) is released from nerve terminals in the median eminence and from nerve terminals of these central axon collaterals (Renaud, 1975). While this may be difficult to demonstrate, there are indirect indications from pharmacological, immunohistochemical and behavioural studies that at least three structurally characterised hypothalamic peptides are active in neural tissue. Electrophysiology has provided a positive contribution through attempts to define the electrical characteristics of tuberoinfundibular neurons, and record the behaviour of neural tissue to exogenously applied TRH, LHRH and somatostatin.

CONCLUSION

The results outlined here would seem sufficient to indicate that peptides are engaged in some aspect of neural function. Peptides selectively modify neuronal activity with respect to the mode and location of action. Studies with analogues of hypophysiotropic peptides are usually in agreement with their potencies on the adenohypophysis. What these peptides do naturally in the brain is still an unsolved question. Their short-term actions can be followed electrophysiologically on microiontophoresis and the evidence suggests that they are good candidates for neurotransmitter agents. Their equimolar potency also appears to be much higher than that of conventional neurotransmitters. However, a survey of the results presented here indicates that there are multiple sites of action and a rigid classification as neurotransmitter or neuromodulator is not yet possible. In fact, further studies may well result in a re-evaluation of the definition of a neurotransmitter or a neuromodulator.

Electrophysiology has played a significant role in the assessment of neuropeptides and their action in neural tissue. It should be evident that a better understanding of centrally acting peptides cannot be accomplished in isolation, and requires the assistance of immunohistochemists, biochemists and pharmacologists as well as electrophysiologists. Such a multidisciplinary approach is required if significant progress is to be made regarding the role of peptides in neural tissue.

ACKNOWLEDGEMENTS

We thank the Biological Council and supporting Societies for the invitation to participate in this Symposium. The authors are grateful to the Canadian Medical Research Council for financial support of the studies conducted in their laboratories, and to the various authors for permission to quote the work conducted in their laboratories. We thank Mrs M. Walker for typing the manuscript, and members of Abbott and Ayerst laboratories, and the Salk Institute for supplies of peptides and peptide analogues.

REFERENCES

Abrash, L., Walter, R. and Marks, N. (1971). *Experientia*, 27, 1352–53
Arnauld, E., Dufy, B. and Vincent, J. D. (1975). *Brain Res.*, 100, 315–25
Auerback, A. A. and Bennett, M. V. L. (1961). *J.gen.Physiol.*, 53, 183–210
Barker, J. L. (1976). *Physiol. Rev.*, 56, 435–52
Barker, J. L. (1977). In *Peptides in Neurobiology* (ed. H. Gainer), Plenum Press, New York (in press)
Barker, J. L., Crayton, J. W. and Nicoll, R. A. (1971). *Brain Res.*, 33, 353–66
Barker, J. L. and Gainer, H. (1975a). *Brain Res.*, 84, 461–77
Barker, J. L. and Gainer, H. (1975b). *Brain Res.*, 84, 479–500
Barker, J. L., Nicoll, R. A. and Padjen, A. (1975). *J. Physiol., Lond.*, 245, 521–36
Barker, J. L. and Smith, T. G. (1977). In *Approaches to the Cell Biology of Neurons* (ed. W. M. Cowan and J. A. Ferrendelli). Society for Neuroscience, Bethesda, Md. pp. 340–73
Bennett, G., Edwardson, J., Holland, D., Jeffcoate, S. and White, N. (1975). *Nature*, 257, 323–25
Bloom, F. E. (1972). *Neurosci.Res.Progm Bull.*, 10, 122–251
Bloom, F. E. (1974). *Life Sci.*, 14, 1819–34
Bloom, F. E. (1975). In *Pre- and Postsynaptic Receptors* (ed. E. Usdin and W. E. Bunney Jr), Marcel Dekker, New York, pp. 67–86
Brazeau, P., Vale, W., Burgus, R., Ling, N., Butcher, M., Rivier, J. and Guillemin, R. (1973). *Science*, 179, 77–79
Breese, G. R., Cott, J. M., Cooper, B. R., Prange, A. J. Jr., Lipton, M. A. and Plotnikoff, N. P. (1975). *J.Pharmac.exp.Ther.*, 193, 11–22
Buranarugsa, P. and Hubbard, J. I. (1976). *Proc.Univ.Otago med.Sch.*, 54, 3–4
Burnstock, G. (1976). *Neuroscience*, 1, 239–48
Burt, D. R. and Snyder, S. H. (1975). *Brain Res.*, 93, 309–28
Calvillo, O., Henry, J. L. and Neuman, R. (1974). *Can.J.Physiol.Pharmac.*, 52, 1207–11
Cross, B. A., Dyball, R. E. J., Dyer, R. G., Jones, C. W., Lincoln, D. W., Morris, J. F. and Pickering, B. T. (1975). *Recent Prog.Horm. Res.*, 31, 243–294
Cross, B. A. and Green, J. D. (1959). *J.Physiol., Lond.*, 148, 554–69
Curtis, D. R. (1964). In *Physical Techniques in Biological Research*, Vol. V. (ed. W. L. Nastuk), Academic Press, New York and London, pp. 144–90
Curtis, D. R., Game, C. J. A., Johnston, G. A. R. and McCulloch, R. M. (1974). *Brain Res.*, 70, 493–99
Dale, H. A. (1935). *Proc. R. Soc. Med.*, 28, 318–32
Davidoff, R. A. and Sears, E. S. (1974). *Neurology*, 24, 957–63

82   Centrally Acting Peptides

Davies, J. and Dray, A. (1976). *Brain Res.*, **107**, 623–27
Davies, J. and Watkins, J. C. (1974). *Brain Res.*, **70**, 501–05
Dreifuss, J. J. and Kelly, J. S. (1972). *J.Physiol., Lond.*, **220**, 87–103
Dyball, R. E. J. (1974). *J.Endocr.*, **60**, 135–43
Dyer, R. G. and Dyball, R. E. J. (1974). *Nature*, **252**, 486–488
Eccles, J. C. (1964). *The Physiology of Synapses.* Springer, Gottingen, Berlin and Heidelberg
Epstein, A. N., Fitzsimmons, J. F. and Johnson, A. F. (1973). *J.Physiol., Lond.*, **230**, 42–43P
Felix, D. and Akert, K. (1974). *Brain Res.*, **76**, 350–53
Fischer–Ferraro, C., Nahmod, V. E., Goldstein, D. J. and Finkielman, S. (1971). *J.exp. Med.*, **133**, 353–61
Fitzsimons, J. T. (1975). In *Hormones, Homeostasis and the Brain; Progress in Brain Research*, Vol. 42, (ed. W. H. Gispen, Tj B. van Wimersma Greidanus, B. Bonus and D. de Wied), Elsevier, Amsterdam, pp. 215–33
Florey, E. (1967). *Fedn Proc.*, **26**, 1164–78
Foote, S. L., Freedman, R. and Oliver, A. P. (1975). *Brain Res.*, **86**, 229–242
Fotherby, K. J., Morrish, N. J. and Ryall, R. W. (1976). *Brain Res.*, **113**, 210–13
Freedman, R., Hoffer, B. J., Puro, D. and Woodward, D. J (1976). *Br.J.Pharmac.*, **57**, 603–05
Fuxe, K., Ganten, D., Hökfelt, T. and Bolme, P. (1976). *Neurosci. Lett.*, **2**, 229–34
Ganten, D., Granger, P., Ganten, U., Boucher, R. and Genest, J. (1972). In *Hypertension* (ed. J. Genest and K. Koiw), Springer-Verlag, Berlin, pp. 423–32
Ganten, D., Hutchinson, J. S., Schelling, J. P., Ganten, U. and Fischer, H. (1976). *Clin.Exp. Pharmac.Physiol.*, **2**, 103–126
Ganten, D., Marques-Julio, A., Granger, P., Hayduk, K., Karminky, K. P., Boucher, R. and Genest, J. (1971*a*). *Am.J.Physiol.*, **221**, 1733–1737
Ganten, D., Minnich, J., Granger, P., Hayduk, K., Brecht, H. M., Barbeau, A., Boucher, R. and Genest, J. (1971*b*). *Science*, **173**, 64–65
Goldstein, D. J., Diaz, A., Finkielman, S., Nahmod, V. E. and Fischer-Ferraro, C. (1972). *J.Neurochem.*, **19**, 2451–52
Green, J. D. (1969). In *The Hypothalamus* (ed. W. Haymaker, E. Anderson and W. J. H. Nauta), Thomas, Springfield, Ill., pp. 276–310
Gronan, R. J. and York, D. H. (1976). *Neurosci.Abstr.*, **2**, 426
Halasz, B. (1969). In *Frontiers in Neuroendocrinology* (ed. W. F. Ganong and L. Martini), Oxford University Press, London, pp. 307–42
Handwerker, H. O., Iggo, A. and Zimmermann, M. (1975). *Pain*, **1**, 147–65
Harris, G. W. (1955). *Neural Control of Pituitary Gland.* Edward Arnold, London
Harris, M. C. and Sanghera, M. (1974). *Brain Res.*, **81**, 401–11
Hayward, J. N. and Jennings, D. P. (1973) *J.Physiol., Lond.*, **232**, 545–72
Henry, J. L., Krnjević, K. and Morris, M. E. (1974). *Fedn Proc.*, **33**, 548
Henry, J. L. (1975). *Neurosci.Abstr.*, **1**, 390
Henry, J. L. (1976). *Brain Res.*, **114**, 439–51
Henry, J. L. (1977). In *Substance P* (ed. U. S. von Euler and B. Pernow), Raven Press, New York, (in press)
Henry, J. L. and Ben-Ari, Y. (1976). *Brain Res.*, **117**, 540–44
Henry, J. L., Krnjević, K. and Morris, M. E. (1975). *Can.J.Physiol.Pharmac.*, **53**, 423–32
Highstein, S. M. and Bennet, M. V. L. (1975). *Brain Res.*, **98**, 229–42
Hökfelt, T., Efendić, S., Hellerström, C., Johansson, O., Luft, R. and Arimura, A. (1975). *Acta Endocr.*, **80**, suppl. 200, 1–40
Hökfelt, T., Elde, R., Johansson, O., Luft, R., Nilsson, G. and Arimura, A. (1976). *Neuroscience*, **1**, 131–36
Hökfelt, T., Fuxe, K., Johansson, O., Jeffcoate, S. L. and White, N. (1975*a*). *Neurosci.Lett.*, **1**, 133–139
Hökfelt, T., Fuxe, K., Johansson, O., Jeffcoate, S. and White, N. (1975*b*). *Eur.J.Pharmac.*, **34**, 389–392
Jackson, I. M. D. and Reichlin, S. (1974). *Endocrinology*, **95**, 854–62
Kandel, E. R. (1964). *J.gen.Physiol.*, **47**, 691–717
Kawakami, M. and Sakuma, Y. (1974). *Endocrinology*, **15**, 290–307
Kawakami, M. and Sakuma, Y. (1976). *Brain Res.*, **101**, 79–94
Kelly, J. S. (1975). In *Handbook of Psychopharmacology*, Vol. II (ed. L. L. Iversen, S. D. Iversen and S. H. Snyder), Plenum Press, New York, pp. 29–67
Kelly, J. S. and Renaud, L. P. (1973). *Br. J. Pharmac.*, **48**, 369–386

Kelly, J. S., Simmonds, M. A. and Straughan, D. W. (1975). In *Methods in Brain Research* (ed. P. B. Bradley), Wiley, New York, pp. 333–77

Kelly, M. J. and Moss, R. M. (1976). *Neuropharmacology*, 15, 325–28

Knuttsson, E., Lindblom, U. and Martensson, A. (1973). *Brain*, 96, 29–46

Koizumi, K. and Yamashita, H. (1972). *J.Physiol., Lond.*, 221, 683–705

Konishi, S. and Otsuka, M. (1974a). *Brain Res.*, 65, 397–410

Konishi, S. and Otsuka, M. (1974b). *Nature*, 252, 734–35

Krnjević, K. (1971). In *Methods in Neurochemistry* (ed. R. Fried), Marcel Dekker, New York, pp. 129–72

Krnjević, K. (1974). *Physiol.Rev.*, 54, 418–540

Krnjević, K. (1977). In *Substance P* (ed. U. S. von Euler and B. Pernow), Raven Press, New York, pp. 217–30

Krnjević, K. and Morris, M. E. (1974). *Can.J.Physiol.Pharmac.*, 52, 736–44

Krnjević, K. and Morris, M. E. (1976). *J.Physiol., Lond.*, 257, 791–815

Krnjević, K., Pumain, R. and Renaud, L. (1971). *J.Physiol., Lond.*, 215, 247–268

Lote, C. J., Gent, J. P., Wolstencroft, J. H. and Szelke, M. (1976). *Nature*, 264, 188–89

Makara, G. B., Harris, M. C. and Spyer, K. M. (1972). *Brain Res.*, 40, 283–90

McLean, A. S., Sirett, N. E., Bray, J. J. and Hubbard, J. I. (1975). *Proc.Univ. Otago med.Sch.*, 53, 19–20

McLellan, D. L. (1973). *J.Neurol.Neurosurg.Psychiat.*, 36, 555–60

Miletić, V., Kovacs, M. S. and Randić, M. (1977). *Fedn Proc.*, 36, 1014, Abstr., 3915

Minnich, J. L., Ganten, D., Barbeau, A. and Genest, J. (1972). In *Hypertension* (ed. J. Genest and K. Koin), Springer-Verlag, Berlin, pp. 432–35

Moss, R. L. (1976). In *Frontiers in Neuroendocrinology*, Vol. 4 (ed. L. Martini and W. F. Ganong), Raven Press, New York, pp. 95–128

Moss, R. L. (1977). *Fedn Proc.*, 36, 1978–83

Moss, R. L., Dudley, C. A. and Kelly, M. J. (1977). *Neuropharmacology*, (in press).

Moss, R. L., Dyball, R. E. J. and Cross, B. A. (1972). *Expl Neurol.*, 14, 95–102

Moss, R. L., Kelly, M. and Riskind, P. (1975). *Brain Res.*, 89, 265–77

Moss, R. L. and McCann, S. M. (1973). *Science*, 181, 177–79

Myers, R. D. (1974). *Handbook of Drug and Chemical Stimulation of the Brain*. Van Nostrand Reinhold Co., New York

Negoro, H., Visessuwan, S. and Holland, R. C. (1973). *Brain Res.*, 57, 479–83

Nicoll, R. A. (1976). In *Neurotransmitters, Hormones and Receptors: Novel approaches* (ed. J. A. Ferrendelli, B. S. McEwan and S. H. Snyder), Society for Neuroscience, Bethesda, Md. pp. 99–122

Nicoll, R. A. (1977). *Nature*, 265, 242–43

Nicoll, R. A. and Barker, J. L. (1971a). *Nature new Biol.*, 233, 172–74

Nicoll, R. A. and Barker, J. L. (1971b). *Brain Res.*, 35, 501–11

Nicoll, R. A., Siggins, G. R., Ling, N., Bloom, F. E. and Guillemin, R. (1977). *Proc. natn Acad. Sci., U.S.A.*, 74, 2584–89

Osborne, M. J., Pooters, N., Angles d'Auriac, G., Epstein, A. N., Worcel, M. and Meyer, P. (1971). *Pflügers Arch.ges.Physiol.*, 326, 101–114

Otsuka, M. and Konishi, S. (1974).

Otsuka, M. and Konishi, S. (1975). *Cold Spring Harb. Symp. quant.Biol.*, 40, 135–43

Otsuka, M. and Konishi, S. (1976). *Nature*, 264, 83–84

Otsuka, M., Konishi, S. and Takahashi, T. (1972a). *Proc.Japan Acad.*, 48, 747

Otsuka, M., Konishi, S. and Takahashi, T. (1972b). *Proc.Japan.Acad.*, 48, 342–46

Otsuka, M. and Konishi, S. (1974). *Nature*, 252, 733

Pfaff, D. W. (1973). *Science*, 182, 1148–49

Phillips, M. I. and Felix, D. (1975). *Neurosci. Abstr.*, 1, 469

Phillips, M. I. and Felix, D. (1976). *Brain Res.*, 531–40

Phillis, J. W. (1976). *Experientia*, 32, 593–94

Phillis, J. W. (1977). In *Approaches to the Cell biology of Neurons* (ed. W. M. Cowan and J. A. Ferrendelli). Society for Neuroscience, Bethesda, Md. pp. 241–264

Phillis, J. W. and Limacher, J. J. (1974a). *Brain Res.*, 69, 158–63

Phillis, J. W. and Limacher, J. J. (1974b). *Expl Neurol.*, 43, 414–23

Pickel, V. M., Reis, D. J. and Leeman, S. E. (1977). *Brain Res.*, 122, 534–40

Pierau, F. K. and Zimmermann, P. (1973). *Brain Res.*, 54, 376–80

Poth, M. M., Heath, R. G. and Ward, M. (1975). *J.Neurochem.*, 25, 83–85

Prange, A. J. Jr., Nemeroff, C. B., Lipton, M. S., Breese, G. R. and Wilson, I. C. (1977). In *Handbook of Psychopharmacology*, Section II (ed. L. L. Iversen, S. D. Iversen and S. H. Snyder), Plenum Press, New York, (in press)

Randić, M. and Miletić, V. (1977). In *Iontophoresis and Transmitter Mechanisms in the Mammalian Central Nervous System* (ed. R. Ryall and J. S. Kelly), Elsevier, Amsterdam, (in press)

Reid, I. A. and Ramsay, D. J. (1975). *Endocrinology*, 97, 536–42

Renaud, L. P. (1975). *Neurosci. Abstr.*, 1, 441

Renaud, L. P. (1976a). *J. Physiol., Lond.*, 254, 20P–21P

Renaud, L. P. (1976b). *Brain Res.*, 105, 59–72

Renaud, L. P. (1976c). *J.Physiol., Lond.*, 260, 237–52

Renaud, L. P. (1977a). *J.Physiol., Lond.*, 264, 541–64

Renaud, L. P. (1977b). In *Approaches to the Cell Biology of Neurons* (ed. W. M. Cowan and J. A. Ferrendelli), Society for Neuroscience, Bethesda, Md. pp. 265–90

Renaud, L. P., Blume, H. W. and Pittman, Q. J. (1978). In *Frontiers in Neuroendocrinology*, Vol. 5 (ed. W. F. Ganong and L. Martini), Raven Press, New York, (in press)

Renaud, L. P. and Martin, J. B. (1975a). In *Anatomical Neuroendocrinology* (ed. W. E. Stumpf and L. E. Grant), Karger, Basel, pp. 354–56

Renaud, L. P. and Martin, J. B. (1975b). *Brain Res.*, 86, 150–54

Renaud, L. P., Martin, J. B. and Brazeau, P. (1975). *Nature*, 255, 233–35

Renaud, L. P., Martin, J. B. and Brazeau, P. (1976). *Pharmac.Biochem.Behav.*, 5, suppl. 1, 171–78

Roth, M., Weitzman, A. F. and Piquilloud, Y. (1969). *Experientia*, 25, 1247

Saito, K., Konishi, S. and Otsuka, M. (1975). *Brain Res.*, 97, 177–80

Sakai, K. K., Marks, B. H., George, J. M. and Koestner, A. (1974a). *J.Pharmac.Exp.Ther.*, 190, 482–91

Sakai, K. K., Marks, B. H., George, J. and Koestner, A. (1974b). *Life Sci.*, 14, 1337–44

Sawaki, Y. and Yagi, K. (1973). *J.Physiol., Lond.*, 230, 75–85

Sawaki, Y. and Yagi, K. (1976). *J.Physiol., Lond.*, 260, 447–460

Schulman, J. A. and Weight, F. F. (1976). *Science*, 194, 1437–39

Severs, W. B. and Daniels-Severs, A. E. (1973). *Pharmac.Rev.*, 25, 415–449

Severs, W. B. and Summy-Long, J. (1975). *Life Sci.*, 17, 1513–26

Simpson, J. B. and Routtenberg, A. (1973). *Science*, 181, 1172–75

Spira, M. E., Model, P. G. and Bennet, M. V. L. (1970). *J.Cell Biol.*, 47, 199–200A

Steinacker, A. and Highstein, S. M. (1976). *Brain Res.*, 114, 128–133

Steiner, F. A. (1975). In *Anatomical Neuroendocrinology* (ed. W. E. Stumpf and L. E. Grant), Karger, Basel, pp. 270–275

Szentágothai, J., Flerkó, B., Mess, B. and Halász, B. (1968). *Hypothalamic Control of the Anterior Pituitary*. Akademiai Kiado, Budapest

Takahashi, T., Konishi, S., Powell, D., Leeman, S. E. and Otsuka, M. (1974). *Brain Res.*, 73, 59–69

Takahashi, T. and Otsuka, M. (1975). *Brain Res.*, 87, 1–11

Tan, A. T., Tsang, D., Renaud, L. P. and Martin, J. B. (1977). *Brain Res.*, 123, 193–96

Vale, W., Ling, N., Rivier, J., Villarreal, J., Rivier, C., Douglas, C. and Brown, M. (1976). *Metabolism*, 25, suppl. 1, 1491–94

von Euler, U. S. and Gaddum, J. H. (1931). *J.Physiol., Lond.*, 72, 74–87

Walker, R. J., Kemp, J. A., Yajima, H., Kitagawa, K. and Woodruff, G. N. (1976). *Experientia*, 32, 214–15

Wayner, M. J., Ono, T. and Nolley, D. (1973). *Pharmac.Biochem.Behav.*, 1, 679–91

Werman, R. (1966). *Comp.Biochem.Physiol.*, 18, 745–66

Weight, F. F. (1974). In *Neurosciences Third Study Program* (ed. S. O. Schmidt and F. G. Worden), MIT Press, Boston, pp. 929–41

Weight, F. F. and Padjen, A. (1973). *Brain Res.*, 55, 219–24

Wilbur, J. F., Montoya, E., Plotnikoff, N., White, W. F., Genrich, R., Renaud, L. and Martin, J. B. (1976). *Recent Progress Horm. Res.*, 32, 117–53

Yagi, K. and Sawaki, Y. (1975). *Brain Res.*, 84, 155–59

Yagi, K. and Sawaki, Y. (1977). *Brain Res.*, 120, 342–346

Yang, H. Y. and Neff, N. H. (1972). *J.Neurochem.*, 19, 2443–50

Yarbrough, G. G. (1976). *Nature*, 263, 523–24

6
Comparative features of enkephalin and neurotensin in the mammalian central nervous system

Solomon H. Snyder, George R. Uhl and Michael J. Kuhar
(Departments of Pharmacology and Experimental Therapeutics,
and Psychiatry and Behavioral Sciences,
Johns Hopkins University School of Medicine,
Baltimore, Maryland 21205, U.S.A.)

INTRODUCTION

Interest in recent years has centred on small peptides as possible neurotransmitters or neuromodulators in the central nervous system (CNS). Some of these, such as somatostatin and thyrotropin releasing hormone (TRH), were first explored as hypothalamic releasing factors for pituitary hormones. Vasoactive intestinal peptide (VIP) and gastrin were identified as gastrointestinal modulators and angiotensin has been well known as a regulator of vascular muscle activity and adrenal gland secretion. Substance P and neurotensin were first characterised as brain peptides of unknown function; enkephalin as the brain's endogenous opioid peptide. Most 'neuropeptides' appear to be associated with cells that have arisen embryologically from neuroectodermal elements and which are localised in the intestinal system or the CNS. Enkephalin and neurotensin will constitute the major focus of this chapter.

ENKEPHALIN: THE BRAIN'S ENDOGENOUS OPIOID PEPTIDE

The dramatic substrate specificity, regional and subcellular localisation of the opiate receptor (Snyder, 1975) suggested that it might normally interact with an endogenous substance. Hughes et al. (1975) identified the two pentapeptide enkephalins, leucine-enkephalin and methionine-enkephalin in pig brain, findings confirmed independently in bovine brain by Simantov and Snyder (1976a). Several opioid peptides have been identified in pituitary extracts, of which at least three, α-endorphin, β-endorphin and γ-endorphin contain amino acid sequences of varying length that incorporate the sequence of methionine-enkephalin at their N-terminals

(Bradbury *et al.*, 1976; Cox *et al.*, 1976; Guillemin *et al.*, 1976). The possible neuronal localisations of enkephalin and the larger opioid peptides, as well as the relative roles that these substances play in the brain, are examined in this chapter.

Biochemical localisation

Radioreceptor assay, bioassay, and radioimmunoassay have all indicated the presence of opioid peptides in the CNS with amino acid sequences longer than those of the enkephalins (Goldstein, 1976; Rossier, Guillemin and Bloom, personal communication). In our laboratory localisation was investigated by comparative radioreceptor assay and radioimmunoassay (Simantov and Snyder, 1976*b*). The radioimmunoassay used in our laboratory recognises and distinguishes met-enkephalin but shows negligible cross-reactivity with any of the pituitary endorphins. In contrast, the radioreceptor assay will identify any opioid substance. If the pituitary endorphins comprise a major portion of brain opioid peptides, then values by radioreceptor assay should greatly exceed levels determined by specific enkephalin radioimmunoassay. In fact, levels are quite similar for the two procedures (Table 6.1). This suggests that the enkephalins are the major opioid peptides of the brain, which fits in well with the findings of Rossier, Guillemin and Bloom (personal communication) that rat brain levels of β-endorphin, apparently the major or sole brain endorphin, are only 10–20 per cent of enkephalin levels. The ratios of met-enkephalin to leu-enkephalin are fairly similar whether determined by radio immunoassay or radioreceptor assay. In several regions of bovine brain, such as caudate, hypothalamus and thalamus, levels of leu-enkephalin exceed those of met-enkephalin, whereas the reverse holds for bovine amygdala. Levels of the two enkephalins are about the same in bovine cerebral cortex (table 6.1). Whereas leu-enkephalin is the major enkephalin in bovine brain, whole rat brain possesses 4–7 times more met-enkephalin than leu-enkephalin. Total opioid peptide activity in the intermediate lobe of the pituitary (which contains virtually all the opioid peptide content of the pituitary (Simantov and Snyder, 1977; Bloom *et al.*, 1977)) is hundreds of times greater than in brain tissue. However, enkephalin is virtually undetectable in the pituitary, whereas β-endorphin appears to be the major or sole opioid peptide present (Simantov *et al.*, 1977*a*; Simantov and Snyder, 1977; Bloom *et al.*, 1977).

Histochemical localisation

Antisera to enkephalins have also been useful in immunohistochemical localisation of enkephalin to specific neuronal systems in the brain (Elde *et al.*, 1976; Simantov *et al.*, 1977*b*). Because of previous studies of the cellular localisation of the opiate receptor by autoradiography (Pert *et al.*, 1976; Atweh and Kuhar, 1977*a, b, c*) we have been able to compare the detailed microscopic localisations of opiate receptors and enkephalin immunofluorescence (Simantov *et al.*, 1977*b*) (figures 6.1 and 6.2). In the spinal cord, both enkephalin and opiate receptors are sharply localised to a dense band in the dorsal grey matter. Because opiate receptor binding declines sharply following dorsal root lesion (LaMotte *et al.*, 1976), opiate receptors in the dorsal grey matter are probably localised to nerve terminals of primary sensory afferent neurons. Thus, enkephalin neurons presumably make axoaxonic synapses on sensory terminals in a similar way to that postulated for the presynaptic inhibitory actions of γ-aminobutyric acid (GABA). A similar

Table 6.1 Opioid peptides in bovine and rat brain and bovine pituitary

Bovine brain region	Opioid peptide levels					
	Radioimmunoassay			Radioreceptor assay		
	Met-enkephalin	Leu-enkephalin	Ratio $\frac{\text{Met-enk}}{\text{Leu-enk}}$	Met-enkephalin	Leu-enkephalin	Ratio $\frac{\text{Met-enk}}{\text{Leu-enk}}$
	(pmol/g wet weight)			(pmol/g wet weight)		
Caudate nucleus	90	450	0.2	35	345	0.1
Amygdala	300	100	3.0	190	110	1.7
Hypothalamus	110	270	0.4	20	200	0.1
Thalamus	60	150	0.4	40	65	0.6
Cerebral cortex	100	100	1.0	40	28	1.4

	Total enkephalin level (pmol/g wet weight leu-enkephalin equivalents)	Total opioid activity (pmol β-endorphin equivalents/g wet weight)
Bovine pituitary-pars intermedia	< 2	210 000

Tissues were extracted by sucrose homogenisation as described by Simantov and Snyder (1976a) and levels were corrected for recovery, assessed with unlabelled enkephalins added to tissue before homogenisation. The radioreceptor assay was performed with and without cyanogen bromide treatment for determination of the relative amounts of met- and leu-enkephalin.

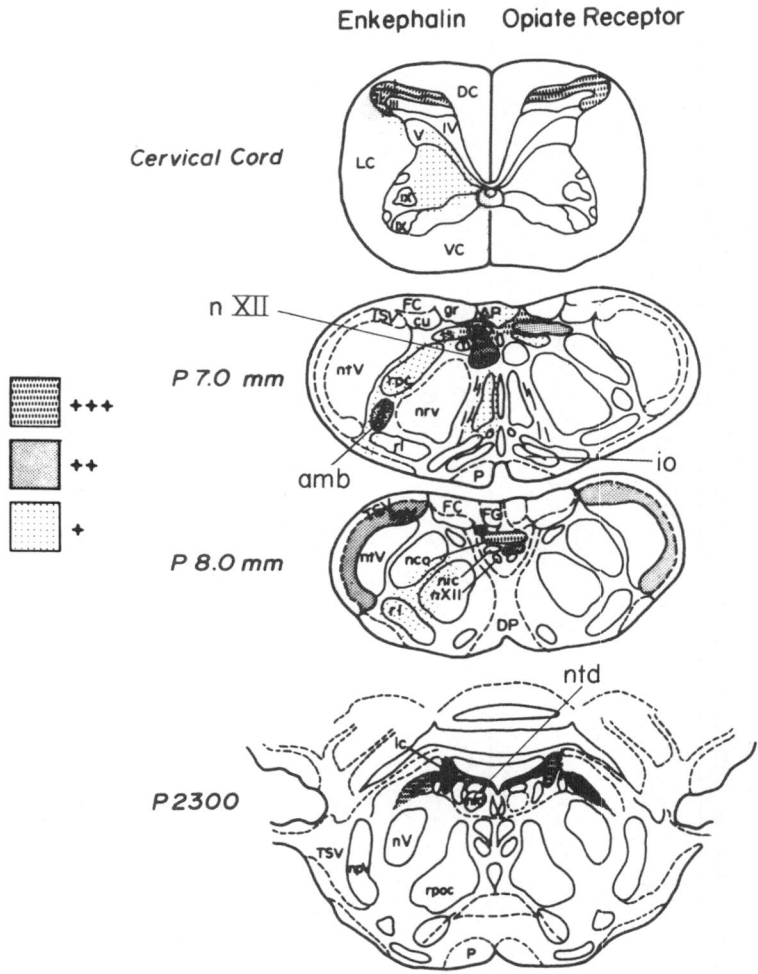

Figure 6.1 Distributions of enkephalin (left) and opiate receptors (right) at different levels of the spinal cord and medulla. Abbreviations are as follows: amb, nucleus ambiguus; AP, area postrema; cu, nucleus cuneatus; DC, dorsal column; DP, pyramidal decussation; FC, fasciculus cuneatus; FG, fasciculus gracilis; gr, nucleus gracilis; io, nucleus olivaris inferior; LC, lateral column; lc, locus coeruleus; nco, nucleus commissuralis; nic, nucleus intercalatus; npV, nucleus principis nerve trigemenii; nrv, nucleus reticularis medullae oblongata pars ventralis; ntd, nucleus tegmenti dorsalis Gudden; nts, nucleus tractus solitarius; ntV, nucleus tractus spinalis nervi trigemini; nV, nucleus originis nervi trigemini; nX, nucleus originis dorsalis vagi; nXII, nucleus originis nervi hypoglossi, P, tractus corticospinalis; rl, nucleus reticularis lateralis; rpc, nucleus reticularis parvocellularis; rpoc, nucleus reticularis pontis caudalis; sgV, substantia gelatinosa trigemini; ts, tractus solitarius; TSV, tractus spinalis nervi trigemini; VC, ventral column. Derived from Simantov *et al.* (1977*b*), Atweh and Kuhar (1977*a*) and Pert *et al.* (1976).

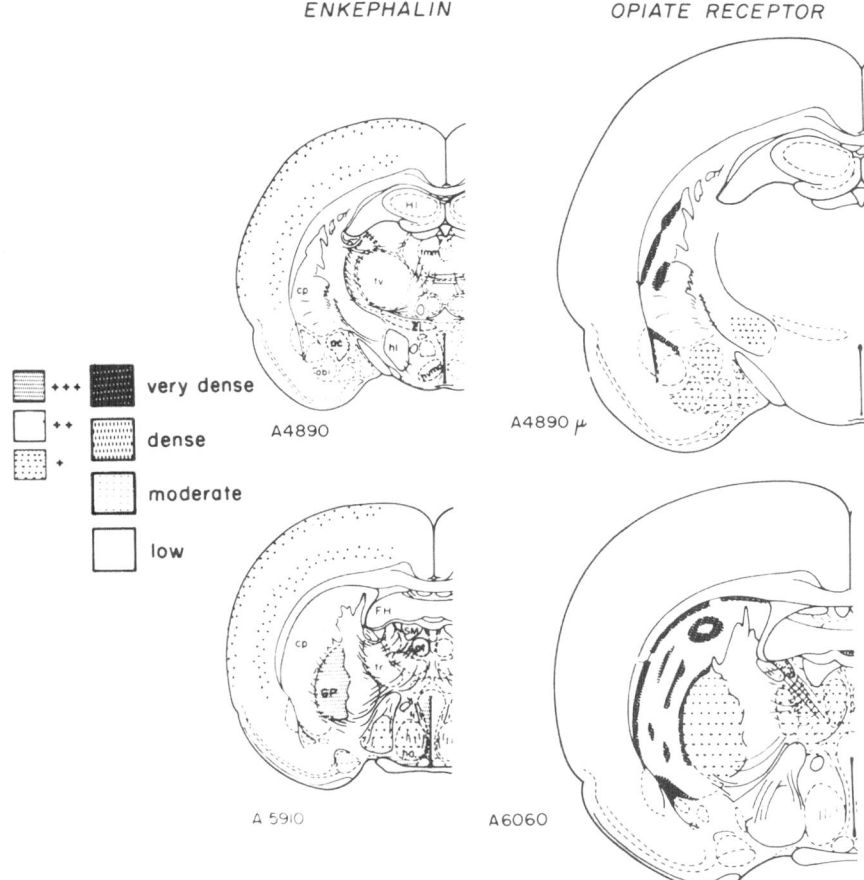

Figure 6.2 Distribution of enkephalin (left) and opiate receptors (right) at certain level of the diencephalon and telencephalon. Abbreviations are as follows: abl, nucleus amygdaloideus basalis, pars lateralis; ac, nucleus amygdaloideus centralis; cp, nucleus caudatus putamen; ha, nucleus anterior (hypothalami); hl, nucleus lateralis (hypothalami); hvma, nucleus ventromedialis (hypothalami), pars anterior; pt, nucleus paratenialis; tmm, nucleus medialis thalami, pars medialis; tr, nucleus reticularis thalami; tv, nucleus ventralis thalami; FH, fimbria hippocampi; GP, globus pallidus; HI, hippocampus; SM, stria medullaris thalami; ZI, zona incerta. Derived from Simantov *et al.* (1977*b*).

'presynaptic' localisation of opiate receptors has been demonstrated through the loss of auto-radiographic opiate receptor grains in vagal nuclei of brainstem following lesion of the vagus nerve in the neck, and a loss of receptors in the accessory optic nuclei after unilateral enucleation (Atweh and Kuhar, 1977*b*, *c*).

In the brainstem both opiate receptors and enkephalin are localised to vagal nuclei and the area postrema, where they might mediate visceral reflexes such as coughing and vomiting that are influenced by opiates (figure 6.1). Opiate receptors in the substantia gelatinosa of the spinal nucleus of the trigeminal are probably

associated with pain perception for facial areas (figure 6.1). Opiate receptors and
enkephalin in the vicinity of the locus coeruleus, which contains exclusively
noradrenergic neurons, may be associated with euphoric actions of opiates. A
similar function might be adduced for the opiate receptors and enkephalin neurons
of the amygdala.

The thalamus, dorsal and medial nuclei are associated with the affective aspects
of pain perception, and they contain high densities of opiate receptors and
enkephalin. In contrast, the lateral portion of the thalamus, which mediates in a
somatotopic fashion the more highly localised, less affective perception of pain
which is not influenced by opiates, is largely devoid of opiate receptors and
enkephalin.

In the hypothalamus enkephalin fluorescence is localised in a band on the
ventral surface and in periventricular areas (figure 6.2). Receptors here may account
for some of the numerous endocrinological actions of opiates.

NEUROTENSIN

Neurotensin is a tridecapeptide whose amino acid sequence is pyroGlu-Leu-Tyr-
Glu-Asn-Lys-Pro-Arg-Arg-Pro-Tyr-Ile-Leu-OH (Carraway and Leeman, 1973,
1975*a*, *b*). Neurotensin was isolated from hypothalamic extracts as a by-product
of the isolation of Substance P (SP). It was detected as a material that, in small
doses, could elicit vasodilation; it also causes increased vascular permeability,
pain sensation, increased haematocrit, cyanosis, stimulation of adrenocorticotropic hormone (ACTH) secretion, increased luteinising hormone (LH) secretion,
increased follicle stimulating hormone (FSH) secretion, and hyperglycaemia.
Neurotensin also displays a variety of smooth muscle effects including contraction
of the rat uterus and guinea pig ileum.

Like enkephalin, somatostatin and SP, neurotensin is localised to the CNS and
gut (Carraway and Leeman, 1976). Radioimmunoassay reveals marked regional
differences in neurotensin levels throughout the brain (Carraway and Leeman,
1976; Uhl and Snyder, 1976; Kobayashi *et al.*, 1977). In calf brain the highest
concentrations occur in the hypothalamus, with much lower values in the
cerebellum and white matter. In the cerebral cortex, there are pronounced
variations with the highest levels in the parahippocampal gyrus. Despite general
similarities, such as high levels in the hypothalamus, there are differences between
rat and calf in regional localisation of neurotensin (Uhl and Snyder, 1976; Kobayashi
et al., 1977). Whereas endogenous neurotensin values are relatively low in calf
amygdala, rat amygdala contains moderate levels of neurotensin. Whereas neurotensin values are high in the calf caudate, levels in the rat striatum (caudate/
putamen) are relatively low. Subcellular fractionation studies indicate a localisation
of neurotensin in synaptosomal fractions which are enriched in pinched-off nerve
terminals (Uhl and Snyder, 1976).

Receptor binding
Using [^{125}I] -neurotensin, we have detected specific neurotensin receptor binding
in brain (Uhl and Snyder, 1977; Uhl *et al.*, 1977). The dissociation constant of
neurotensin for binding sites is about 3 nM. There is a single population of binding
sites with a Hill coefficient of 1.2, indicating the absence of co-operativity. The
density of binding sites in rat cerebral cortex is about 3 pmol per g wet weight,

similar to that of several neurotransmitter receptors. Marked regional variations exist in neurotensin binding which parallel, in part, variations in endogenous neurotensin. The highest binding is found in the dorsomedial thalamus, para-hippocampal cerebral cortex (which possesses the highest endogenous levels of the cerebral gyri) and hypothalamus. The lowest levels occur in the cerebellum and brainstem. The strongest evidence that the binding sites involve physiological neurotensin receptors emerges from examining the relative abilities of five partial sequence fragments of neurotensin to compete for binding. Their relative potencies correspond fairly well to their relative activities in peripheral systems. The 2-13 and 4-13 fragments have about the same potency as neurotensin in displacing [^{125}I]-neurotensin binding, while the 6-13 fragment is 50 per cent, the 8-13 fragment is 10 per cent, and the 9-13 fragment is 0.5 per cent as potent as neurotensin itself. Numerous other peptides and non-peptides have negligible affinity for neurotensin binding sites.

Histochemical localisation
The existence of specific receptor sites and a synaptosomal localisation favour a neurotransmitter or neuromodulator role for neurotensin. This possibility is strengthened by our recent immunohistochemical localisation of neurotensin to specific neuronal systems in the brain (Uhl, Kuhar and Snyder, 1977 (figures 6.3-6.7).

Utilising antisera of high specificity and high titre we have detected neurotensin by immunohistofluorescence examination of rat brain. Neurotensin immunoreactivity varies throughout the CNS in parallel with variations of endogenous levels of neurotensin reported in the rat (Kobayashi *et al.*, 1977). Control experiments, using antibody-free serum or serum antibodies previously incubated with neurotensin, show negligible fluorescence.

Immunoreactive neurotensin appears to be localised to fibre-like structures and small dots throughout the rat CNS. These resemble patterns of neuronal processes and nerve terminal varicosities associated with the histofluorescence for other brain peptides and biogenic amines (figure 6.6). Such structures appear to reflect the localisation of neurotensin to neuronal fibres and terminals. In the midbrain tegmentum and in several areas of the hypothalamus, a limited number of fluorescent cell bodies have been identified as has been the case for enkephalin (Simantov *et al.*, 1977*b*).

In the spinal cord, the distribution of neurotensin resembles that of enkephalin as well as SP and somatostatin (Hökfelt *et al*, 1976). Fluorescence occurs in lamina II and to a lesser degree in lamina I, whereas enkephalin is concentrated in both laminae I and II (figure 6.3). As with enkephalin, neurotensin-containing fibres are also scattered throughout the rest of the dorsal horn with a few in the ventral horn as well. Neurotensin fluorescence is observed to a limited extent in the fasciculus proprius in the white matter closely lateral to laminae I and II, as is the case with enkephalin.

In the medulla oblongata, just as with enkephalin, neurotensin occurs most densely in the 'substantia gelatinosa' of the caudal nucleus of the trigeminal (figure 6.3). Several components associated with the vagus nerve also display both enkephalin and neurotensin fluorescence. The nucleus tractus solitarius has

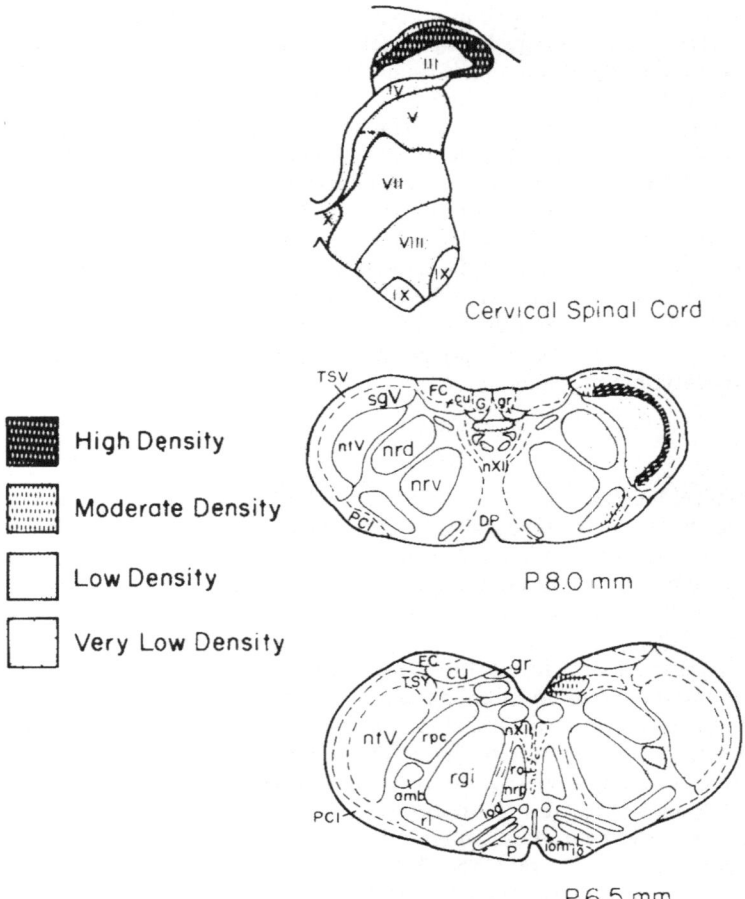

Figure 6.3 Distribution of neurotensin immunofluorescence. Abbreviations are as follows: amb, nucleus ambiguus; cu, nucleus cuneatus; DP, pyramidal decussation; FC, fasciculus cuneatus; G, fasciculus gracilis; gr, nucleus gracilis; io, nucleus olivaris inferior; iod, nucleus accessorius olivaris dorsalis; iom, nucleus accessorius olivaris medialis; nrd, nucleus reticularis medullae oblongata, pars dorsalis; nrp, nucleus reticularis paramedianus; nrv, nucleus reticularis medullae oblongata pars ventralis; ntV, nucleus tractus spinalis nervi trigemini; nXII, nucleus originis nervi hypoglossi; P, pyramid; PCI, inferior cerebellar peduncle; rl, nucleus reticularis lateralis; ro, nucleus raphe obscurans; sgV, 'substantia gelatinosa' trigemini; TSV, tractus spinalis nervi trigemini. Levels P 8.0 and P 6.5 from Palkovitz and Jakobwitz (1974), cervical spinal cord from Steiner and Turner (1972).

moderate fluorescence as does the nucleus commisuralis, formed by the confluence of the solitary nuclei from the two sides. Sparse fluorescence is also observed in the area of the nucleus ambiguus.

The most intense fluorescence in the mesencephalon occurs in the midbrain tegmentum just dorsal to the interpeduncular nucleus (figure 6.4). The zona compacta of the substantia nigra displays limited fluorescence whereas none is

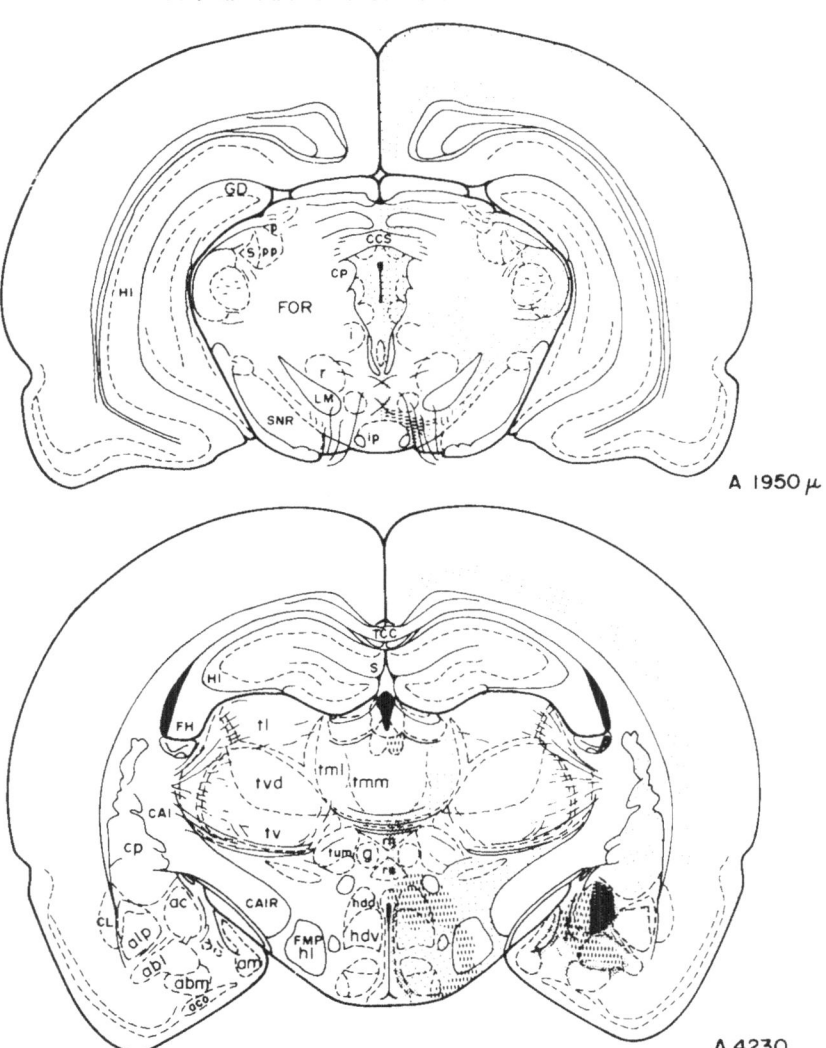

A 1950 μ

A 4230

Figure 6.4   Distribution of neurotensin immunofluorescence. Abbreviations are as follows: abl, nucleus amygdaloideus basalis, pars lateralis; abm, nucleus amygdaloideus basalis, pars medialis; ac, nucleus amygdaloideus centralis; aco, nucleus amygdaloideus corticalis; alp, nucleus amygdaloideus lateralis, pars posterior; am, nucleus amygdaloideus medialis; CAI, capsula interna; CAIR, capsula interna, pars retrolenticularis; CCS, commissura colliculorum superior; CL, claustrum; CP, nucleus caudatus putamen; FH, fimbria hippocampus; FMP, medial forebrain bundle; FOR, reticular formation; g, nucleus gelatinosus; GD, gyrus dentatus; hdd, nucleus dorsomedialis hypothalami, pars dorsalis; hdv, nucleus dorsomedialis hypothalami, pars ventralis; HI, hippocampus; hl, nucleus lateralis hypothalami; i, nucleus interstitialis (Cajal); ip, nucleus interpedunculairs; LM, Lemniscus medialis; re, nucleus reuniens; rh, nucleus rhomboideus; S, nucleus suprageniculatus; T, tapetum; TCC, truncus corpus callosi; tml, nucleus medialis thalami, pars lateralis; tmm, nucleus medialis thalami, pars medialis; tv, nucleus ventralis thalami; tvd, nucleus ventralis thalami, pars dorsomedialis, tvm, nucleus ventralis thalami, pars magnocellularis. Derived from Konig and Klippel (1963).

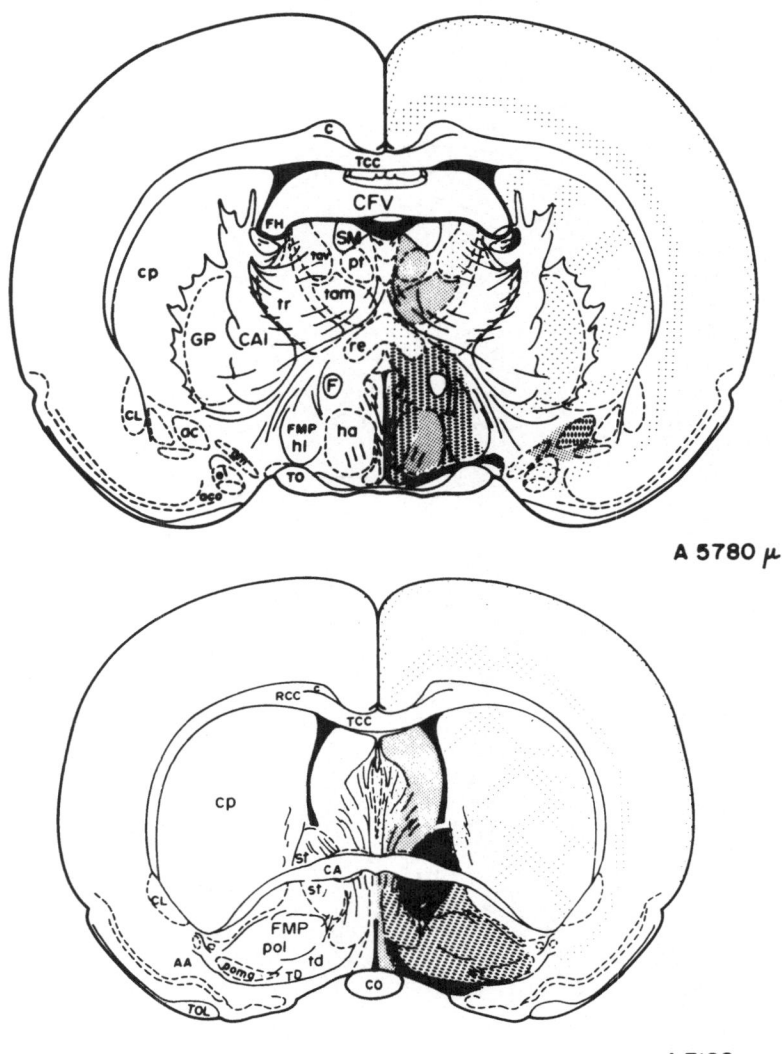

A 5780 μ

A 7190 μ

Figure 6.5 Distribution of neurotensin immunofluorescence. Abbreviations are as follows: AA, anterior amygdala; ac, nucleus amygdaloideus centralis; aco, nucleus amygdaloideus corticalis; am, nucleus amygdaloideus medialis; C, cingulum; CA, anterior commissure; CAI, internal capsule; CFV, commissura fornicis ventrailis; CL, claustrum; CO, optic chiasma, cp, nucleus caudatus putamen; F, fornix; FH, fimbria hippocampi; FMP, medial forebrain bundle; GP, globus pallidus; ha, nucleus anterior hypothalami; hl, nucleus lateralis hypothalami; pol, nucleus preopticus lateralis; poma, nucleus preopticus magnocellularis; pt, nucleus paratenialis; re, nucleus reuniens; RCC, radiatio corpus callosi; SM, stria medullaris thalami; st, nucleus interstitialis striae terminalis; tav, nucleus anterior ventralis thalami; tam, nucleus anterior medialis thalami; TCC, Truncus corpus callosi; td, nucleus tractus diagonalis (Broca); TD, tractus diagonalis (Broca); TO, tractus opticus; TOL, tractus olfactorius lateralis; tr, nucleus reticularis thalami. From Konig and Klippel (1963).

observed in the zona reticulata of this structure. Some fluorescence is observed throughout the periaqueductal grey.

Anterior areas of the hypothalamus display more fluorescence than posterior areas (figure 6.5). As with enkephalin, dense hypothalamic fluorescence is observed in a band along the ventral hypothalamic border, most pronounced in the anterior zones. Preoptic areas display fluorescence almost as dense as that of this ventral band. Unlike enkephalin, neurotensin is not highly concentrated in periventricular areas. Areas of the hypothalamus with little or no neurotensin include the fornix, the mamillothalamic pathway, the anterior commissure, optic chiasma and mamillary bodies.

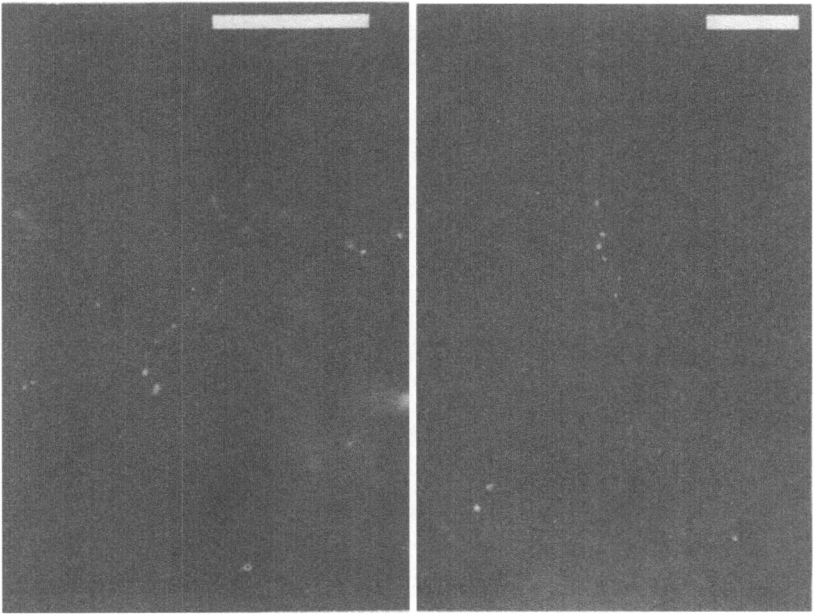

Figure 6.6 Photomicrograph of neurotensin immunohistofluorescence. Left, lateral posterior amygdala; right, zona incerta. Bars = 25 μm.

In the thalamus (figure 6.6) neurotensin, like enkephalin, is enriched in medial periventricular zones. Medial thalamic nuclei with concentrations of neurotensin fluorescence include the rhomboid, parataenial and reuniens nuclei, while the lateral complex of thalamic nuclei are virtually devoid of fluorescence.

Among the most densely fluorescent areas of the telencephalon is the central nucleus of the amygdala which is also highly enriched with enkephalin (figures 6.5 and 6.7). The lateral portion of the central nucleus is more fluorescent than the medial area and fluorescence decreases progressively in more ventral and lateral portions of the amygdala. The stria terminalis is enriched in both enkephalin and

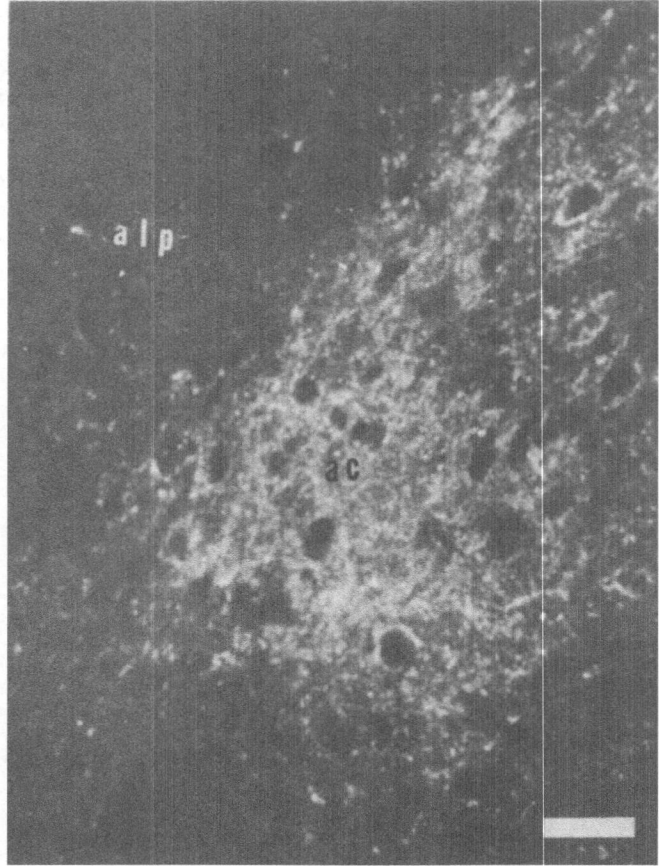

Figure 6.7 Photomicrograph of neurotensin immunohistofluorescence. ac, central amygdaloid
nucleus; alp, lateral posterior amygdala. Bar = 25 μm.

neurotensin fluorescence as is the interstitial nucleus of the stria terminalis. In the
corpus striatum, a very sparse neurotensin fluorescence occurs in patches throughout the caudate-putamen with a more even distribution in the globus pallidus
(figure 6.5).

SUMMARY

In summary, enkephalins and neurotensin share numerous properties that appear
to be common to certain neuropeptides. They are localised in the gut and in the
CNS. They interact with specific receptor sites. They are both contained within
discrete neuronal systems, and share several specific localisations. The concentration of enkephalin neurons in regions associated with pain perception and
emotional integration accords with our knowledge of opiate actions. The function
of neurotensin neurons is as yet unclear.

Despite certain similarities in the localisation of various neuropeptides there are definite differences in distribution from one to another. For instance, Substance P and somatostatin have been identified in dorsal root ganglia and are depleted from the dorsal grey of the spinal cord following dorsal rhizotomy. Thus these two peptides appear to be sensory transmitters. Enkephalin has not been detected in dorsal root ganglia and rhizotomy fails to alter spinal cord enkephalin immunofluorescence (LaMotte, Uhl, Kuhar and Snyder, in preparation). It is more likely that spinal cord enkephalin is contained within interneurons that modulate the release of sensory transmitters, perhaps by a process akin to presynaptic inhibition.

REFERENCES

Atweh, S. and Kuhar, M. J. (1977a). *Brain Res.,* **124**, 53–67

Atweh, S. and Kuhar, M. J. (1977b). *Brain Res.,* (in press)

Atweh, S. and Kuhar, M. J. (1977c). *Brain Res.,* (in press)

Bloom, F. E., Battenberg, E., Rossier, J., Ling, N., Leppaluoto, J., Vargo, T. and Guillemin, R. (1977). *Life Sci.,* **20**, 43–48

Bradbury, A. F., Smith, D. G., Snell, C. R., Birdsall, N. J. M. and Hulme, E. C. (1976). *Nature,* **260**, 793–95

Carraway, R. and Leeman, S. (1973). *J. biol. Chem.,* **248**, 6854–61

Carraway, R. and Leeman, S. (1975a): *J. biol. Chem.,* **250**, 1907–11

Carraway, R. and Leeman, S. (1975b): *J. biol. Chem.,* **250**, 1912–18

Carraway, R. and Leeman, S. (1976). *J. biol. Chem.,* **251**, 7045–52

Cox, B. M., Goldstein, A. and Li, C. H. (1976). *Proc. natn. Acad. Sci. U.S.A.,* **73**, 1821–23

Elde, R., Hökfelt, T., Johannson, O. and Terenius, L. (1976). *Neuro-sciences,* **1**, 349–55

Goldstein, A. (1976): *Science,* **193**, 1081–86

Guillemin, R., Ling, N. and Burgus, R. (1976). *C. r. hebd. Séanc. Acad. Sci., Paris D.* **282**, 783–85

Hökfelt, T., Elde, R., Johannson, O., Luft, R., Nilsson, G. and Arimura, A. (1976). *Neuroscience,* **1**, 131–36

Hughes, J., Smith, T. W., Kosterlitz, H. W., Fothergill, L., Morgan, B. A. and Morris, H. R. (1975). *Nature,* **258**, 577–79

Kobayashi, R., Brown, M. and Vale, W. (1977). *Brain Res.,* **126**, 584–88

Konig, J. and Klippel, R. (1963). *The Rat Brain.* Williams and Wilkins, Baltimore.

LaMotte, C., Pert, C. B. and Snyder, S. H. (1976). *Brain Res.,* **112**, 407–12

Palkovitz, M. and Jakobwitz, L. (1974). *J. comp. Neurol.,* **157**, 29–42

Pert, C. B., Kuhar, M. J. and Snyder, S. H. (1976). *Proc. natn. Acad. Sci. U.S.A.,* **73** 3729–33

Simantov, R. and Snyder, S. H. (1976a). *Proc. natn. Acad. Sci. U.S.A.,* **73**, 2515–19

Simantov, R. and Snyder, S. H. (1976b). In *Opiates and Endogenous Opioid Peptides* (ed. H. W. Kosterlitz), North Holland, Amsterdam, pp. 41–48

Simantov, R. and Snyder, S. H. (1977). *Brain Res.,* **124**, 178–84

Simantov, R., Childers, S. R. and Snyder, S. H. (1977a). *Brain Res.,* **135**, 358–67

Simantov, R., Kuhar, M. J., Uhl, G. R. and Snyder, S. H. (1977b). *Proc. natn. Acad, Sci, U.S.A.,* **74**, 2167–71

Snyder, S. H. (1975). *Nature,* **257**, 185–89

Steiner, T. and Turner, L. (1972). *J. Physiol., Lond.,* **222**, 123–25

Uhl, G. R. and Snyder, S. H. (1976). *Life Sci.,* **19**, 1827–32

Uhl, G. R. and Snyder, S. H. (1977). *Eur. J. Pharmac.,* **41**, 89–91

Uhl, G. R., Bennett, J. P. and Snyder, S. H. (1977). *Brain Res.,* **130**, 299–313

Uhl, G. R., Kuhar, M. J. and Snyder, S. H. (1977). *Proc. natn. Acad. Sci. U.S.A.,* **74**, 4059–63

7

Behavioural effects
of hypothalamic peptides

Arthur J. Prange, Jr, Charles B. Nemeroff and Peter T. Loosen
(Department of Psychiatry, The Neurobiology Program,
Biological Sciences Research Center,
University of North Carolina School of Medicine,
Chapel Hill, North Carolina 27514, U.S.A.)

Hypothalamic peptides appear to serve a variety of functions. While many exert classical endocrine effects, certain of them appear also to cause alterations in behaviour, alterations in behavioural responses to drugs, or alterations in systems that themselves affect behaviour. Such actions are the subject of this chapter.

We have limited our subject to those peptides that have been chemically characterised. One should recognise, however, that hypothalamic peptides not yet fully identified have been isolated. These include the elusive corticotropin releasing factor (CRF), prolactin releasing factor (PRF), and growth hormone releasing factor (GHRF).

BASIC STUDIES ON HYPOTHALAMIC RELEASING AND RELEASE-INHIBITING HORMONES

According to the portal vessel–chemotransmitter hypothesis of anterior pituitary regulation, which is now generally accepted (Knowles, 1974), after appropriate stimulation substances produced in the hypothalamus are released into the hypophyseal portal system. These hypothalamic hypophysiotropic hormones are peptides (Vale and Rivier, 1975; Vale *et al.*, 1977). They bind to specific membrane receptors in the anterior pituitary and cause (or inhibit) the release of adenohypophyseal hormones. Several reviews of the endocrine actions of the hypothalamic hormones are available (McCann *et al.*, 1974; Reichlin *et al.*, 1976; Vale *et al.*, 1977).

Thyrotropin releasing hormone (TRH)
TRH, a tripeptide (pyroGlu–His–Pro–NH$_2$), was the first hypothalamic hormone to be chemically identified (Vale and Rivier, 1975). Radioimmunoassay procedures for the measurement of this tripeptide are now available. Immunoreactive TRH is present in both hypothalamic and extrahypothalamic brain areas (Brownstein *et al.*, 1974; Jackson and Reichlin, 1974*a*, *b*; Winokur and Utiger, 1974; Olivier *et al.*,

1974), a result not predicted by the portal vessel–chemotransmitter hypothesis. The presence of immunoreactive TRH in extrahypothalamic brain areas and its low acute toxicity in the rat (LD_{50} = 2500 mg/kg i.v.) (Piva and Steiner, 1972) have heightened interest in studies of its behavioural effects.

TRH potentiates the excitation caused by L–DOPA in pargyline-pretreated mice and rats (Plotnikoff *et al.*, 1972), a test devised by Everett (1966) for screening drugs for antidepressant properties. This effect of TRH is apparently independent of its action on the pituitary–thyroid axis since TRH also acts in hypophysectomised and thyroidectomised rats (Plotnikoff *et al.*, 1974*a*, *b*). Moreover, neither thyrotropin (TSH) nor thyroid hormones are active in this paradigm (Plotnikoff *et al.*, 1974*a*, *b*). In similar studies on the interaction of TRH and serotonergic systems, the tripeptide was found to potentiate the behavioural effects of administered serotonin (5-hydroxytryptamine, 5-HT) precursors (Green and Grahame–Smith, 1974; Huidobro–Toro *et al.*, 1974, 1975).

TRH, administered peripherally or centrally, antagonises the sedation and hypothermia induced by barbiturates, ethanol and several other central depressants (Prange *et al.*, 1974; Breese *et al.*, 1974*a*, *b*; Bissette *et al.*, 1976*a*).

Table 7.1   Effect of intracisternal (i.c.) and intraperitoneal (i.p.) injection of TRH on pentobarbitone-induced sedation in the mouse

Regain righting time (% of control ± s.e.m.)	
Control (saline)	TRH
i.c.* (N = 240) 100.0 ± 1.66	(N = 240) 49.98 ± 0.93‡
i.p.† (N = 430) 100.0 ± 1.67	(N = 430) 46.21 ± 0.82‡

*All mice received 55 mg/kg sodium pentobarbitone (i.p.) 10 min before injection (i.c.) of 10 μl vehicle (0.9% NaCl) or 10 μg TRH in 10 μl vehicle.
†All mice received 55 mg/kg sodium pentobarbitone (i.p.) 2 min after pretreatment with 1 mg/kg TRH or vehicle (i.p.).
‡P < 0.001 one-way ANOVA, Duncan's multiple range test.

Data were collected from a series of studies of TRH congeners, in which saline-treated and TRH-treated animals were included for comparison.

Table 7.1 summarises the results of several experiments conducted in our laboratory. TRH is clearly a potent barbiturate antagonist. This analeptic effect has been found in rats, mice, hamsters, rabbits, gerbils, guinea pigs and monkeys (Breese *et al.*, 1975; Brown and Vale, 1975*a*; Kraemer *et al.*, 1976). It appears to occur independently of the classical endocrine actions of TRH, that is, (1) it is not modified by hypophysectomy (Breese *et al.*, 1975); (2) it is not produced by treatment with thyrotropin (TSH) or thyroid hormones (Breese *et al.*, 1975); (3) analogues of TRH show a dissociation of analeptic potency and TSH releasing potency (Prange *et al.*, 1975*a*). Thus, pyrazolyl TRH is approximately equipotent with TRH in reversing the sedative properties of barbiturates, although it possesses only 5 per cent of the thyrotropin-releasing potency of the parent compound. TRH does not exert its analeptic action by altering the metabolism of barbiturates (Breese *et al.*, 1975). Although the mechanism by which TRH exerts its analeptic effects is uncertain, some studies suggest the involvement of cholinergic (Cott *et al.*, 1976) and gabaminergic systems in the central nervous system (CNS) (Cott and Engel, 1977).

TRH and pentobarbitone appear to interact with other systems. Brown and Vale (1975*b*) reported that the tripeptide inhibits the release of growth hormone (GH) *in vivo* induced by pentobarbitone or morphine. Collu *et al.* (1975) reported that TRH also blocks pentobarbitone-induced prolactin (PRL) release.

Since we had demonstrated that TRH *antagonises* the sedative effects of barbiturates, we were interested in the effect of TRH on the anticonvulsant properties of these drugs. Accordingly, we examined the effect of the tripeptide and a linear tetrapeptide analogue (pyroGlu-His-Pro-β-Ala-NH$_2$) on the action of phenobarbitone in a laboratory model of grand mal epilepsy. Both peptides, surprisingly, *potentiated* the antiepileptic action of phenobarbitone in mice, an effect not mimicked by TSH or thyroid hormones (Nemeroff *et al.*, 1975). Thus, TRH or its tetrapeptide analogue may deserve consideration as adjuncts to the phenobarbitone treatment of grand mal epilepsy. In contrast, TRH had no effect on the anticonvulsant potency of trimethadione (Nemeroff *et al.*, 1977*a*). Amphetamine, like TRH, antagonises barbiturate sedation and enhances barbiturate efficacy in the treatment of epilepsy. The two substances share other properties as well (Simone *et al.*, 1975; Lipton *et al.*, 1977) (table 7.2).

Table 7.2   Pharmaco-behavioural comparison of TRH and amphetamine

	TRH	(+)-Amphetamine
(1) Antagonism of pentobarbitone-induced narcosis and hypothermia	+	+
(2) Induces hyperactivity (locomotor)	+	+
(3) Anorexic activity	+	+
(4) Ability to act as a discriminative stimulus	+	+
(5) Induction of stereotypic behaviour	−	+
(6) Antagonism of ethanol-induced sedation	+	−
(7) Antagonism of chlorpromazine-induced muscle relaxation in mice	+	+
(8) Actions antagonised by 6-hydroxydopamine pretreatment	−	+
(9) Inhibition of isolation-induced fighting in mice	+	−

(Adapted from Prange *et al.*, 1977 and Lipton *et al.*, 1977. See text for references)
+ = activity
− = no activity

Apart from its interactions with other drugs, TRH appears to exert independent effects in the untreated animal. Intravenous (i.v.) injection of the tripeptide causes muscle tremor, piloerection, tail lifting and 'wet dog' shakes (Schenkel–Hulliger, 1974). Similar results are produced by direct injection into specific brain regions (Wei *et al.*, 1975). Administered i.v. or centrally, it induces hyperthermia and excitation in rabbits (Horita and Carino, 1975). On the other hand, TRH is a potent hypothermic agent in cats (Metcalf, 1974). Segal and Mandell (1974) found that intracerebral infusion of TRH to the ventricles (i.c.v), increases locomotor activity of rats. Our group has shown that the tripeptide produces dose-related decreases in food consumption and food-reinforced, fixed ratio, bar-press responding. No effects on electrical self-stimulation or on active avoidance conditioning have been observed (Barlow *et al.*, 1975). Jones and her colleagues (1975, and personal communication), working in our laboratory, found that rats distinguish between TRH (administered peripherally or centrally) and saline conditions when tested repeatedly in a state-dependent learning situation. Malick (1976) has reported that small doses of the peptide antagonise isolation-induced aggression in mice. The i.c.v. administration of $200\mu g$ TRH to cats has been reported to lead to changes in the sleep–wakefulness cycle. Total time awake increased and both slow wave and REM sleep were inhibited (King, 1975). The central administration of 200 μg TRH to curarised rabbits leads to an activated electroencephalogram (EEG) profile (White and Beale, 1975). These findings are consistent with the notion that TRH is a physiological excitant.

Several groups have studied the neurochemical effects of TRH. TRH administration does not appear to change brain levels of monoamine neurotransmitters (Breese *et al.*, 1975; Plotnikoff *et al.*, 1975; Reigle *et al.*, 1974). However, both biochemical and histochemical evidence suggests that it enhances brain noradrenaline (NA) turnover (Keller *et al.*, 1974; Constantinidis *et al.*, 1974). We found that chronic TRH administration has no effect on regional rat brain tyrosine hydroxylase activity (Nemeroff *et al.*, 1977b). TRH does, however, cause an enhanced release of NA and dopamine (DA) from presynaptic nerve endings (Horst and Spirt, 1974; Reigle *et al.*, 1974).

Melanocyte stimulating hormone release-inhibiting factor (MSH–IF)

The identification of an endogenous hypothalamic peptide that inhibits the release of melanocyte stimulating hormone (MSH) from the intermediate lobe of the pituitary is a controversial issue in neuroendocrinology (Vale *et al.*, 1977). Two peptides that inhibit MSH release have been reported (Nair *et al.*, 1971a, b, 1972): MSH-IF I (Pro–Leu–Gly–NH$_2$) and MSH-IF II (Pro–His–Arg–Gly–NH$_2$). Whether these factors are indeed the physiological mediators of MSH secretion is not yet certain.

MSH-IF I has been tested extensively for behavioural effects. It markedly potentiates the stimulant properties of L-DOPA (Plotnikoff *et al.*, 1971), and it is active after oral or parenteral administration in intact and hypophysectomised animals (Plotnikoff *et al.*, 1974c). These findings naturally led to studies of the effect of MSH-IF I on brain DA systems. The tripeptide has been reported to cause a dose-related increase in striatal DA synthesis (Friedman *et al.*, 1974). Spirtes *et al.* (1975) reported that DA and NA levels in brain are increased in animals receiving MSH-IF I during the DOPA potentiation test. In addition, the tripeptide abolishes oxotremorine–induced tremors in mice and hormone–induced tremors in rabbits (Plotnikoff and Kastin, 1974, 1976). MSH-IF I has no effect on the cardiovascular system, possesses no anticonvulsant properties, and does not affect locomotor activity in intact or methamphetamine-treated animals. It does appear to activate the EEG and to possess weak analgesic activity (Plotnikoff and Kastin, 1974).

MSH-IF I has no effect on pentobarbitone-induced sedation (Prange *et al.*, 1974, 1975a) or on the anticonvulsant properties of phenobarbitone (Nemeroff *et al.*, 1975). The pharmacological properties of MSH-IF I suggests interactions with brain DA systems: DOPA potentiation, increased DA biosynthesis, oxotremorine and harmine antagonism. Moreover, the tripeptide induces dose-dependent stereotypic behaviour in cats (North *et al.*, 1973). The behaviour is comparable to that observed after L-DOPA, amphetamine or other treatments known to enhance dopaminergic transmission. Further support for this concept comes from studies which indicate that MSH-IF I administration results in increased mounting behaviour in apomorphine-treated male rats. A recent report (Barbeau *et al.*, 1975) found that MSH-IF I potentiates the increase in motor activity in apomorphine-treated rats. These data have led to clinical trials of MSH-IF I in Parkinson's disease, and preliminary results appear promising (see below).

Somatotropin release-inhibiting factor (SRIF)
The hormone that inhibits the secretion of GH is a tetradecapeptide (H-Ala-Gly-Cys-Lys-Ala-Phe-Phe-Trp-Lys-Thr-Phe-Thr-Ser-Cys-OH). It was characterised by Brazeau *et al.* (1973). Radioimmunoassay and immunohistochemical techniques have shown immunoreactive SRIF in hypothalamus, extrahypothalamic brain (Hökfelt *et al.*, 1974; Brownstein *et al.*, 1975), pancreas and stomach (Arimura *et al.*, 1975; Rufener *et al.*, 1975).

In addition to its inhibition of GH, SRIF inhibits the release of insulin, glucagon, renin, vasoactive intestinal peptide and gastrin. It also blocks TRH-induced TSH release in man and other animals (see Vale *et al.*, 1977 for review). Thus comparisons of the behavioural actions of TRH and SRIF are of special interest. Segal and Mandell (1974) noted that after i.c.v. infusion in rats TRH and SRIF exerted contrasting effects on locomotor activity. TRH caused an increase, SRIF a decrease. We studied the influence of SRIF on pentobarbitone-induced sedation. In contrast to TRH, SRIF, given intraperitoneally (i.p.), caused a slight but definite extension of barbiturate-induced sleeping time (Prange *et al.*, 1975*a*, *b*; Brown and Vale, 1975*a*). However, small doses (25 µg subcutaneously) of SRIF block the rise in GH seen after sodium thiamylal or morphine (Brown and Vale, 1975*b*).

Cohn and Cohn (1975) found that the i.c.v. injection of 5–45 µg of SRIF in rats causes sedation and hypothermia. They also observed 'barrel-rolling' behaviour (animals roll laterally). Rezek *et al.* (1976) investigated changes in behaviour, motor co-ordination, sleep–wakefulness cycle and electrical activity after SRIF infusion into two areas of the hippocampus. They noted stereotyped behaviour, changes in the sleep–wakefulness cycle and dissociation between EEG activity and behavioural state. Rezek and his colleagues suggested that the hippocampus is at least one of the anatomical substrates mediating the central, non-endocrine actions of SRIF.

Plotnikoff *et al.* (1974*d*) reported that SRIF is active in the DOPA-potentiation test in mice. They found it inactive in altering oxotremorine-induced tremors, the 5-hydroxytryptophan (5-HTP) potentiation test, audiogenic seizures and footshock-induced fighting in mice.

A recent study has suggested a role for SRIF in behavioural homeostasis. Terry *et al.* (1976) reported that the administration of antisera to SRIF blocks the fall in GH which otherwise occurs in stressed rats. Arimura *et al.* (1976) have confirmed these findings.

Luteinising hormone-releasing hormone (LHRH)
LHRH is a decapeptide (pyroGlu-His-Trp-Ser-Tyr-Gly-Leu-Arg-Pro-Gly-NH$_2$) (Matsuo *et al.*, 1971; Monahan *et al.*, 1971) that releases the gonadotropins, luteinising hormone (LH) and follicle stimulating hormone (FSH), from the adenohypophysis. The vast majority (80 per cent) of LHRH is localised in the hypothalamus with the remaining 20 per cent in circumventricular organs (see Brownstein *et al.*, 1976*a* for review). LHRH has been reported in extrahypothalamic brain areas (Wilber *et al.*, 1976), but this finding is controversial. The induction of lordosis behaviour

by very small quantities (150–500 ng subcutaneously (s.c.)) of LHRH in ovariecto-mised (Moss and McCann, 1973) or ovariectomsed, hypophysectomised (Pfaff, 1973) oestrogen-primed female rats is one of the most impressive demonstrations of a direct effect of a peptide on the CNS. Other peptides are inactive. LHRH does not exert this action by releasing adrenal progesterone, since Moss (1975) showed that the phenomenon occurs in adrenalectomised rats. Selective injection of LHRH into the hypothalamus produces similar effects (Moss and Foreman, 1976). LHRH also affects sexual behaviour in the male; in intact, or castrated, testosterone-primed male rats, 500 ng of LHRH accelerates ejaculation (Moss *et al.*, 1975).

Other effects of LHRH have been described. Although peripherally administered LHRH has no effect on pentobarbitone-induced sedation, the central administration of the decapeptide markedly reduces barbiturate-induced sleeping time (Bissette *et al.*, 1976a). In the pole-jumping (active avoidance) paradigm utilised by de Wied and his colleagues (1975) LHRH is as potent as $ACTH_{4-10}$ in inhibiting the extinc-tion of the response.

Like MSH-IF I and TRH, LHRH (4–8 mg/kg, oral or i.p.) has been reported to potentiate the stimulant properties of L-DOPA in mice pretreated with pargyline (Plotnikoff *et al.*, 1976). LHRH (2 and 8 mg/kg i.p.) also enhanced the stimulant properties of 5-HTP, a serotonin precursor, in pargyline-pretreated mice. The decapeptide had no effect on oxotremorine-induced tremors, aggressive behaviour or audiogenic seizures in mice.

OTHER HYPOTHALAMIC PEPTIDES

Substance P

This undecapeptide (H-Arg-Pro-Lys-Pro-Gln-Gln-Phe-Phe-Gly-Leu-Met-NH$_2$) was characterised (Chang *et al.*, 1971) 40 years after von Euler and Gaddum (1931) discovered a potent hypotensive substance in alcoholic extracts of equine brain and intestine. Substance P (SP) is present in intestinal and nervous tissue of several verte-brate species. In agreement with bioassay and radioimmunoassay data (Brownstein *et al.*, 1976b), immunohistochemical studies have revealed a heterogeneous distri-bution of SP in brain and spinal cord. Recently, SP-like immunoreactivity has been observed in nerve endings in the human cortex (Hökfelt *et al.*, 1976). This differential distribution of SP, its synaptosomal localisation and the presence of enzymes in ner-vous tissue that inactivate it suggest that SP is a viable meurotransmitter candidate (see Leeman and Carraway 1977 for review).

Intramuscularly administered SP (0.5 mg/kg) abolishes the abstinence syndrome in morphin-treated mice and also tranquillises aggressive mice, but has no effect on strychnine-induced seizures (Stern and Hadzovic, 1973). In addition, intracistern-ally (i.c.)-administered SP exerts a significant analeptic action (Bissette, Nemeroff, Loosen, and Prange, unpublished observations).

Neurotensin

This tridecapeptide (pyroGlu-Leu-Tyr-Glu-Asn-Lys-Pro-Arg-Arg-Pro-Tyr-Ile-Leu-OH) was isolated from hypothalamus and later characterised. It is distributed heterogereously in brain (Carraway and Leeman, 1973, 1975; Kobayashi *et al.*, 1977). I is also present in the gastrointestinal tract of dogs (Orci *et al.*, 1976). It is classifieu as a kinin because of its effects on peripheral vasculature and smooth muscle preparations *in vitro* (Carraway and Leeman, 1973, 1975). When administered peripherally, neurotensin induces hypotension, cyanosis, hyperglycaemia, hypoin-sulinaemia and hyperglucagonaemia (Carraway and Leeman, 1975; Brown *et al.*, 1976;

Nagai and Frohman, 1976). These effects of neurotensin on blood sugar homeostasis occur after peripheral but not after central injection (Nagai and Frohman, 1976). Controversy exists concerning the effects of neurotensin on pituitary hormone secretion. Makino *et al.* (1973) reported that in the rat the peptide induces release of corticotropin (ACTH) and gonadotropins. Rivier and her colleagues (1977) reported that neurotensin is a potent releaser of GH and PRL in the rat.

Initial studies in our laboratory demonstrated that i.c. injection of neurotensin markedly potentiates pentobarbitone-induced narcosis and lethality (Nemeroff *et al.*, 1976, 1977*c*). Further studies utilising radiolabelled pentobarbitone revealed that centrally administered neurotensin decreased the rate of metabolic degradation of the barbiturate (Nemeroff *et al.*, 1977*d*). These observations led us to the examination of the effects of centrally administered neurotensin on thermoregulatory

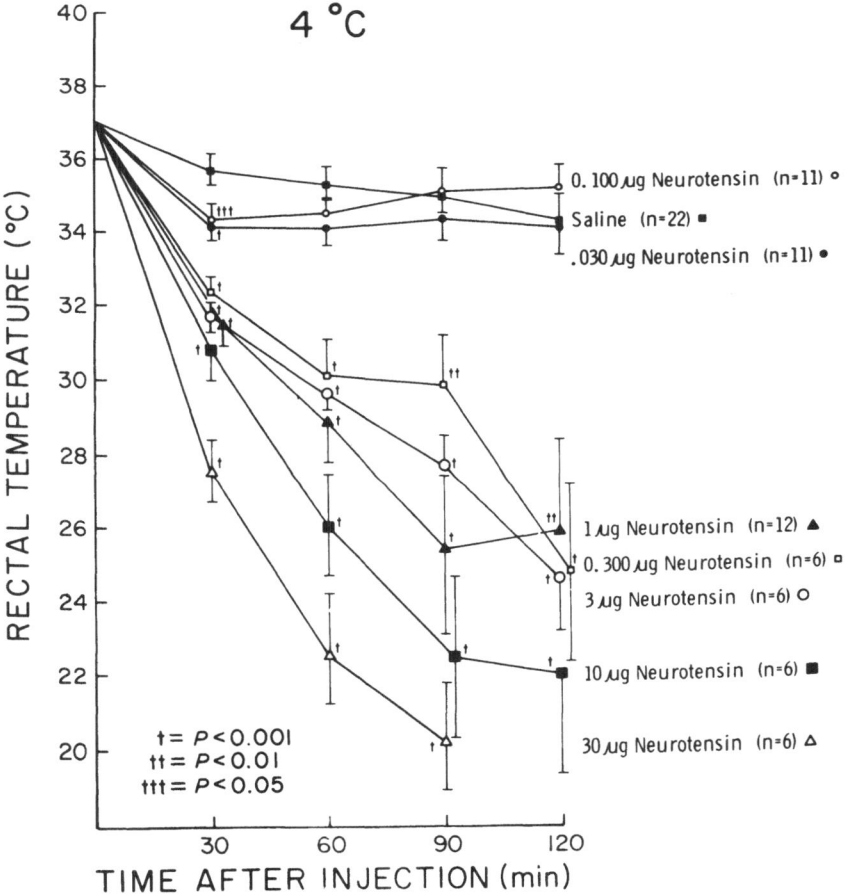

Figure 7.1 Effect of i.c. neurotensin on the core temperature of mice subjected to a cold environment (4 °C). Data are expressed as °C ± s.e.m. Experimental values were compared with those of controls by Student's *t* test (two-tailed).

processes in laboratory animals. It induces a marked dose-related hypothermia (Bissette *et al.*, 1976*b*), an effect exaggerated when the animals are placed in an ambient temperature of 4°C (figure 7.1). Peripherally administered neurotensin, even in doses as high as 10 mg/kg, has no such effect. Centrally administered neurotensin (30 μg) also results in a short-lived decrease in spontaneous motor activity (Nemeroff *et al.*, 1977*d*). The hypothermic effect of neurotensin is strikingly specific; other hypothalamic peptides (TRH, LHRH, MSH-IF, SP) do not possess this activity (Bissette *et al.*, 1976*b*). The central administration of the tridecapeptide results in hypothermia in several different mammalian species including the rat, mouse, gerbil, hamster and monkey (Nemeroff *et al.*, 1977*d*). The results suggest that neurotensin may play a physiological role in temperature regulation. This notion is supported by the demonstration of brain receptor binding of radiolabelled neurotensin (Uhl *et al.*, 1977). Neurotensin or an active congener may find clinical utility in modifying hyperpyrexia.

CLINICAL STUDIES ON HYPOTHALAMIC RELEASING AND RELEASE-INHIBITING HORMONES

TRH: Behavioural actions

In early studies of depressed patients we examined interactions between imipramine and L-triiodothyronine (Prange *et al.*, 1969; Wilson *et al.*, 1970) and between imipramine and TSH (Prange *et al.*, 1970). More recently we have explored the possible antidepressant effects of TRH. Our interest in this area was stimulated by the finding that TRH is active in the pargyline–DOPA test, a screening procedure for putative antidepressant substances.

In a double blind, crossover study of ten women with unipolar depression TRH 0.6 mg given as an i.v. bolus, exerted a rapid, though partial and brief, beneficial effect. Patients improved measurably within a few hours and showed the greatest improvement the following day. One week after treatment the patients had relapsed to pretreatment severity (Prange and Wilson, 1972; Prange *et al.*, 1972) (figure 7.2).

Other investigators have addressed themselves to the question of whether TRH could be used as an effective clinical treatment for depression. The results are disappointing (see Prange *et al.*, 1978 for review), though it is difficult to generalise about these trials because frequency and size of dose and route of administration have varied substantially. Furlong *et al.* (1976) have suggested that the disparity between findings might be resolved by attention to patterns of endocrine disturbances in depressed patients.

Since some treatments useful in depression are also active in mania, it was of interest to examine TRH for possible antimanic effects. In a double-blind, placebo-controlled study of five manic men, Huey *et al.* (1975) found reliable advantages for TRH as against saline.

Depression is a common clinical feature in alcoholism, including alcoholic withdrawal (Weingold *et al.*, 1968; Woodruff *et al.*, 1973; Tyndel, 1974). For this reason we examined the behavioural effects of TRH in the latter condition. In a double-blind study of 33 men undergoing alcohol withdrawal we compared TRH with two placebo injections, nicotinic acid and saline (Loosen *et al.*, in preparation). A reliable beneficial effect of TRH was found on factor I of the Hamilton rating scale for depression, a measure concerned mainly with retardation and depressed mood, but only in the first few hours after injection (figure 7.3). At other times during the following week the differences between the groups were statistically insignificant.

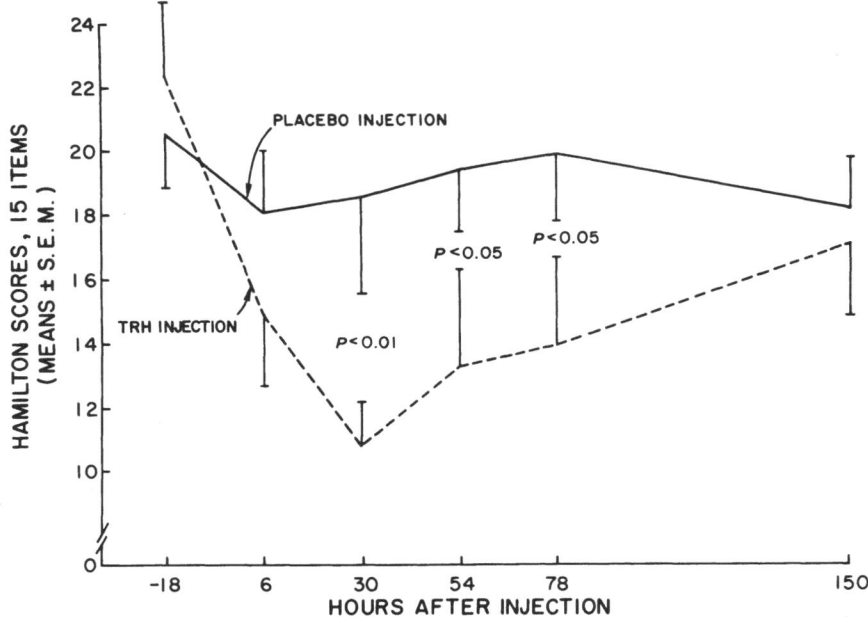

Figure 7.2 Ten women with primary unipolar depression were given TRH, 0.6 mg i.v., and then saline 1 week later, or were given the substances in reverse order according to a random sequence. Improvement after TRH was rapid, but partial, and relapse to baseline occurred within a week. Mean values ± s.e.m. are plotted.

Huey *et al.* (1975), also using a single i.v. injection of TRH treated three patients in a state of pre-delirium tremens. They observed no beneficial effects. However, two patients in milder stages of withdrawal showed improvement in their sense of well-being and in increased relaxation.

After our preliminary study of four schizophrenic patients (Wilson *et al.*, 1973), we performed a double-blind trial, controlled by use of i.v. nicotinic acid to mimic the side effects of TRH (Wilson *et al.*, in preparation). Figure 7.4 shows that in these ten patients, with the prominent features of process schizophrenia, TRH exerted a clearly beneficial effect. As in depression, the extent of improvement was about 50 per cent. Relapse to baseline severity was complete in about two weeks, though patients showed considerable variability in this regard.

After reviewing our own work and seven other reports we tried to find a consensus about the effects of TRH in schizophrenia (Wilson *et al.*, in preparation). TRH aggravates the mental state of patients diagnosed as paranoid while it benefits patients who display prominent social withdrawal and anhedonia. In the latter group, improvement is not limited to the sphere of affectivity but pertains to all aspects of psychosis, including thinking disorders. In a large multi-hospital, controlled study, Inanaga *et al.* (1975*a*, *b*) confirmed this generalisation as it applies to withdrawn patients. Since TRH may exert dopaminergic properties, these findings traduce the relevance of the DA hypothesis (Snyder, 1976) for some subgroups of schizophrenic patients.

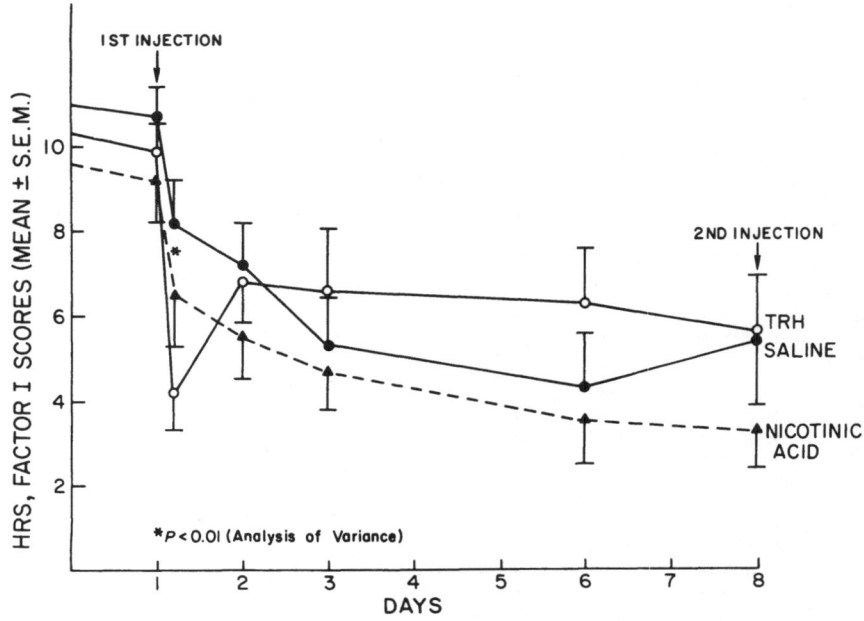

Figure 7.3 Ten patients in acute alcohol withdrawal were given TRH, 0.5 mg i.v., while an equal number were given saline and an equal number nicotinic acid, 1.0 mg i.v. TRH showed a reliable advantage a few hours after injection but thereafter the treatment groups were not statistically different. HRS, Hamilton Rating Scale for Depression. Factor I is concerned largely with depressed mood and psychomotor retardation. Mean values ± s.e.m. are plotted.

Tiwary *et al.* (1975) reported that TRH exerts a therapeutic effect in hyperkinetic children. Campbell found that TRH reduced hyperactivity in five of seven autistic children (see Prange *et al.*, 1978, for review).

Since TRH may enhance brain DA activity, several investigators have assessed the action of the tripeptide in Parkinson's disease (Chase *et al.*, 1974; Lakke *et al.*, 1974; McCaul *et al.*, 1974). No consistent effects have been found. However, it is interesting to note that McCaul *et al.* (1974) reported that two of three Parkinsonian patients taking L-DOPA, when given TRH, experienced a 'dramatic improvement in well-being, including enhanced clarity of thought'.

TRH: endocrine responses

Administration of TRH presents the opportunity to observe simultaneously both behavioural changes and endocrine responses. To identify deviant endocrine responses in our patients, we noted the lowest and highest individual TSH peak response shown by normal control subjects, matched with patients for sex and age. We then inspected the TSH response data for each patient and classified them accordingly. The results of this classification are shown in table 7.3.

No member of any patient population showed a TSH response greater than the greatest response shown by a control subject. Diminished TSH responses were

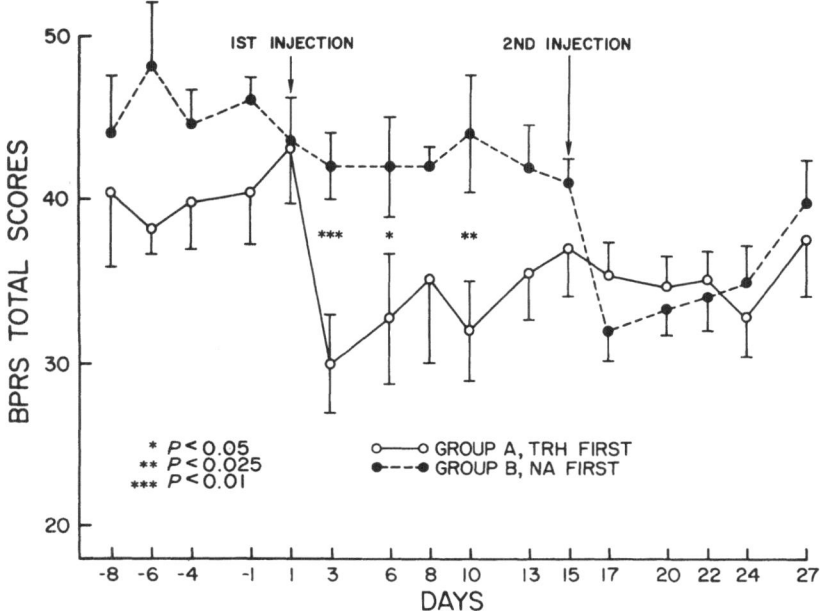

Figure 7.4 Five schizophrenic patients were given TRH, 0.5 mg i.v., and two weeks later nicotinic acid, 2.0 mg i.v., while five others were given the substances in reverse order according to a random sequence. As in depressed patients, improvement was rapid, but partial. Its duration was quite variable. Severity was assessed by use of the Brief Psychiatric Rating Scale (BPRS). Mean values ± s.e.m. are plotted.

observed in depressed and alcoholic patients, but not in schizophrenics. In the alcoholic patients, TSH blunting was observed in both the withdrawal and the post-withdrawal state.

Considering our own findings and those of others, we attempted to establish the context of TSH blunting in depression (Loosen *et al.*, 1976; Prange, 1977). The defect does not seem to be related to previous drug intake, diagnosis, sex, age or severity of illness. In a similar way, there seems to be no clear relationship between TSH blunting and behavioural response to TRH injection. Some patients, both depressive and alcoholics, who demonstrate this fault during illness do not correct it on recovery (Coppen *et al.*, 1974; Kirkegaard *et al.*, 1975; Maeda *et al.*, 1975; Loosen *et al.*, 1977).

The phenomenon of a blunted TSH response to TRH must also be scrutinised on endocrinologic grounds. Hyperthyroidism is the most frequent cause (Hollander *et al.*, 1972). In addition TSH blunting has been observed in Klinefelter's syndrome (Ozawa and Shishiba, 1975), 36 hour starvation (Vinik *et al.*, 1975) and chronic renal failure (Czernichow *et al.*, 1976). However, these disorders were excluded in our patients. Some degree of thyroid activation is sometimes seen in depression (Whybrow *et al.*, 1972), and it is possible that heightened thyroid state even within the normal range could depress TSH response through enhanced negative feedback.

Table 7.3   Distribution of abnormal peak TSH responses to TRH injection
in psychiatric patients

	Depression $n = 23$	Schizophrenia $n = 17$	Severe alcohol intoxication	
			Withdrawal $n = 12$	Post-withdrawal $n = 14$
TSH ↑	0	0	0	0
TSH ↓	6	0	4	3
TSH normal	17	17	8	11

However, all authors who have studied this phenomenon in depression have
found no correlation between thyroid indices and TSH response. In fact, Takahashi
et al. (1974) found that patients with *low* free thyroxine indices often showed the
most diminished TSH responses. In our alcoholic patients we found moderate thyroid activation in early withdrawal but not in the post-withdrawal state (Loosen
et al., in preparation). Thus thyroid activation appears unlikely to account for TSH
blunting observed in the latter condition.

Another possible explanation for TSH blunting, also based on endocrine changes,
requires consideration. Several authors have shown that the administration of gluco-
corticoids diminishes the TSH response to TRH (Otsuki *et al.*, 1973; Re *et al.*,
1976). Accordingly, we studied the relationship between baseline serum cortisol
levels and TRH-induced TSH release in our patients. These results are summarised

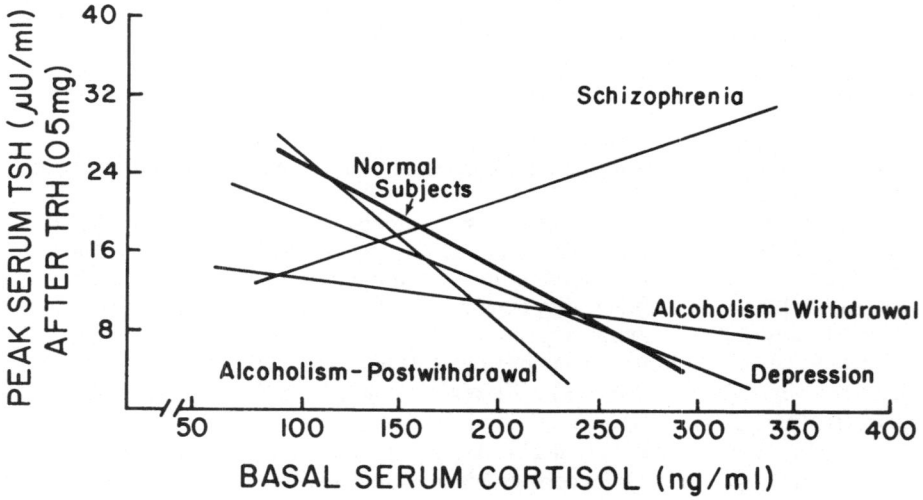

Figure 7.5 For the various clinical groups shown and for normal subjects cortisol at baseline
before injection was plotted against peak TSH response to TRH injection. All patient groups,
except schizophrenics, tended to show the same inverse relationship between variables as did
normals. For alcoholics the relationship more closely resembled the normal one after withdrawal than during it. Schizophrenics showed loss of integration between the two endocrine
systems. 'Best' lines were computed for each group.

in figure 7.5, wherein correlations are expressed by interslope lines. Basal cortisol and TSH peak response showed a significant inverse relationship in normal subjects and depressed patients. An inverse though not statistically significant, relationship between these parameters was also found in alcoholic patients. The relationship more nearly resembled that of normals in the post-withdrawal state than in early withdrawal. Unlike all other groups, schizophrenic patients tended to show a *positive* correlation between the two variables.

The foregoing data suggest that cortisol elevation, which is known to occur in some depressed patients, accounts for TSH blunting in this condition. However, we have noted, as have others (Coppen *et al.*, 1974; Kirkegaard *et al.*, 1975; Maeda *et al.*, 1975), that the blunted TSH response in depression may persist after remission. In such patients the abnormality probably cannot be attributed to elevated cortisol nor to any other aspect of the state of depression Thus, in some patients a blunted TSH response may be related to the trait of depression rather than to the state of being depressed. This possibility runs parallel to an observation we have made concerning alcoholic patients. In this population, blunted TSH responses were found in acute withdrawal and, less frequently, in the post-withdrawal state as well. Thus it is possible that in alcoholism, as in depression, some instances of TSH blunting represent traits of patients suffering from these disorders rather than states of the disorders themselves.

MSH-IF I

Ehrensing and Kastin (1974) studied 18 depressed women in a double-blind trial of oral MSH-IF I. Six patients received 60 mg per day; six received 150 mg; six received placebo capsules. The authors reported a marked beneficial effect from the lower dose of the peptide.

While both TRH and MSH-IF I potentiate the behavioural effects of L-DOPA in mice, MSH-IF I is clearly more potent (Huidobro-Toro *et al.*, 1975). This suggests that MSH-IF I might exert a greater benefit than TRH in Parkinson's disease and this appears to be the case. All trials of MSH-IF I in Parkinson's disease have produced positive results (Kastin and Barbeau, 1972; Kastin *et al.*, 1976). Large doses of MSH-IF I have been used, a tactic that can be used since the pituitary secretions affected by MSH-IF I, unlike those affected by TRH, lack a discrete target organ. MSH-IF I, of course, inhibits the release of MSH, which has been reported to be elevated in patients with Parkinson's disease; administration of MSH also exacerbates Parkinsonian symptoms (see Kastin *et al.*, 1976 for review). Whether the benefits of MSH-IF I in Parkinson's disease are related to dopamine activation or to MSH inhibition is uncertain.

LHRH

Three groups have given LHRH, all by different routes, to impotent men with or without evidence of endocrine disorder (Mortimer *et al.*, 1974; Benkert, 1975; Schwartzstein *et al.*, 1975), and quite variable results have been obtained. Mortimer *et al.* (1974) suggested that LHRH may exert a behavioural effect independent of its endocrine effects in selected patients: 'In six of the seven adult patients there was an early increase in potency, 7 to 14 days after starting therapy, which was maintained despite circulating 17-β-hydroxyandrogen levels well below the lower limit of the normal male range.'

Benkert (1975) gave LHRH or TRH (0.5 mg i.v.) or saline, each on one occasion, to a small series of depressed patients. Both hormones were somewhat more effective antidepressants than saline, though the effect of neither was statistically superior to the effect of saline.

Brambilla *et al.* (1976) performed LHRH challenge tests in chronic schizophrenic patients. They noted increases in both FSH and LH which were greater than those of control subjects. A recent study has demonstrated a marked increase in the concentration of immunoreactive LHRH in the median eminence of women suffering from Huntington's chorea (Bird *et al.*, 1976). The significance of this finding is at present unknown.

Other hypothalmic peptides
We do not know of any reports that might suggest that Substance P or neurotensin exert behavioural effects in man.

DISCUSSION
We have stated elsewhere what we have termed a generic hypothesis (Prange *et al.*, 1975*b*, 1978; Lipton *et al.*, 1977). 'Hypothalamic releasing hormones may exert brain effects apart from their actions on the anterior pituitary gland. These brain effects may have neurologic or behavioural consequences, and in clinical disorders these consequences may manifest themselves as benefits or as aggravations.' Martin and his colleagues (1975) have stated a similar view. Of course, the concept that a substance may serve both an endocrine role and a non-endocrine role has precedent. Adrenaline of adrenal medullary origin, for example, acts as a circulating hormone and as a neurotransmitter in the CNS (Hökfelt *et al.*, 1973). Moreover, de Wied (chapter 16) and his colleagues have argued that both anterior and posterior pituitary hormones exert both endocrine effects and behavioural effects. One manner in which a messenger might accrue additional functions has been proposed by Wallis (1975). A change in the target of a substance could occur without any change in the substance itself or in its receptor. The receptor genome might simply be expressed in a tissue in which it had previously been repressed.

The position we have taken is directly supported by the behavioural events described above and indirectly by plausibility. First, some peptides we have considered are hormones, and there appear to be few if any hormones that lack behavioural effects (McEwen, 1976). Other points regarding plausibility are best illustrated by reference to TRH. The tripeptide occurs in species in which it lacks pituitary–thyroid effects (Jackson and Reichlin, 1974*a*); it is found in extra-hypothalamic brain areas (Brownstein *et al.*, 1976). Certain of its behavioural effects, as noted, occur in animals that have been deprived of the possibility of pituitary–thyroid responses (Plotnikoff *et al.*, 1974*a*).

The notion of non-endocrine, behaviourally directed effects of releasing hormones may be refined. There may exist a principle of harmony between the behavioural actions of the component members of a hypothalamus–pituitary–target gland axis. Thus, for example, the behavioural actions of LHRH are not in the realm, say, of food seeking but in the realm of sexual activity as are the effects of other members of the hypothalamus–pituitary–gonad axis, LH, FSH, and the gonadal steroids. We have illustrated this notion in more detail elsewhere and also have cited possible exceptions (Prange *et al.*, 1978).

If many peptides exert behavioural effects that cannot be explained as consequences of hormonal changes, there remains the question of whether these effects are an aspect of pharmacology or of physiology. To argue for a physiological role one must set aside peptide-induced alterations in responses to drugs and cite only behavioural effects consequent to the administration of a peptide to an otherwise untreated animal. TRH, again as an example, exerts several such effects. It induces anorexia (Barlow *et al.*, 1975); it increases motor activity (Segal and Mandell, 1974); it diminishes isolation induced aggression in male mice (Malick, 1976). Magnitude of dose must also be considered, and this is a complex matter. When TRH is given peripherally, what constitutes a large or small dose? Comparison with doses that release TSH from the anterior pituitary gland may not be a useful guide, for the pituitary is outside the blood–brain barrier. On the other hand, the circumventricular organs of the brain are also outside the barrier. It may be instructive to realise that if the total TRH content of rat brain is about 15 ng (Winokur and Utiger, 1974), then penetration by 1 per cent of a peripherally administered dose of 1 mg/kg would result in about a 130-fold increase in brain levels of the tripeptide. A central difficulty is that we do not know if TRH (or other peptides) must penetrate the blood–brain barrier to exert its behavioural effects. When TRH is infused into brain it appears at least as potent in its behavioural effects as noradrenaline. In the rat the effective i.c.v. dose per minute required to increase locomotor activity is about 4 nmol TRH (Segal and Mandell, 1974) or 6 nmol NA (Emlen *et al.*, 1972). The problem of assessing the magnitude of the dose might be helped by the description of behavioural systems in which peptides are more potent. Recently, progress has been made in this direction regarding TRH. In rats, 20 mg/kg i.p. may be needed to induce anorexia (Barlow *et al.*, 1975), but less than 0.04 mg/kg i.p. of TRH may be needed to antagonise isolation-induced aggression in the mouse (Malick, 1976).

Another puzzling aspect of peptide pharmacodynamics arises from the fact that these substances have short half-lives. The half-life of TRH in plasma, for example, is only about 2 min (Nair *et al.*, 1971*b*). Why, then, is the onset of its behavioural action so slow and how can it persist so long? TRH, for example, may be given as long as 30 min before pentobarbitone and still antagonise sedation (Breese *et al.*, 1975). LHRH-induced lordosis behaviour peaks at about 5 hours after injection and may then persist for another 3 hours (Moss *et al.*, 1975). Although peptides attach to membranes (Posner, 1975) and are probably destroyed quickly, they may initiate a train of events that gradually reach threshold for expression. Latency and duration of action are probably related, and possibly they can be rationalised by neuroanatomical considerations. Nearly a century ago Ramon y Cajal (see Rakic, 1975 for review) indicated that the majority of brain neurons are not members of projecting fibre tracts, but rather are components of local, redundant circuits. The proportion of these neurons increases in higher animals, and they may be concerned with modulation and integration of projected messages. We suggest that organisms require responses that are neither instantaneous, as subserved by nerve tracts, nor delayed, as subserved by hormones. Local redundant circuits may serve a temporally intermediate function and peptides may play a role in their activity.

Another observation concerning the behavioural effects of peptides poses a problem in interpretation but may also suggest a principle in this new area of neurobiology. In their behavioural actions peptides seem largely to lack specificity. The production of lordosis behaviour by LHRH appears to be the most discrete behavioural action of a peptide. Nevertheless, one other peptide, TRH, is active in the paradigm; it tends to block the action of LHRH (Moss, personal communication). Many substances delay the extinction of a conditioned avoidance response, and these include ACTH fragments, LHRH, TRH, and vasopressin (de Wied and Gispen, 1977). These substances, of course, vary in potency, and the phenomenon of overlapping actions of peptides should not be overstated. TRH and MSH-IF I, for example, share activity in some tests, but the former is potent in antagonising pentobarbitone while the latter is inactive (Prange *et al.*, 1975*a*). It would not be surprising if the biological principle of redundancy applies to functions served by peptides as it does to other functions.

Another observation about peptides can only be suggested as a stimulus for research. Immunoassay techniques show that some peptides exist both in the gut and the nervous system. These include SRIF (Rufener *et al.*, 1975), Substance P (Leeman and Carraway, 1977), neurotensin (Orci *et al.*, 1976), gastrin (Vanderhaegen *et al.*, 1975), and VIP (Said and Rosenberg, 1976). The implications are unclear, but Wurtman (personal communication) has made an interesting observation. In primitive forms, the environment is often sampled by swallowing it. Thus, the gut may serve as a sensory organ, a role later assumed largely by the nervous system.

The time course of some peptide actions suggest that they are not to be classified with classical neurotransmitter substances (Prange *et al.*, 1978). Peptides, however, need not be transmitters to be determinants of neuronal activity, and it is possible that we may eventually be able to formulate behavioural states in terms of peptide changes as aptly as in terms, say, of biogenic amine changes. By providing an entering wedge, the studies we have cited have illuminated the possibility that peptides play a role in behavioural homeostasis. Their mechanism of action will require further study.

ACKNOWLEDGEMENTS

This work was supported in part by a U.S. Public Health Service Career Scientist Award (MH-22536) to A. J. P.; an NINCDS postdoctoral fellowship (1-F32-NSO5722-01) to C. B. N.; and USPHS grants MH-15631 and HD-03110.

REFERENCES

Arimura, A., Sato, H., Dupont, A., Nishi, N. and Schally, A. V. (1975). *Science*, **189**, 1007–09
Arimura, A., Smith, W. D. and Schally, A. V. (1976). *Endocrinology*, **98**, 540–43
Barbeau, A., Burnett, C., Strother, E. and Butterworth, R. F. (1975). *Clin. Res.*, **23**, 641A
Barlow, T. S., Cooper, B. R., Breese, G. R., Prange, A. J., Jr and Lipton, M. A. (1975) *Neurosci Abst*, **1**, 334
Benkert, O. (1975). In *Hormones, Homeostasis and the Brain* (ed. W. H. Gispen, Tj. B. van Wimersma Greidanus, B. Bohus and D. de Wied), Progress in Brain Research 42, Elsevier, Amsterdam
Bird, E. D., Chiappa, S. A. and Fink, G. (1976). *Nature*, **260**, 536–38
Bissette, G., Nemeroff, C. B., Loosen, P. T., Prange, A. J., Jr and Lipton, M. A. (1976*a*). *Pharmac. Biochem. Behav.*, **5** (suppl. 1), 135–38
Bissette, G., Nemeroff, C. B., Loosen, P. T., Prange, A. J., Jr and Lipton, M. A. (1976*b*). *Nature*, **262**, 607–09
Brambilla, F., Rovere, C., Guastalla, A., Guerrini, A. and Riggi, F. (1976). *Acta psychiat. scand.*, **54**, 131–45
Brazeau, P., Vale, W., Burgus, R., Ling, N., Butcher, M., Rivier, J. and Guillemin, R. (1973) *Science*, **179**, 77–79
Breese, G. R., Cott, J. M., Cooper, B. R., Prange, A. J., Jr and Lipton, M. A. (1974*a*). *Life Sci.*, **14**, 1053–63
Breese, G. R., Cooper, B. R., Prange, A. J., Jr, Cott, J. M., and Lipton, M. A. (1974*b*). In *The Thyroid Axis, Drugs and Behavior*, (ed. A. J. Prange, Jr) Raven, Press, New York
Breese, G. R., Cott, J. M., Cooper, B. R., Prange, A. J., Jr, Lipton, M. A. and Plotnikoff, N. P. (1975). *J. Pharmac. exp. Ther.*, **193**, 11–22
Brown, M. and Vale, W. (1975*a*). *Endocrinology*, **90**, 1333–36

Brown, M. and Vale, W. (1975b). *Endocrinology*, 97, 1151–1156
Brown, M., Villarreal, J. and Vale, W. (1976) *Metabolism*, 25 (suppl. 1), 1459–63
Brownstein, M. J., Palkovits, M., Saavedra, J. M., Bassiri, R. M. and Utiger, R. D. (1974). *Science*, 185, 267–69
Brownstein, M. J., Arimura, A., Sato, H., Schally, A. V. and Kizer, J. S. (1975) *Endocrinology*, 96, 1456–61
Brownstein, M. J., Palkovits, M., Saavedra, J. M. and Kizer, J. S. (1976a). In *Frontiers in Neuroendocrinology*, Vol. 4 (ed. L. Martini and W. F. Ganong), Raven Press, New York, pp. 1–24
Brownstein, M. J., Mroz, E. A., Kizer, J. A., Palkovits, M. and Leeman, S. E. (1976b). *Brain Res.*, 116, 299–305
Carraway, R. and Leeman, S. E. (1973). *J. biol. Chem.*, 248, 6851–54
Carraway, R. and Leeman, S. E. (1975). *J. biol. Chem.*, 250, 1907–18
Chang, M. M., Leeman, S. E. and Niall, H. D. (1971). *Nature new Biol.*, 232, 86–89
Chase, T. N., Woods, A. C., Lipton, M. A. and Morris, C. E. (1974). *Archs Neurol.*, 31, 55–56
Cohn, M. L. and Cohn, M. (1975): *Fedn Proc.*, 34, 738
Collu, R., Clermont, M. J., Letarte, J., Leboeuf, F. and Ducharme, J. R. (1975). In *Hypothalamus and Endocrine Functions* (ed. F. Labrie, J. Meites and G. Pelletier), Plenum Press, New York, p. 479
Constantinidis, J., Geissbuhler, F., Gaillard, J. M., Hovaguimian, T. H. and Tissot, R. (1974). *Experientia*, 30, 1182
Coppen, A., Peet, M., Montgomery, S. and Bailey, J. (1974). *Lancet*, i, 433–35
Cott, J. M., Breese, G. R., Cooper, B. R., Barlow, T. S. and Prange, A. J., Jr. (1976): *J. Pharmac. exp. Ther.*, 196, 594–604
Cott, J. M. and Engel, J. (1977). *Psychopharmacology*, (in press).
Czernichow, P., Dauzet, M. C., Broyer, M. and Rappoport, R. (1976). *J. clin. Endocr. Metab.*, 43, 630–37
de Wied, D. (1974). In *Hormones and Brain Function* (ed. K. Lissak), Plenum Press, New York
de Wied, D., Witter, A. and Greven, H. M. (1975). *Biochem. Pharmac.*, 24, 1463–68
de Wied, D., and Gispen, W. H. (1977). In *Peptides in Neurobiology* (ed. H. Gainor), Plenum Press, New York, (in press)
Ehrensing, R. H. and Kastin, A. J. (1974). *Archs. gen. Psychiat.*, 30, 63–65
Emlen, W., Segal, D. S. and Mandell, A. J. (1972). *Science*, 175, 79–82
Everett, G. M. (1966). *Excerpta Medica Int. Congr. Series*, No. 122, Milan
Friedman, E., Friedman, J. and Gershon, S. (1974). *Science*, 182, 831–32
Furlong, F. W., Brown, G. M. and Beeching, M. F. (1976). *Am. J. Psychiat.*, 133, 1187–90
Green, A. R. and Grahame-Smith, D. C. (1974). *Nature*, 251, 524–26
Hökfelt, T., Fuxe, K., Goldstein, M. and Johansson, O. (1973). *Acta physiol. scand.*, 89, 286–88
Hökfelt, T., Efendić, S., Johansson, O., Luft, R., and Arimura, A. (1974). *Brain Res.*, 80, 165–69
Hökfelt, T., Meyerson, B., Nilsson, G., Pernow, B. and Sachs, C. (1976). *Brain Res.*, 104, 181–86
Hollander, C. S., Mitsuma, T. and Nilhei, N. (1972). *Lancet*, i, 609
Horita, A. and Carino, M. A. (1975). *Psychopharmac. Commun.*, 1, 403–14
Horst, W. D. and Spirt, N. (1974). *Life Sci.*, 15, 1073–82
Huey, L. Y., Janowsky, D. S., Mandell, A. J., Judd, L. L. and Pendery, M. (1975). *Psychopharmac. Bull.*, 11(1), 24–27
Huidobro-Toro, J. P., Scotti de Carolis, A. and Longo, V. G. (1974). *Pharmac. Biochem. Behav.*, 2, 105–09
Huidobro-Toro, J. P., Scotti de Carolis, A. and Longo, V. G. (1975). *Pharmac. Biochem. Behav.*, 3, 235–42
Inanaga, K., Nakano, T., Nagata, T. and Tanaka, M. (1975a). *Kurume med. J.*, 22, 159–68
Inanaga, K., Ohshima, M., Nagata, T. and Yamauchi, I. (1975b). *Folia psychiat. neurol. jap.*, 29, 197–205
Jackson, I. M. D. and Reichlin, S. (1974a). *Endocrinology*, 95, 816–24
Jackson, I. M. D. and Reichlin, S. (1974b). *Life Sci.*, 14, 2259–66
Jones, C. N., Grant, L. D., Prange, A. J., Jr and Breese, G. R. (1975). *Neurosci. Abstr.*, 1, 246
Kastin, A. J. and Barbeau, A. (1972). *Can. med. Ass. J.*, 107, 1079–81
Kastin, A. J., Plotnikoff, N. P., Schally, A. V. and Sandman, C. A. (1976). In *Reviews of Neuroscience 2*, (ed. S. Ehrempreis and I. J. Kopin), Raven Press, New York
Keller, H. H., Bartholini, G. and Pletscher, A. (1974). *Nature*, 248, 528–29

King, C. D. (1975). *Pharmacologist,* 17, 211

Kirkegaard, C., Norlem, N., Lauridsen, B. U., Bjorum, N. and Christiansen, C. (1975). *Archs. gen. Psychiat.,* 32, 1115–18

Knowles, F. (1974). In *Essays on the Nervous System* (ed. R. Bellaris and E. G. Gray), Clarendon Press, Oxford

Kobayashi, R. M., Brown, M. and Vale, W. (1977). *Brain Res.,* (in press)

Kraemer, G. W., Mueller, R., Breese, G. R., Prange, A. J., Jr, Lewis, J. K., Morrison, H. and McKinney, W. T., Jr (1976). *Pharmac. Biochem. Behav.,* 4, 709–12

Lakke, J. P. W. F., van Praag, H. M., van Twist, R., Doorenbos, H. and Witt, F. G. J. (1974). *Clin. Neurol. Neurosurg.,* 3 (4), 1–5

Leeman, S. E. and Carraway, R. E. (1977). In *Peptides in Neurobiology* (ed. H. E. Gainor) Plenum Press, New York, (in press)

Lipton, M. A., Prange, A. J., Jr., Nemeroff, C. B., Breese, G. R. and Wilson, I. C. (1977). In *Neuroregulators and Hypotheses of Psychiatric Disorders* (ed. E. Usdin, D. A. Hamburg and J. Barchas), Oxford University Press, New York

Loosen, P. T., Prange, A. J., Jr, Wilson, I. C. and Lara, P. P. (1976). *Pharmac. Biochem. Behav.,* 5 (suppl. 1), 95–101

Loosen, P. T., Prange, A. J., Jr, Wilson, I. C., Lara, P. P. and Pettus, C. (1977). *Psychoneuroendocrinology,* (in press).

Maeda, K., Kato, Y., Ohgo, S., Chihara, K., Yashimoto, Y., Yamaguchi, N., Korumaru, S., and Imura, H. (1975). *J. clin. Endocr. Metab.,* 40, 501–05

Makino, T., Carraway, R., Leeman, S. E. and Greep, R. O. (1973). *Soc. Study Repro.,* 26

Malick, J. B. (1976). *Pharmac. Biochem. Behav.,* 5, 665–69

Martin, J. B., Renaud, L. P. and Brazeau, P. (1975). *Lancet,* i, 393–95

Matsuo, H., Baba, Y., Nair, R. M. C., Arimura, A. and Schally, A. V. (1971). *Biochem. biophys. Res. Commun.,* 43, 1334–39

McCann, S. M., Fawcett, C. P. and Krulich, L. (1974). In *MTP International Review of Science Series I Physiology, Vol. 5* (ed. S. M. McCann), Butterworth, London

McCaul, J. S., Cassell, K. J. and Stern, G. M. (1974). *Lancet,* ii, 735

McEwen, B. S. (1976). In *Basic Neurochemistry* (ed. G. J. Siegel, R. W. Albers, R. Katzman and B. W. Agranoff), Little, Brown & Co., Boston

Metcalf, G. (1974). *Nature,* 252, 310–11

Monahan, M., Rivier, J., Burgus, R., Amoss, M., Blackwell, R., Vale, W. and Güillmein, R. (1971). *C. r. Séanc. Soc. Biol.,* 273. 508–10

Mortimer, G. H., McMeilly, A. S., Fisher, R. A., Murray, M. A. F. and Besser, G. M. (1974). *Br. med. J.,* 4, 617–21

Moss, R. L. (1975). In *Reproductive Behaviour* (ed. W. A. Sadler and W. Montagna), Plenum Press, New York

Moss, R. L. and McCann, S. M. (1973) *Science,* 181, 177–79

Moss, R. L., McCann, S. M. and Dudley, C. A. (1975). In *Homeostatsis, and the Brain* (ed. W. H. Gispen, Tj. B. van Wimersma Greidanus, B. Bohus and D. de Wied), Elsevier, Amsterdam

Moss, R. W. and Foreman, M. M. (1976). *Neuroendocrinology,* 20, 176–81

Nagai, K. and Frohman, L. A. (1976). *Life Sci.,* 19, 273–80

Nair, R. M. G., Kastin, A. J. and Schally, A. V. (1971a). *Biochem. biophys. Res. Commun.,* 43, 1376–81

Nair, R. M. G., Redding, T. W. and Schally, A. V. (1971b). *Biochemistry,* 10, 3621–24

Nair, R. M. G., Kastin, A. J. and Schally, A. V. (1972) *Biochem. biophys. Res. Commun.,* 47, 1420–25

Nemeroff, C. B., Prange, A. J., Jr., Bissette, G., Breese, G. R. and Lipton, M. A. (1975) *Psychopharmac. Commun.,* 1, 305–17

Nemeroff, C. B., Bissette, G., Prange, A. J., Jr, Loosen, P. T. and Lipton, M. A. (1976). *Proc. Endo. Soc. A. Mtg,* 312

Nemeroff, C. B., Bissette, G., Prange, A. J., Jr and Lipton, M. A. (1977a). *Proc. 34th Ann. Meeting Psychosomatic Med. Soc. Abstr.*

Nemeroff, C. B., Diez, J. A., Bissette, G., Harrell, L. E., Prange, A. J., Jr and Lipton, M. A. (1977b): *Pharmac. Biochem. Behav.,* 6, 467–69

Nemeroff, C. B., Bissette, G., Loosen, P. T., Barlow, T. S., Prange, A. J., Jr and Lipton, M. A. (1977c): *Brain Res.,* 128, 485–98

Nemeroff, C. B., Bissette, G., Loosen, P. T., Burnett, G. B., Prange, A. J., Jr and Lipton, M. A. (1977*d*). Int. Soc. Psychoneuroendocrinology, Eighth Annual Meeting, Atlanta (abstract)
North, R. B., Harik, S. I., and Snyder, S. H. (1973). *Brain Res.*, 63, 435–39
Olivier, C., Eskay, R. L., Ben-Jonathan, N. and Porter, J. C. (1974). *Endocrinology*, 96, 540–46
Orci, L., Baetens, O., Rufener, C., Brown, M., Vale, W. and Guillemin, R. (1976). *Life Sci.*, 19, 559–62
Otsuki, M., Dakoda, M. and Baba, S. (1973). *J. clin. Endocr. Metab.*, 36, 95–102
Ozawa, J. and Shishiba, J. (1975). *Endocr. Japon*, 22, 269–75
Pfaff, D. S. (1973). *Science*, 182, 1148–49
Piva, F. and Steiner, H. (1972). *Front. Horm. Res.*, 1, 11–21
Plotnikoff, N. P., Kastin, A. J., Anderson, M. S. and Schally, A. V. (1971). *Life Sci.*, 10, 1279–83
Plotnikoff, N. P., Prange, A. J., Jr, Breese, G. R., Anderson, M. S. and Wilson, I. C. (1972). *Science*, 178, 417–18
Plotnikoff, N. P. and Kastin, A. J. (1974). *Archs int. Pharmacodyn. Thér.*, 211, 211–14
Plotnikoff, N. P., Prange, A. J., Jr, Breese, G. R. and Wilson, I. C. (1974*a*). *Life Sci.*, 14, 1271–78
Plotnikoff, N. P., Prange, A. J., Jr, Breese, G. R., Anderson, M. S., and Wilson, I. C. (1974*b*). In *The Thyroid Axis, Drugs, and Behavior* (ed. A. J. Prange, Jr), Raven Press, New York
Plotnikoff, N. P., Minard, F. N., and Kastin, A. J. (1974*c*). *Neuroendocrinology*, 14, 271–79
Plotnikoff, N. P., Kastin, A. J. and Schally, A. V. (1974*d*). *Pharmac. Biochem. Behav.*, 2, 693–96
Plotnikoff, N. P., Breese, G. R. and Prange, A. J., Jr (1975). *Pharmac. Biochem Behav.*, 3, 665–670
Plotnikoff, N. P. and Kastin, A. J. (1976). *Life Sci.*, 18, 1217–22
Plotnikoff, N. P., White, W. F., Kastin, A. J. and Schally, A. V. (1976): *Life Sci.*, 17, 1685–92
Posner, B. L. (1975). *Can. J. Physiol. Pharmac.*, 53, 689–703
Prange, A. J., Jr (1977). In *Phenomenology and Treatment of Depression* (ed. W. E. Fann, L. Karacan, A. D. Porkorny and R. L. Williams), Spectrum, New York, pp. 1–15
Prange, A. J. Jr, Wilson, I. C., Rabon, A. M. and Lipton, M. A. (1969): *Amer. J. Psychiat.*, 126 457–69
Prange, A. J., Jr, Wilson, I. C., Knox, A., McClane, T. K., and Lipton, M. A. (1970). *Amer. J. Psychiat.*, 127, 190–99
Prange, A. J., Jr and Wilson, I. C. (1972). *Psychopharmacologia*, 26, 82
Prange, A. J., Jr, Wilson, I. C., Lara, P. P., Alltop, L. B. and Breese, G. R. (1972). *Lancet*, ii, 999–1002
Prange, A. J., Jr, Breese, G. R., Cott, J. M., Martin, B. R., Cooper, B. R., Wilson, I. C. and Plotnikoff, N. P. (1974). *Life Sci.*, 14, 447–55
Prange, A. J., Jr, Breese, G. R., Jahnke, G. D., Martin, B. R., Cooper, B. R., Cott, J. M., Wilson, I. C., Alltop, L. B., Lipton, M. A., Bissette, G., Nemeroff, C. B. and Loosen, P. T. (1975*a*) *Life Sci.*, 16, 1907–14
Prange, A. J., Jr, Breese, G. R., Wilson, I. C. and Lipton, M. A. (1975*b*). In *Anatomical Neuroendocrinology* (ed. W. E. Stumpf and L. D. Grant), S. Karger, Basel, pp. 357–66
Prange, A. J., Jr, Nemeroff, C. B., Lipton, M. A., Breese, G. R. and Wilson, I. C. (1978). In *Handbook of Psychopharmacology* (ed. L. L. Iversen, S. D. Iversen and S. H. Snyder), Plenum Press, New York, pp. 1–117
Rakic, P. (1975) *Neurosci. Res. Prog.*, 13(3)
Re, R. N., Kourides, I. A., Ridgeway, E. C., Weintraub, B. D., and Maloot, F. (1976). *J. Clin. Endocr. Metab.*, 43, 338–46
Reichlin, S., Saperstein, R., Jackson, I. M. D., Boyd, A. E. and Patel, Y. (1976). *A. Rev. Physiol* 38, 389–424
Reigle, T. J., Avni, J., Platz, P., Schidkraut, J. J. and Plotnikoff, N. P. (1974). *Psychopharmacologia*, 37, 1–6
Rezek, M., Havlicek, V., Hughes, K. R. and Friesen, H. (1976). *Neuropharmacology*, 15, 499–504
Rivier, C., Brown, M. and Vale, W. (1977). *Endocrinology*, 100, 751–54
Rufener, C., DuBois, M. P., Malaisse Lagae, F. and Orci, L. (1975). *Diabetologia*, 11, 321–24
Said, S. T. and Rosenberg, R. N. (1976). *Science*, 192, 907–08

Schenkel-Hulliger, L., Koella, W. P., Hartmann, A. and Maltre, L. (1974). *Experientia*, **30**, 1168–70
Schwarzstein, L., Aparico, N. J., Turner, D., Calamera, J. C., Mancini, R. and Schally, A. V. (1975). *Fert. Steril,* **26**(4), 331–36
Segal, D. S. and Mandell, A. J. (1974). In *The Thyroid Axis, Drugs, and Behavior* (ed. A. J. Prange, Jr) Raven Press, New York
Simon, P., Goujet, M. A. and Boissier, J. R. (1975). *Therapie*, **30**, 485–97
Snyder, S. H. (1976). *Am. J. Psychiat.*, **133**, 197–202
Spirtes, M. A., Kostrezewa, R. M., Plotnikoff, N. P. and Kastin, A. J. (1975). *Neurosci. Abstr.*, **1**, 241
Stern, P. and Hadzovic, J. (1973). *Archs. int. Pharmacodyn. Thér.*, **202**, 259–62
Takahashi, S., Kondo, H., Yoshimura, M. and Ochi, Y. (1974). *Folia psychiat. neurol. jap.* **28**(4), 355–65
Terry, L. C., Willoughby, J. O., Brazeau, P., Martin, J. B. and Patel, Y. (1976). *Science*, **192**, 565–66
Tiwary, C. N., Rosenbloom, A. L., Robertson, M. F. and Parker, J. C. (1975). *Pediatrics*, **56**, 119–21
Tyndel, M. (1974). *Can. Psychiat. Asso. J.*, **19**, 21–24
Uhl, G., Bennett, J. P. and Snyder, S. H. (1977). *Brain Res.*, **130**, 299–314
Vale, W. and Rivier, C. (1975). In *Handbook of Psychopharmacology* (ed. L. L. Iversen, S. D. Iversen and S. H. Snyder), Plenum Press, New York
Vale, W., Rivier, C. and Brown, M. (1977). *A. Rev. Physiol.*, **39**, 473–527
Vanderhaegen, J. J., Signeay, J. C. and Gepts, W. (1975) *Nature*, **257**, 604
Vinik, A. I., Kalk, W. J., McLaren, H., Hendricks, S., and Pimstone, B. L. (1975). *J. Clin. Endocr. Metab.*, **40**, 509–11
von Euler, U.S. and Gaddem, J. H. (1931). *J. Physiol , Lond,* **72**, 74–87
Wallis, M. (1975). *Biol. Rev.*, **50**, 35–98
Wei, E., Sigel, S., Loh, H. and Way, E. L. (1975). *Nature*, **253**, 739–40
Weingold, L. P., Lachin, I. M., Bell, A. H. and Coxe, R. C. (1968). *J. abnorm. Psychol,* **72**, 195–97
White, R. P. and Beale, J. S. (1975). *Neurosci. Abstr.*, **1**, 727
Whybrow, P. C., Coppen, A., Prange, A. J., Jr, Noguera, R. and Bailey, J. R. (1972). *Archs. gen. Psychiat.*, **26**, 242–45
Wilber, J. F., Montoya, E., Plotnikoff, N. P., White, W. F., Gendrich, R., Renaud, L. P. and Martin, J. B. (1976). *Recent Prog. Horm. Res.*, 117–53
Wilson, I. C., Prange, A. J., Jr, McClane, T. K., Rabon, M. A. and Lipton, M. A. (1970). *New Eng. J. Med.,* **282**, 1063–1067
Wilson, I. C., Lara P. P. and Prange, A. J., Jr (1973). *Lancet*, **ii**, 43–44
Winokur, A. and Utiger, R. D. (1974). *Science*, **185**, 265–66
Woodruff, R. A., Jr, Guze, S. B., Clayton, P. J. and Carr, D. (1973). *Archs. gen. Psychiat.*, **28**, 97–100

8
Substance P:
a historical account

F. Lembeck (Institut für experimentelle und klinische Pharmakologie
der Universität Graz, Universitätsplatz 4, A-8010 Graz, Austria)

Substance P (SP) was first detected (von Euler and Gaddum, 1931) during a search for acetylcholine in alcoholic extracts of brain and intestine of the rabbit. These extracts contracted the isolated intestine and lowered the rabbit's blood pressure, but contrary to expectation these actions were resistant to atropine. It soon became clear that the extracts owed their biological activity not to acetylcholine, histamine or derivatives of adenylic acid, but to material of polypeptide nature (von Euler, 1936). The name 'Substance P' appeared first in a paper by Gaddum and Schild (1934).

Not much attention was paid to SP for the next 20 years. Then Pernow (1953) extensively studied the actions of SP on smooth muscles and its occurrence in tissues. He reached a purification of 3000 units per mg from the laboratory standard preparation containing 10 U/mg.

At the same time I investigated the distribution of SP in the nervous system. The discovery of 'at least 10 times more SP' in the dorsal than in the ventral roots was the most exciting and 'opened the possibility that SP was a transmitter of the primary sensory neuron' (Lembeck, 1953). When Otsuka et al. (1975) found a ratio of SP in dorsal and ventral roots of 9:1 in crude extracts and of 27:1 in purified extracts I was very pleased, the more so as my laboratory facilities in 1953 were fairly primitive. The isolated organ bath to suspend a guinea pig ileum was home made, the guinea pigs were home bred and the kymograph was an heirloom of Otto Loewi's time. The only really excellent equipment was the experience on bioassay which I collected in Sir John Gaddum's department shortly beforehand.

The investigations on SP in the following years were mainly concerned with its distribution in the central nervous system (CNS) and performed by Kopera and Lazarini (1953), Amin et al. (1954), Zetler and Schlosser (1955) and Paasonen and Vogt (1956). Despite the well-known influence on the bioassays of other biologically active material in the tissue (Cleugh et al., 1964), the distribution of SP in brain regions was, with a few exceptions, confirmed by the recently developed radioimmunoassay.

These studies on the distribution of SP stimulated several investigations on possible actions of SP in the CNS. They were initiated by Zetler (1956, 1959). Some experiments were performed with the aim of studying the content of SP in the brain under various pharmacological treatments. Many of the results obtained must now be regarded of limited value unless confirmed by the use of synthetic SP. It has to be remembered that even a very pure extract of 1000 U/mg contained 99.9 per cent impurities at that time. An example of the limited significance of experiments concerning the action of unpurified 'SP extracts' is the following. Lembeck (1957) observed a stimulation of paravascular pain receptors by such a SP preparation. This observation was confirmed by Juan and Lembeck (1974). Recent findings have shown however that synthetic SP is completely devoid of such an algesic effect, the effect of the 'SP extracts' could be explained by contamination with bradykinin, potassium chloride or ammonium sulphate, depending on the source and preparation of the extract (Lembeck and Gamse, 1977). It seems to be very likely that many other older findings of the actions of SP on the nervous system are of equally limited value.

The development of peptide chemistry between 1950 and 1960, leading to the isolation of oxytocin and bradykinin, finally stimulated vigorous attempts of three research groups in Basel (see Haefely and Hurlimann, 1962 for a review) and Meinardi and Craig (1966) to isolate SP. A purification up to about 100 000 U/mg was achieved at that time. At a symposium in 1961 in Sarajevo, organised by Pavel Stern, who promoted the research on SP with many stimulating ideas, and attended by U.S. von Euler and Sir John Gaddum, it was generally agreed in the closing discussion that further research on SP should be slowed down since the imminent isolation and synthesis would open the door to more exact studies.

But the door still remained closed for several years. I think the next important step emerged from the Mediterranian Sea, like Aphrodite, but in the shape of a mollusc, *Eledone moschata*. Erspamer discovered eledoisin in its salivary glands and among so many other interesting peptides, shortly thereafter, physalaemin in the skin of a South American frog. The isolation and careful pharmacological analysis of these peptides led to a statement by Erspamer in 1965 (Bertaccini *et al.*, 1965) which is worth quoting:

> It may be seen that as many as ten of the amino acids found in SP are also present in eledoisin, and eight in physalaemin. The only amino acid lacking in the molecule of substance P is apparently the tremendously important methioninamide. The suspicion seems to be justified that the labile methioninamide residue has escaped the attention of research workers, who have isolated and studied substance P. We would suggest that this possibility be checked.

The close similarity between SP and the other 'tachykinins' could at that time also be shown by crossed tachyphylaxis (Lembeck and Fischer, 1967).

Following an observation by Vogler *et al.* (1963) that their highly purified SP stimulated salivation, we investigated this action of SP in several species and used these experiments for some kinetic studies of SP (Lembeck and Starke, 1968; Lembeck *et al.*, 1968*a*; Lembeck *et al.*, 1968*b*; Lembeck *et al.*, 1968*c*).

At that time, Leeman and Hammerschlag (1967) traced the 'salivary secretion stimulating factor' in peptide extracts from the hypothalamus. They discovered its similarity with SP and finally succeeded in its isolation and its subsequent synthesis.

The activity still exceeded all earlier assumptions: 1 mg of synthetic SP corresponds to about 2×10^6 units.

The present situation of research on SP, based on the results of the Symposium on SP held in Stockholm last year, is mainly based and shaped by the following achievements:

(1) The isolation and synthesis of SP (Chang and Leeman, 1970, Chang and Leeman, 1971, Tregear *et al.*, 1971, Studer *et al.*, 1973).

(2) The development of a specific radioimmunoassay for SP by Powell *et al.* (1973).

(3) The introduction of the elegant immunohistochemical techniques by Hökfelt *et al.* (1975) to localise SP in the various tissues.

(4) The electrophysiological investigations of synthetic SP by Otsuka *et al.* (1975), Henry (1976), Phillis and Limacher (1974), Davies and Dray (1976).

The achievements reached in these years can be seen by comparing the statement of Gaddum in 1960

These studies on the distribution of SP suggest that it must have some physiological function, but do not show what this function is.

with that of Leeman and Mroz in 1975

Well-defined studies on the physiological roles of Substance P have only been possible since Substance P has received a chemical definition and been available in pure form. The next few years should continue to yield much information on the functional role of this interesting neural peptide.

POSSIBLE PHYSIOLOGICAL FUNCTIONS OF SP

(1) The most interesting and also most likely physiological function of SP is certainly that of an excitatory central transmitter (see Cuello *et al.* chapter 9). In addition several other possible functions should probably be considered as well.

(2) Dale (1935) suggested that the antidromic vasodilation of the so-called 'axon reflex' might be produced by release of the same substance that mediates the centripetal transmission of the impulse. SP has indeed been shown to produce vasodilation and protein extravasation. Experiments by Lembeck *et al.* (1977a) showed that the effect of SP on vascular permeability in the rat paw on intra-arterial injection is 30 times that produced with bradykinin. Release of SP at the peripheral nerve endings has not yet been demonstrated, however. The release of SP in the CNS might also result in a localised vasodilation or an increase in vascular permeability. The existence of a factor which increases the capillary permeability in different regions of the brain could be closely correlated with the amount of SP in these regions (Lembeck and Starke, 1963) and it has now been identified as SP (Heizmann and Lembeck, 1966).

(3) Endocrine actions of SP seem also to be worth further investigation, since it has recently been shown that SP releases growth hormone and prolactin (Kato *et al.*, 1976). This should stimulate further investigations on a function of SP in relation to other hypothalamic releasing hormones. A glucogenetic effect of SP has been found by Brown and Vale (1976).

(4) The distribution of SP in the spinal cord and certain other regions of the CNS parallels the localisation of opiate receptors (Lamotte *et al.*, 1976). Consequently, the possibility of certain interactions between SP and endorphins should be considered, especially since an analgesic action of SP has been reported recently (Stewart *et al.*, 1976; Malick and Goldstein, 1977). Consequently, possible interactions between SP and enkephalin are under investigation (see Cuello *et al.*, chapter 9). As some evidence indicates that the enkephalins act as neurotransmitters, any form of such an interaction returns to the model of SP as a neurotransmitter.

(5) The occurrence of SP within the intestinal tract and its histochemical localisation in the cell bodies and processes of Auerbach's and Meissner's plexuses (Pearse and Polak, 1976), as well as its action on the gut, deserve consideration for possible physiological functions. The stimulation of intestinal smooth muscle led to its discovery, and may well be of physiological significance. The powerful stimulation of the salivary secretion which was the guide line for its isolation has been found in most laboratory animals, though not in cats and rabbits (Lembeck and Starke, 1968). Furthermore, SP in doses much smaller than those which stimulate salivation bring about a contraction of the gall bladder and an increase of flow through the sphincter of Oddi (Lembeck and Juan, 1972).

(6) Since the physiological actions of other peptide hormones are sometimes found to undergo considerable changes during the phylogenetic development, such possibilities have to be kept in mind when discussing certain effects of SP with regard to their role in physiology.

TRANSMITTER FUNCTION OF SP

A considerable amount of recent work is in favour of a transmitter function for SP. However, questions such as the following remain to be answered before the possible role of SP can be regarded as fully established.

Where and how is SP released within the CNS?

Otsuka and Konishi (1976) found a release of SP from the isolated rat spinal cord stimulated electrically or by increased $[K^+]$. A release of SP from synaptosomes by electrical stimulation (Lembeck *et al.*, 1977*b*) and by increased $[K^+]$ has also been described (Jessel, 1977; Schenker *et al.*, 1976; see Cuello *et al.*, chapter 9). The thin afferent fibres emerging from the spinal neuron, the trigeminus fibres and a part of the striato-nigral pathway (Davies and Dray, 1976; Hong *et al.*, 1977; see Cuello *et al.*, chapter 9) contain SP and can be regarded as sites where a release of SP seems to be very likely.

Is SP released at peripheral nerve endings?

No results dealing with this question have been published. The evidence of a release of SP may be complicated by the inactivation of SP by the endothelium (Johnson and Erdös, 1977).

How are the receptors of SP on neurons to be defined?

No details are known at present. The antagonism between SP and baclofen (Saito *et al.*, 1975) may be the first step to elucidate this mechanism.

The biosynthesis and degradation of SP in the nervous system

This is still unknown. A centripetal flux of SP from the cell body to the nerve endings has already been shown by Holton (1960) and confirmed and extended recently by Takahashi and Otsuka (1975). Some experiments on the storage of SP in nervous tissue have been performed recently and are described in the second half of this chapter.

REFERENCES

Amin, A. H., Crawford, T. B. B. and Gaddum, J. H. (1954). *J. Physiol., Lond.*, **126**, 596–617
Bertaccini, G., Cei, J. M. and Erspamer, V. (1965). *Br. J. Pharmac.*, **25**, 380–91
Brown, M. and Vale, W. (1976). *Endocrinology*, **98**, 819–22
Chang, M. M. and Leeman, S. E. (1970). *J. biol. Chem.*, **245**, 4784–90
Chang, M. M. and Leeman, S. E. (1971). *Nature new biol.*, **232**, 86–87
Cleugh, J., Gaddum, J. H., Mitchell, A. A., Smith, M. W. and Whittaker, V. P. (1964). *J. Physiol., Lond.*, **170**, 69–85
Dale, H. M. (1935). *Proc. R. Soc. Med.*, **28**, 319–32
Davies, J. and Dray, A. (1976). *Brain Res.*, **107**, 623–27
Euler, U. S. von (1936). *Arch. exp. Path. Pharmak.*, **181**, 181–97
Euler, U. S. von and Gaddum, J. H. (1931). *J. Physiol., Lond.*, **72**, 74–87
Gaddum, J. H. (1960). In *Polypeptides which affect Smooth Muscles and Blood Vessels* (ed. M. Schacter), Pergamon Press, Oxford, 163–70
Gaddum, J. H. and Schild, H. (1934). *J. Physiol., Lond.*, **83**, 1–14
Haefely, W. and Hürlimann, A. (1962). *Experientia*, **18**, 297–344
Heizmann, A. and Lembeck, F. (1966). *Arch. exp. Path. Pharmak.*, **253**, 260–64
Henry, J. L. (1976). *Br. J. Pharmac.*, **57**, 435P
Hökfelt, T., Kellerth, J. O., Nilsson, G. and Pernow, B. (1975). *Brain Res.*, **100**, 235–52
Holton, P. (1960). In *Polypeptides which affect Smooth Muscles and Blood Vessels* (ed. M. Schachter), Pergamon Press, Oxford, 192–94
Hong, J. S., Yang, H. Y.T. and Costa, E. (1977). *Fedn Proc.*, **36**, 394
Jessell, T. M. (1977). *Br. J. Pharmac.*, **59**, 486P
Johnson, A. R. and Erdös, E. G. (1977). *Fedn Proc.*, **36**, 1014
Juan, H. and Lembeck, F. (1974). *Naunyn-Schmiedeberg's Arch. exp. Path. Pharmac.*, **283**, 151–64
Kato, Y., Chihara, K., Ohgo, Y., Iwasaki, Y., Abe, H. and Imura, H. (1976). *Life Sci.*, **19**, 441–46
Kopera, H. and Lazarini, W. (1953). *Arch. Exp. Path. Pharmak.*, **219**, 214–22
Lamotte, C., Pert, C. B. and Snyder, S. H. (1976). *Brain Res.*, **112**, 407–12
Leeman, S. E. and Hammerschlag, R. (1967). *Endocrinology*, **81**, 803–09
Leeman, S. E. and Mroz, E. A. (1975). *Life Sci.*, **15**, 2033–44
Lembeck, F. (1953). *Arch. exp. Path. Pharmak.*, **219**, 197–213
Lembeck, F. (1957). *Arch. exp. Path. Pharmak.*, **230**, 1–9
Lembeck, F. and Fischer, G. (1967). *Arch. exp. Path. Pharmak.*, **258**, 452–56
Lembeck, F. and Gamse, R. (1977). *Naunyn-Schmiedeberg's Arch. exp. Path. Pharmac.*, (in **299**, 295–303
Lembeck, F., Gamse, R. and Juan, H. (1977a). In *Substance P* (ed. U. S. von Euler and B. Pernow), Raven Press, New York, (in press)
Lembeck, F., Geipert, F. and Starke, K. (1968a). *Arch. exp. Path. Pharmak.*, **261**, 422–33
Lembeck, F. and Juan, H. (1972). In *Vasopeptides* (ed. N. Back and J. Sicuteri), Plenum Press, New York
Lembeck, F., Mayer, N., Schindler, G. (1977b). *Naunyn-Schmiedeberg's Arch. exp. Path. Pharmac.* (in press)
Lembeck, F., Oberdorf, A., Starke, K. and Hettich, R. (1968b). *Arch. exp. Path. Pharmak.*, **261**, 338–45
Lembeck, F. and Starke, K. (1963). *Nature*, **199**, 1295–96
Lembeck, F. and Starke, K. (1968). *Arch. exp. Path. Pharmak.*, **259**, 375–85
Lembeck, F., Starke, K. and Weiss, U. (1968c). *Arch. exp. Path. Pharmak.*, **261**, 329–37
Malick, J. B. and Goldstein, J. M. (1977). *Fedn Proc.*, **36**, 994

Meinardi, H. and Craig, L. C. (1966). In *Hypotensive Peptides* (ed. E. G. Erdös, N. Back, F. Sicuteri and A. Wilde), Springer-Verlag, New York, 594–607
Otsuka, M. and Konishi, S. (1976). *Nature*, 264, 83–84
Otsuka, M., Konishi, S. and Takahashi, T. (1975). *Fedn Proc.*, 34, 1922–28
Paasonen, M. K. and Vogt, M. (1956). *J. Physiol., Lond.*, 131, 617–26
Pearse, A. G. E. and Polak, J. M. (1976). *Histochemistry*, 41, 373–75
Pernow, B. (1953). *Acta physiol. scand.*, 29, suppl. 105, 1–90
Phillis, J. W. and Limacher, J. J. (1974). *Brain Res.*, 69, 158–63
Powell, D., Leeman, S., Tregear, G. W., Niall, H. D. and Potts, J. T. (1973). *Nature*, 241, 252–54
Saito, K., Konishi, S. and Otsuka, M. (1975). *Brain Res.*, 97, 177–80
Schenker, C., Mroz, E. A. and Leeman, S. E. (1976). *Nature*, 264, 790–92
Stewart, J. M., Getto, C. J., Neldner, K., Reeve, E. B., Krovoy, W. A. and Zimmermann, E. (1976). *Nature*, 262, 784–85
Studer, R. O., Trzeciak, A. and Lergier, W. (1973). *Helv. chim. Acta*, 56, 860–66
Takahashi, T. and Otsuka, M. (1975). *Brain Res.*, 87, 1–11
Tregear, G. W., Niall, H. D., Potts, J. T., Leeman, S. E. and Chang, M. M. (1971). *Nature new Biol.*, 232, 87–88
Vogler, K., Haefely, W., Hürlimann, A., Studer, R. O., Lergier, W., Strässle, R. and Berneis, K. H. (1963). *Ann. N.Y. Acad. Sci.*, 104, 378–89
Zetler, G. (1956). *Arch. exp. Path. Pharmak.*, 228, 513–38
Zetler, G. (1959). *Arch. exp. Path. Pharmak.*, 237, 11–16
Zetler, G. and Schlosser, L. (1955). *Arch. exp. Path. Pharmak.*, 224, 159–75

REVIEWS

Euler, U. S. von and Pernow, B. (eds) (1977). *Substance P.* Raven Press, New York
Dix Christensen, H. and Haley, T. J. (1966). *J. Pharm. Sci.*, 55, 747–57
Haefely, W. and Hürlimann, A. (1962). *Experientia*, 18, 297–303
Leeman, S. E. and Mroz, E. A. (1975). *Life Sci.*, 15, 2033–44
Leeman, S. E. *et al.* (1977). In *Peptides in Neurobiology* (ed. H. Gainer), Plenum Press, New York and London, p. 99
Lembeck, F. and Zetler, G. (1962). *Int. Rev. Neurobiol.*, 4, 159–215
Lembeck, F. and Zetler, G. (1971). *Int. Encyclop. Pharmac. Ther.*, Section 72, 1, 28–71
Otsuke, M. and Takahashi, T. (1977). Putative peptide neurotransmitters, *A. Rev. Pharmacol. Tox.*, 17, 425
Stern, P. (1961). *Proc. Sci. Soc. Bosnia Herzegovina*, 1

Substance P: binding to lipids in brain

INTRODUCTION

Heizmann *et al.* (1966) found that a brain homogenate completely loses its SP activity after previous extraction of the total lipids by choloroform/methanol (for details of the method see Folch *et al.*, 1957). By suspension of the total lipid extract of brain in 0.1 N HCl and boiling for a few minutes, SP could be quantitatively transferred from the lipid extract to the aqueous phase. SP was identified by all the bioassays and enzymatic methods available in 1966. Parallel extraction of several regions of brain by the SP-extraction method of Gaddum (1964) and by the lipid extraction method resulted in identical recoveries of SP by both methods (figure 8.1). It was concluded that SP is bound to a lipid in brain tissue.

The availability of synthetic SP made it necessary to repeat and extend these experiments in order:

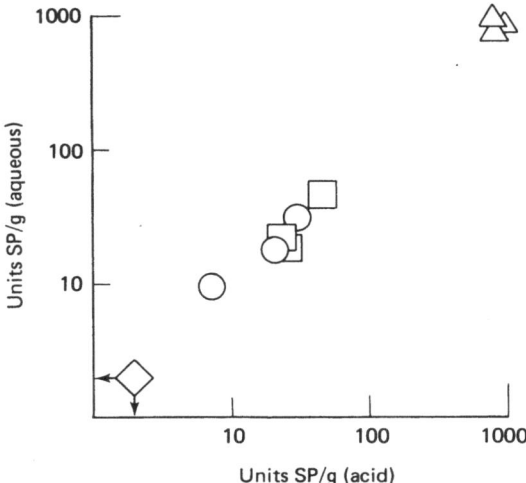

Figure 8.1 Extraction of SP from substantia nigra (△), brain stem (□), cortex (○) and cerebellum (◇) by two methods (from Heizmann *et al.*, 1966; ©1966 Springer-Verlag). Abscissa: units SP/g separated by acid treatment from a total lipid extract prepared according to the method of Folch *et al.* (1957). Ordinate: units SP/g obtained by aqueous extraction with 0.1 N HCl according to the method of Gaddum (1964).

(a) to confirm the identity of the lipid extractable biologically active fraction as SP

(b) to investigate the mode of separation of SP from the lipid into the aqueous phase

(c) to define the SP combining lipid, and

(d) to try a recombination of SP to the total lipid extract or an isolated lipid.

These results will be published in detail shortly (Lembeck, Mayer and Schindler, in preparation).

EXPERIMENTS

Preparation of the total lipid extract (TLE) from brainTLE) FROM BRAIN

First, 50 g brain were homogenised in 500 ml chloroform/methanol (2:1). Then 100 ml 0.034 per cent $MgCl_2$ solution was added and stirred for 10 min to transfer phosphatidylinositol and cerebrosides into the chloroform phase. The chloroform phase was separated by centrifugation (1000*g*, 15 min). This fraction contains all the lipids and proteolipids. The TLE of 20 g rat brain resulted in about 200 ml chloroform solution, from which 3100 ng (that is, 155 ng/g) SP could be recovered. Since proteolipid can be precipitated mainly by ether (Soto *et al.*, 1969), 40 ml ethyl ether were added slowly to 10 ml TLE with stirring for 20 min at 0 °C. The precipitate formed was separated by centrifugation (2000*g*, 20 min), the ether-soluble fraction was brought to dryness *in vacuo* and used for chromatographic separations. The ether precipitate of 100 ml TLE contained 8.9 mg phosphatides, from which 68 ng SP could be extracted. The ether soluble fraction contained 219 mg phosphatides with 684 ng extractable SP.

Separation of SP from TLE
The TLE was brought to dryness *in vacuo*, and 10 ml 0.1 N HCl were added to 1 g
of this dry extract and a fine suspension prepared by sonication. Thereafter,
50 ml acetone were added and the suspension stirred over night at 4 °C. The
resulting protein precipitate was removed by centrifugation. Thereafter, the lipids
were extracted twice with 20 ml petroleum ether. Acetone was then removed
in vacuo. The remaining aqueous solution was either freeze dried or, after adjust-
ment of pH to 6.0 by addition of $NaHCO_3$, used for bioassays.

Identification of SP in the aqueous extract of TLE
The aqueous extract obtained from bovine TLE as described above was used for
chromatography on a Sephadex G-25 column (25 × 1000 mm). Elution was carried
out with 0.5 ml/min 0.1 M acetic acid. Fractions of 5 ml bovine serum albumin
were used to determine the void volume. In two separate runs 10 μg SP and a
freeze-dried aqueous extract prepared from 7.9 g TLE (dry weight) were fractionated. All fractions were first bioassayed on the guinea pig ileum. All active fractions
from this preparation were pooled, freeze dried and used for a parallel quantitative
bioassay on the guinea pig ileum (acetylcholine (ACh), histamine and serotonin
(5-hydroxytryptamine) were excluded by addition of antagonists), the rat salivation
and the rabbit blood pressure tests (table 8.1).

Table 8.1   Identification of SP in aqueous extract of TLE
by gel chromatography and parallel bioassay

Sephadex G-25 column	V_e/V_0
Active fraction in the 0.1 N HCl extract prepared from TLE	1.67
Synthetic substance P	1.63
Parallel bioassay of active fraction from Sephadex G-25 column	
	ng SP/ml
Guinea pig ileum	418 ± 9(5)
Rat salivation	430 ± 15(4)
Rabbit blood pressure	440 ± 44(3)

Mean ± s.e.m. (*n*)
V_e/V_0 = ratio of the elution volume to the void volume of the column.

The relative elution volume was found to be identical with that of synthetic SP.
The parallel bioassays revealed exactly the same activity expressed as SP in all three
preparations. No other biologically active fractions were found in the eluate from
the column. Therefore the activity found in the aqueous extract prepared from
TLE was identified as SP.

Separation of SP from TLE at different pH

Samples of 3.0 ml TLE were evaporated, redissolved in 10 ml chloroform/methanol
(2:1) and 0.6 ml buffer solutions added (0.1 M glycine-HCl buffer in the range of
pH 1.2–3.6, phosphate buffer in the higher range of pH). After shaking for 30 min
at 30 °C, 3 ml of the same buffer were added and the phases were separated by
centrifugation. The aqueous phase was bioassayed on the guinea pig ileum (figure
8.2). The release of SP from bovine or rat TLE was found to be maximal at pH 2.0.
No release of SP from bovine TLE was seen at pH 6.0, and none from rat TLE at a
pH higher than 3.0.

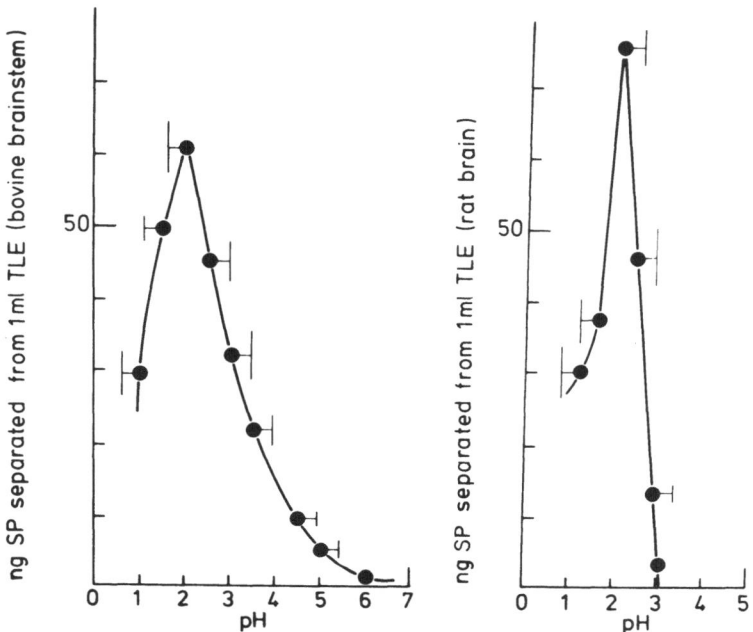

Figure 8.2 Transfer of SP from TLE of bovine and rat brain into the aqueous phase at different
pH. The ordinate represents the amount of SP found in the aqueous phase. Details of the method
are in the text.

The experiments show that the 'dissociation' of SP from TLE is pH dependent
within a very narrow range. At a pH lower than 2.0, less SP was found in the
aqueous phase than at pH 2.0. It was first suspected that SP undergoes destruction
at this low pH, but control experiments showed that SP remained stable at pH 1.2
for the duration of the experiment. It seems possible that SP at a pH lower than
2.0 is again 'bound' apparently by some other mechanism.

Recombination of SP to a SP-deprived TLE and its release

Samples of 3 ml TLE were dried *in vacuo* and redissolved in 10 ml chloroform/
methanol (2:1). To this was added 0.6 ml 0.1 M glycine-HCl buffer (pH 2.0) and
the mixture incubated for 30 min at 30 °C in a shaking water bath. After further

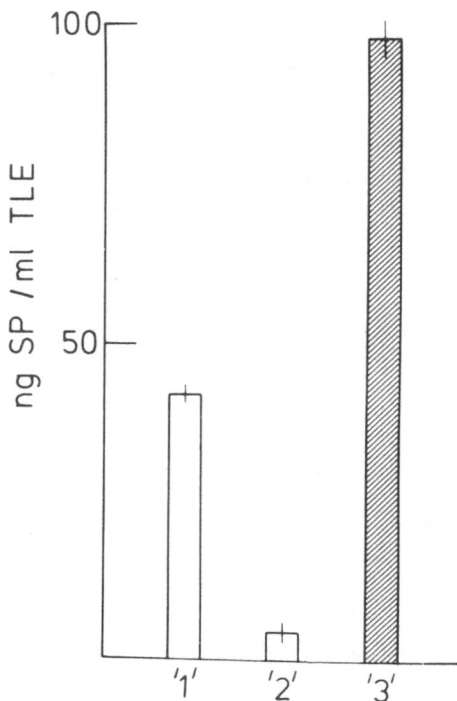

Figure 8.3 Recombination of SP to a SP-deficient TLE and its release. '1', release of SP from TLE at pH 2.0; '2', SP recovered from the aqueous phase after addition of 100 ng PS at pH 5.5 to TLE; '3', release of 100 ng SP which was previously recombined to TLE at 5.5 by separation at pH 2.0.

addition of 3 ml glycine-HCl buffer (pH 2.0) the aqueous phase '1' was removed by centrifugation, neutralised and bioassayed on the guinea pig ileum. This treatment resulted, as known from previous experiments, in a complete release of the 'lipid bound' SP into the aqueous phase and released 45 ng SP per ml TLE.

The chloroform phase from which the SP had been extracted was again dried and the residue redissolved in 10 ml chloroform/methanol (2:1). Then 0.6 ml 0.03 M phosphate buffer (pH 5.5) were added in which 100 ng synthetic SP was dissolved. After incubation as above, another 3 ml phosphate buffer (pH 5.5) were added and the aqueous phase '2' was separated by centrifugation and bioassayed. It was seen that the SP added to the buffer solution had disappeared from the aqueous phase.

The chloroform phase was thereafter treated like the original sample of TLE by addition of the same amount of glycine-HCl buffer (pH 2.0). The resulting aqueous phase '3' was separated in the same way, and contained the total amount of SP that was added previously. This shows that SP is not only released from TLE at pH 2.0 but also recombines to TLE at pH 5.5 (figure 8.3).

The capacity of TLE to bind added SP was investigated in a further series of samples of TLE, from which SP had been removed. From samples of 3 ml TLE about 40 ng SP could be released by acid treatment. The recombination of SP

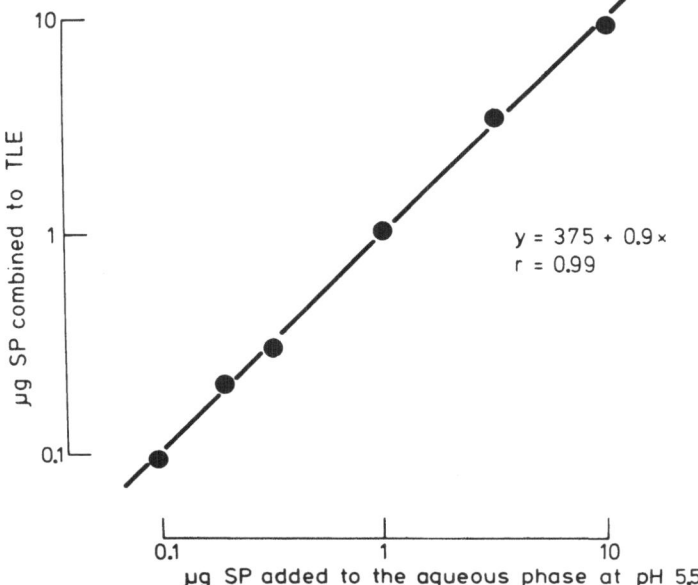

$$y = 375 + 0.9 \times$$
$$r = 0.99$$

Figure 8.4 Capacity of TLE to bind SP from the aqueous phase. Details of the method are in the text.

added to the buffer solution at pH 5.5 showed an enormous capacity of the TLE to take up SP; the uptake was still 100 per cent, even when 10 μg SP (more than 200 times the amount of SP released initially) were added (figure 8.4).

Fractionation of the soluble fraction of TLE on Sephadex LH 20

The binding of SP to TLE raised the question as to which type of lipid binds the SP. Therefore TLE was fractionated on Sephadex LH 20 by the method of Soto *et al.* (1969).

Initial experiments showed that trioleate and phospholipids are eluted by chloroform, cholesterol is eluted with chloroform/methanol (15:1) and phrenoside does not appear before changing to elution with chloroform/methanol (6:1).

The phospholipid-containing fractions of a TLE of rat brain were treated in the same way as the samples of TLE from which SP was extracted. The bioassay of the resulting aqueous extracts revealed that the peak of SP corresponded well with the peak of the phospholipids (figure 8.5).

In another experiment fractions 48–58 obtained from identical runs on Sephadex LH 20 were pooled and evaporated. The content of phospholipids in this pool was investigated by two-dimensional thin-layer chromatography according to the method of Rouser (1967), and compared with reference phospholipids.

The phosphatides found in the extract were mainly phosphatidylserine ($\sim$ 40 per cent), phosphatidylethanolamine ($\sim$ 20 per cent) and others including phosphoinositol, phosphatidic acid and cardiolipine ($\sim$ 10 per cent each). Since the

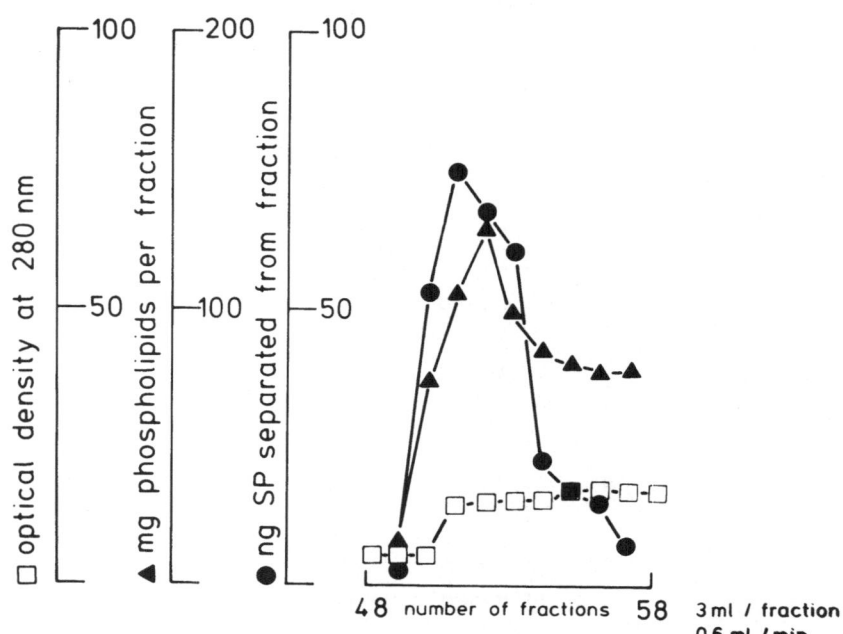

Figure 8.5 Fractionation of the ether soluble fraction of TLE on a Sephadex LH 20 column (25 × 1000 mm). Elution with 0.6 ml/min chloroform, fractions of 3 ml. □, Optical density at 280 nm; ▲, phospholipids, mg per fraction (ammonium molybdate estimation); ●, SP separated from the chloroform phase.

amounts of phosphatidylserine and phosphatidylethanolamine were found to be predominant, the following experiments were essentially repetitions of those previously performed with TLE, but now done with phosphatidylserine.

Combination of SP to phosphatidylserine and phosphatidyl–ethanolamine

These experiments were performed to investigate if one of the two main phosphatides would bind SP in a similar manner to TLE. In the first step, the binding of SP to the two phosphatides was compared. In a second step the binding of SP to phosphatidylserine over a wide range of pH was compared with its binding to TLE. The third step was the investigation of the dissociation of SP previously combined to phosphatidylserine.

The control experiments without addition of a phospholipid to the chloroform phase showed that more than 80 per cent of added SP remained in the aqueous phase at pH 2.0 and at pH 5.5. A high proportion of SP remained in the aqueous phase when shaken with chloroform containing phosphatidylethanolamine at both pH values, and also remained in the aqueous phase at pH 2.0 when shaken with the phosphatidylserine-containing chloroform phase. When SP was added to a buffer of pH 5.5 more than 90 per cent was transferred to the phosphatidylserine-containing chloroform phase. The binding of SP to phosphatidylserine and to TLE was com-

Table 8.2 Combination of Substance P to phospholipids

Solution A 5.0 ml chloroform/methanol (2 : 1) with addition of:	Solution B Buffer with added substance P (μg)	Percentage of Substance P remaining in aqueous phase at	
		pH 2.0	pH 5.5
0	1	93	80
1.0 mg phosphatidylethanolamine	1	86	80
1.0 mg phosphatidylethanolamine	5	92	60
1.0 mg phosphatidylserine	1	62	8
1.0 mg phosphatidylserine	5	96	2.4

Solution A (5 ml chloroform/methánol (2 : 1) was added to solution B (0.3 ml of either phosphate buffer (pH 5.5) or glycine-HCl buffer (pH 2.0) in which SP was dissolved) and shaken for 30 min at 30°C. Following this, another 0.5 ml of the same buffer was added, the phases separated by centrifugation (1000g, 10 min), and the aqueous phase was neutralised with NaHCO$_3$ and bioassayed on the guinea pig ileum.

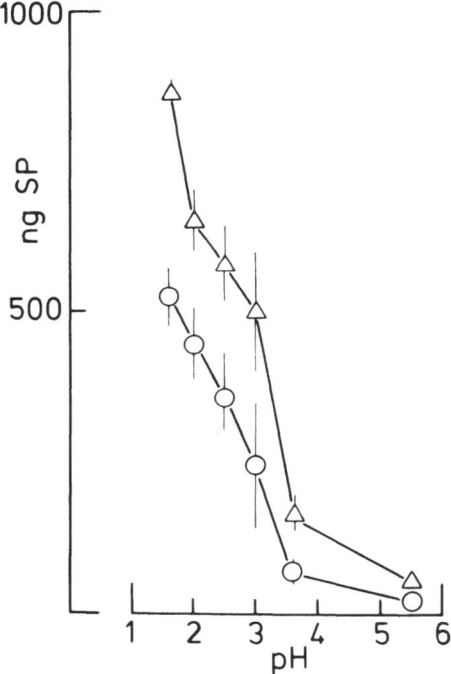

Figure 8.6 Combination of SP to TLE (o) and to phosphatidylserine (△) at different pH. Details of the method are in the text. Abscissa: ng SP (of added 1000 ng) remaining in the aqueous phase.

pared at different pH values between 1.6 and 5.5 using 1000 ng SP dissolved in 0.3 ml buffer solution. For the range from pH 1.6 to 3.6, 0.1 M glycine-HCl buffer was used. For higher pH values 0.03 M phosphate buffer was used. These solutions were added to solutions of 1 mg phosphatidylserine or 1 ml TLE in 5.0 ml chloroform/methanol (2:1). After shaking for 30 min at 30 °C 0.5 ml of the same buffer solution was added and the phases separated by centrifugation (1000*g*, 10 min). The aqueous phase was brought to dryness *in vacuo*, dissolved in 0.4 ml saline solution, the pH adjusted to 7.0 by addition of $NaHCO_3$ and the samples bioassayed on the guinea pig ileum. It can be seen that the binding of SP to phosphatidylserine and to TLE at different pH values is reasonably parallel (figure 8.6).

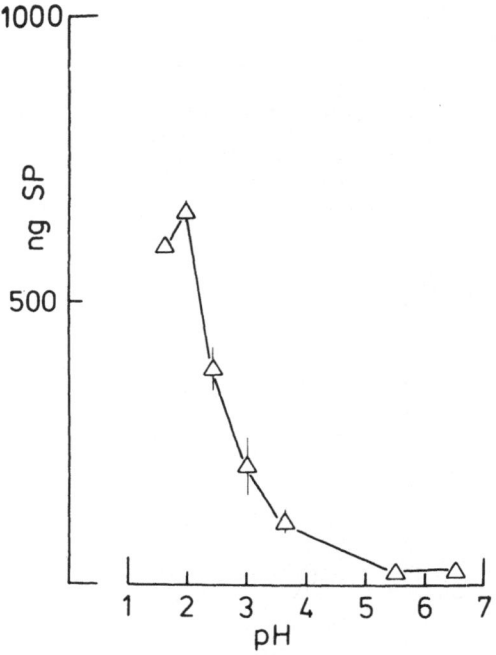

Figure 8.7 Separation of SP previously combined to phosphatidylserine at different pH values. Abscissa: ng SP released into the aqueous phase of a total 1000 ng SP previously combined to 1 mg phosphatidylserine.

The dissociation of SP bound to phosphatidylserine was investigated in experiments similar to those described for TLE. SP (1000 ng) was dissolved in 0.3 ml phosphate buffer (pH 5.5) and added to 5 ml chloroform/methanol (2:1) solution containing 1 mg phosphatidylserine and shaken for 30 min at 30 °C which resulted, as in the previous experiment, in a binding of SP to the phosphatide. Then, 0.5 ml of glycine or phosphate buffer solution between pH 1.6 and 7.0 was added to the samples and the mixture shaken for 30 min at 30 °C. The separation of the aqueous phase and the bioassay was performed as in the other experiments. It could be shown that SP can be dissociated from phosphatidylserine, as it can from TLE, and passes into the aqueous phase (figure 8.7).

Separation of lipid-bound and free SP on a Sephadex LH 20 column

Chromatographic fractionation of the ether-soluble fraction of a TLE from rat brain, of SP bound to phosphatidylserine and of free SP was performed in three identical runs on a Sephadex LH 20 column. It was found that free SP was not eluted with chloroform but appeared in the eluate when the elution was changed after fraction no. 125 to chloroform/methanol (1:1) (figure 8.8).

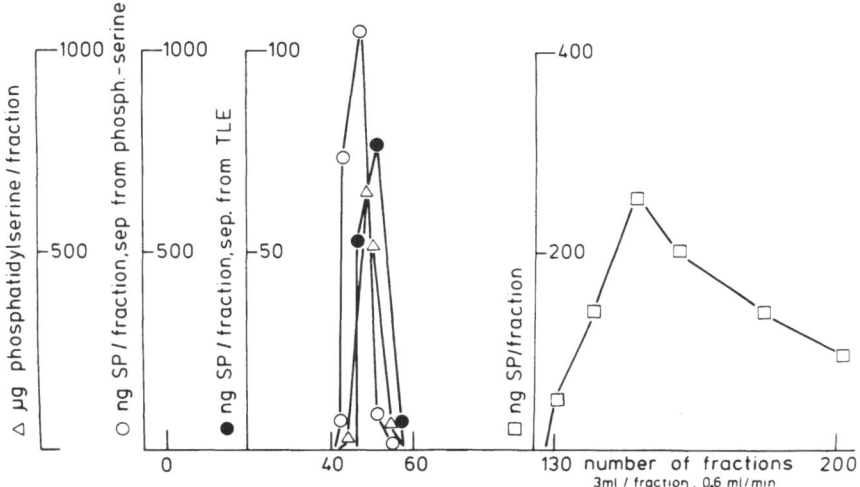

Figure 8.8 Sephadex LH 20 column (25 × 1000 mm). Elution with 0.6 ml/min chloroform until fraction no. 125, thereafter with chloroform/methanol (1: 1). Fractions of 3 ml. □, 10 µg synthetic SP; △, phosphatidylserine; ○, SP split from the fractions containing phosphatidylserine; ●, SP split from the fractions of TLE.

In the second experiment 10 µg SP combined with 2 mg phosphatidylserine was fractionated. Each fraction was divided, one part (1 ml) was used for the estimation of phosphatidylserine by ammonium molybdate reagent, the other part (2 ml) was taken and the SP split off by acid treatment as described in the previous experiments and SP determined by bioassay in the aqueous phase. The peak of phosphatidylserine and of the SP released from it were identical.

In a third experiment the ether soluble fraction of 100 ml TLE was eluted in the same conditions and SP was split from the fractions as in the experiments described above. SP present in the TLE was shown to be eluted in the same fractions as SP associated with phosphatidylserine.

DISCUSSION

The experiments indicate that brain SP is bound to a phosphatide, presumably phosphatidylserine. Some preliminary, unpublished results suggest the same type of binding of SP occurs also in the intestine. Since the total amount of SP present in the brain tissue could be removed by means of a total lipid extraction, the lipid-bound SP can very likely be regarded as the storage form of this peptide. It was noted by Heizmann *et al.* (1966) and confirmed by recent experiments that SP dissociated from the total lipid extract is not contaminated by other biologically active compounds from the brain. However, more investigations are required, particularly to find out how specific such a lipid binding is for SP.

The link of SP to a phosphatidylserine could imply that SP, like other biologically active peptides, is first produced by cleavage from a larger precursor peptide and then stored by attachment to the phosphatide. This storage site of SP would most likely be where phosphatidylserine is predominantly located, that is, in neuronal sites.

The release of SP from the phosphatide and also its uptake *in vitro* has been shown to be highly dependent on the pH. The high isoelectric point (pI) of 10.5 for SP (Geipert *et al.*, 1969) and the carboxyl group of the phosphatidylserine suggest an ionic binding. In the physiological range of pH the binding of SP in tissue would, according to the results *in vitro*, have priority. This would also fit into a model of a lipid storage form of SP. The release of SP from the phosphatide occurs *in vitro* at such a low pH that it seems to be unrealistic that such conditions could exist *in vivo*. The release of SP from the phosphatide could, however, be brought about by a competitive mechanism with another compound or an ion at the binding site of SP.

Recent attempts to characterise the opiate receptor have led to the finding that morphine binds to phosphatidylserine in a 1:1 molar ratio that is stereospecific (Abood and Hoss, 1975). Since catecholamines, serotonin and some alkaloids have also been shown to bind to acidic lipids the specificity of such a binding still needs further elucidation. Phosphatidylserine is the major phospholipid in neural membranes, hence the binding of SP to it has to be seen in connection with its role as a central transmitter (Otsuka *et al.*, 1975). This includes the possibility that the receptor for SP could also be a phosphatide. However, our present experiments do not provide enough information to allow us to distinguish this possibility.

SUMMARY

(1) SP can be extracted from brain with the total lipids by chloroform/methanol.

(2) SP can be transferred from the total lipid extract into the aqueous phase at low pH (2.0–3.0), but not at pH 5.5.

(3) SP can be recombined to a SP-deprived total lipid extract from an aqueous phase of pH 5.5. The binding capacity by far exceeds the amount of SP bound in brain extracts.

(4) Phosphatidylserine has been found to be the lipid in the total lipid extract of brain which binds and releases SP in a pH-dependent manner.

(5) SP bound to phosphatidylserine is regarded at present as the storage form of SP in brain from which it is released by a still unknown mechanism.

REFERENCES

Abood, L. G. and Hoss, W. P. (1975). *Psychopharmac. Commun.*, 1, 29–35
Folch, J., Lees, M. and Stanley, G. H. S. (1957). *J. biol. Chem.*, 226, 497–509
Gaddum, J. H. (1964). *J. Physiol., Lond.*, 172, 207
Geipert, F., Lembeck, F. and Sprössler, B. (1969). *Arch. exp. Path. Pharmak.*, 265, 225–32
Heizmann, A., Lembeck, F. and Seidel, G. (1966). *Arch. exp. Path. Pharmak.*, 253, 265–79
Otsuka, M., Konishi, S. and Takahashi, T. (1975). *Fedn Proc.*, 34, 1922–28
Rouser, G. (1967). *Meth. Enzymol* 10, 402–30
Soto, E. F., Pasquini, J. M., Placido, R. and La Torre, J. L. (1969). *J. Chromatog.*, 41, 400–09

9

Distribution and release of Substance P in the central nervous system

A. C. Cuello, P. Emson, M. del Fiacco, J. Gale, L. L. Iversen, T. M. Jessell,
I. Kanazawa, G. Paxinos and M. Quik (MRC Neurochemical Pharmacology Unit,
Department of Pharmacology, Medical School, Hills Road,
Cambridge CB2 2QD, U.K.)

INTRODUCTION

Following the discovery of Substance P (SP) in extracts of equine gut and brain by von Euler and Gaddum (1931), a considerable number of studies have been carried out on the distribution and pharmacology of this substance in various mammalian tissues. The earlier work has been well reviewed by Lembeck and Zetler (1962). SP was found to be present in relatively high concentrations in various nervous tissues, particularly in sensory nerve pathways. The idea that SP might correspond to the 'Sensory Transmitter Factor' studied by Hellauer and Umrath (1947, 1948) in extracts of dorsal root was also suggested some 25 years ago by Lembeck (1953), and has been championed particularly vigorously since then by this author (Lembeck and Zetler, 1962 and Lembeck this volume, chapter 8). Latterly the work of Otsuka and his colleagues has provided strong support to the view that SP functions as a sensory transmitter in the spinal cord (Otsuka and Konishi, 1975). Earlier studies on SP were hampered by the limited availability of purified material, and by the lack of a simple, specific and sensitive assay method. This situation has changed dramatically since the elucidation by Leeman and her colleagues and by Studer et al. (1973) of the chemical structure of SP as an undecapeptide, (Chang and Leeman, 1970; Chang et al., 1971; Leeman and Mroz, 1974) and by the chemical synthesis of the peptide (Tregear et al., 1971) and the development of a radioimmunoassay technique (Powell et al., 1973). More recently, an immuno-histochemical technique for visualising the cellular localisation of SP has been developed and applied by Nilsson et al. (1974).

With these new and powerful tools available, it has now become possible to obtain detailed information on the quantitative distribution of SP in the mammalian central nervous system (CNS), to begin mapping its precise localisation in specific neuronal pathways, and to investigate its stimulus-evoked release from nervous tissue, and the factors that may modulate this. This chapter will review research on these topics currently in progress in our laboratory, and can only be regarded as a 'progress report' in this actively developing area.

DISTRIBUTION OF SP IN THE CNS

Radioimmunoassay in rat brain and spinal cord

For these studies samples of various regions of rat brain or spinal cord were dissected, usually from 500 μm thick sections of fresh chilled tissue prepared with a McIlwain tissue chopper according to the method developed in this laboratory by Zigmond and Ben-Ari (1976). Frozen tissue samples were extracted with acetone/ hydrochloric acid and assayed for SP by the radioimmunoassay technique of Powell *et al.* (1973). The specificity of the SP-antisera prepared in guinea pigs was checked with the related peptide eledoisin and with somatostatin, met-enkephalin and β-endorphin, all of which failed to cross-react with SP even when present at 1000-fold molar excess. The 9 and 10 amino acid fragments of the C-terminal sequence of SP were found to possess immunoreactivity equal to that of the complete peptide. The sensitivity of the radioimmunoassay technique allows the accurate measurement of as little as 8 fmol per sample, using antiserum dilutions in the range 1:50 000 to 1:250 000. In experiments to study the post-mortem stability of

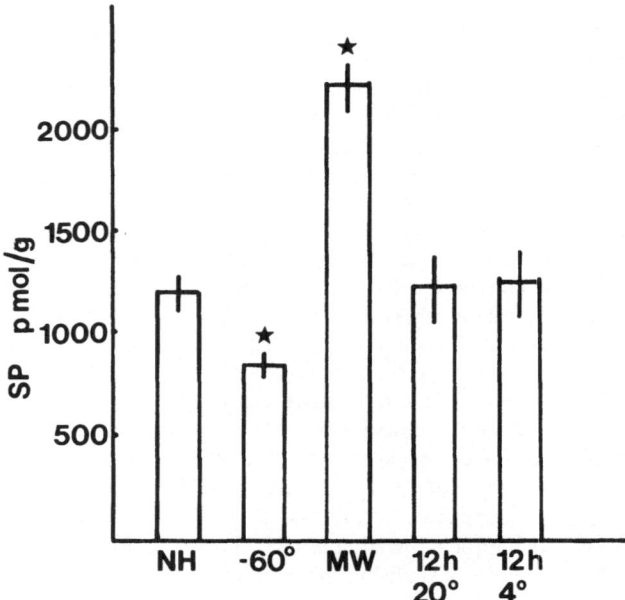

Figure 9.1 Post-mortem stability of SP in mouse hypothalamus. Substance P concentration is expressed as pmol/g (wet weight). Each point is the mean (± s.e.m.) of at least eight determinations. NH, normal handling; −60 °C, Freezing in isopentane; MW, microwave irradiation; 12 h 20 °C, levels 12 hours *post mortem*, brain kept at 20 °C; 12 h 4 °C, levels 12 hours *post mortem*, brain kept at 4 °C. *$P<0.01$.

SP in the CNS, significantly higher concentrations of the peptide were found in hypothalamic and cerebral cortical tissue from mice killed by decapitation at room temperature than in animals killed by rapid freezing in isopentane/dry ice (−60 °C), and the extent of this post-mortem increase in SP was greater in cerebral cortex than in hypothalamus (Kanazawa and Jessell, 1976) (figure 9.1). However, although a rapid post-mortem increase in SP does appear to occur, the concentrations of SP measured in mouse hypothalamus did not alter significantly following decapitation and storage of the brain for up to 12 hours at either 4 °C or at room temperature. Thus, following an initial rise, SP appears to be remarkably stable in brain tissue *post mortem*, as described previously by Zetler and Schlosser (1955). Surprisingly, the effect of rapid fixation of the brain by microwave irradiation was to cause a large increase in the concentration of immunoreactive material, (figure 9.1).

The regional distribution of SP in samples dissected from chilled sections of tissue from animals killed by decapitation is summarised in figure 9.2. These results are similar in most respects to the parallel data reported by Brownstein *et al.* (1976), and are in general agreement with the results of similar studies performed earlier with bioassay techniques (see Lembeck and Zetler, 1962). The highest concentra-

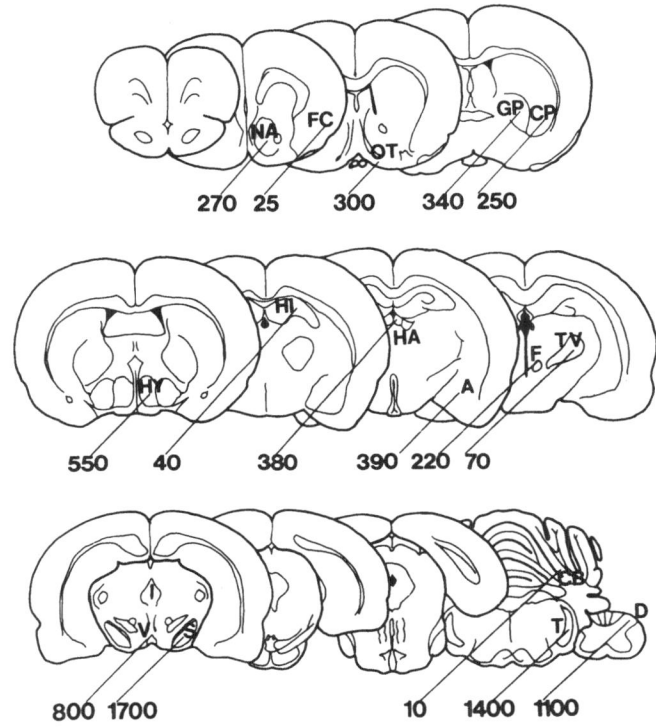

Figure 9.2 Regional distribution of SP in the rat CNS. Values are expressed as pmol/g (wet weight). Each point is the mean (± s.e.m.) of at least four determinations. NA, nucleus accumbens; FC, frontal cortex; OT, tuberculum olfactorium; GP, globus pallidus; CP, caudate putamen; HY, hypothalamus; HI, hippocampus; HA, nucleus habenulae medialis; A, amygdala; F, fasciculus retroflexus; TV, nucleus ventralis thalami; V, area ventralis tegmenti; S, substantia nigra; CB, cerebellum; T, nucleus tractus spinalis trigemini; D, dorsal horn.

tions of SP were found in the superficial layers of the dorsal horn in spinal cord, in the trigeminal nerve nucleus and in the substantia nigra. Moderately high concentrations were present also in the hypothalamus, striatum, globus pallidus, amygdala, habenula, olfactory tubercle, interpeduncular nucleus and septum. Only very low concentrations were present in cerebral or cerebellar cortex, or in white matter. Within the rat substantia nigra, sampled at various rostrocaudal levels, high levels of SP were found in both the zona compacta and zona reticulata with particularly high concentrations in the caudal regions of zona reticulata (Jessell and Cuello, 1977). Brownstein *et al.* (1976), using a frozen slice punch technique, in contrast reported considerably higher levels of SP in zona reticulata than in compacta. Micro-dissection of human post-mortem brain tissue also showed that SP was present in high concentration in the substantia nigra, although the concentration in zona reticulata was higher than in zona compacta (Kanazawa *et al.*, 1977*a*).

In some important respects the present results differ from those reported previously using bioassay techniques, in dog, pig and human brain (Lembeck and Zetler, 1962). Thus, although there is agreement that SP is present in high concentrations in dorsal roots, substantia nigra and other regions of the basal ganglia, the radioimmunoassay results do not confirm the relatively high concentrations of SP reported previously in retina or in the dorsal columns of spinal cord and in the cuneate and gracilis nuclei in brainstem. The latter is a particularly important point, since the present findings—which are supported also by negative immuno-histochemical data for these nuclei—suggest that SP is not likely to represent the sensory transmitter substance released from the terminals of the large diameter myelinated nerve fibres that terminate in these sensory nuclei. SP is instead apparently confined to small diameter sensory nerves that terminate predominantly in the superficial layers of the dorsal horn, or in the superficial grey in the sensory nuclei of cranial nerves such as the trigeminal.

In those regions in which SP is most abundant (dorsal grey, substantia nigra) the absolute concentration (pmol per g wet weight) is similar to that found for other putative transmitter substances such as the catecholamines, noradrenaline and dopamine in other regions of the CNS.

Immunohistochemical localisation of SP

We have developed an immunohistochemical technique for SP, using an indirect fluorescent technique essentially as described by Hökfelt *et al.* (1975*a*). Using this technique, with appropriate controls, a survey of SP distribution has been made in 10 μm thick serial sections taken at 300 μm intervals along the longitudinal axis of the rat brain and in sections of spinal cord. These results are described in detail elsewhere (Cuello and Kanazawa, 1977). As reported by Nilsson *et al.* (1974) and Hökfelt *et al.* (1975*a, b, c*), SP-immunoreactive material was highly localised in nerve terminals in the superficial layers of substantia gelatinosa in the spinal cord. SP-containing terminals were also abundant in sensory nuclei such as the trigeminal nerve nucleus (figure 9.3). High densities of SP-containing nerve terminals were present in the substantia nigra, ventral tegmental area, nucleus amygdaloidus medialis and in various regions of the thalamus and hypothalamus. An interesting observation was that SP-containing terminals appeared to be present in moderate or high density in the various brain regions known to contain the cell bodies of monoamine-containing neurons; for example, in dorsal raphe nuclei, locus coeruleus, substantia nigra, A10 dopamine cell bodies and arcuate nucleus.

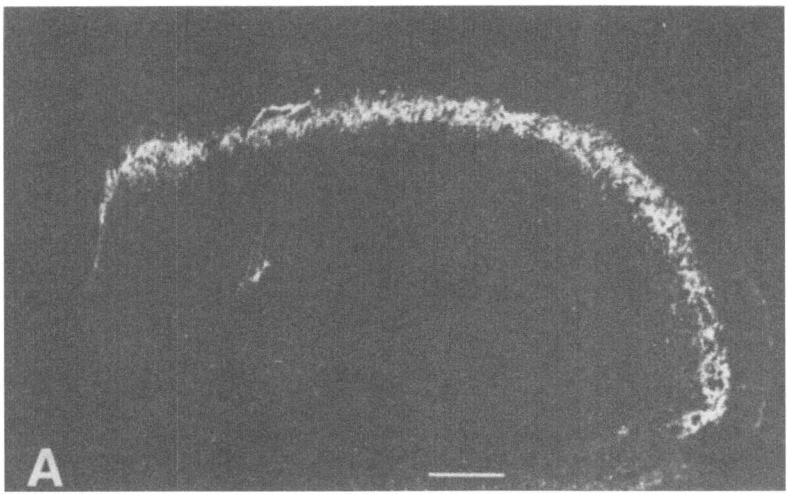

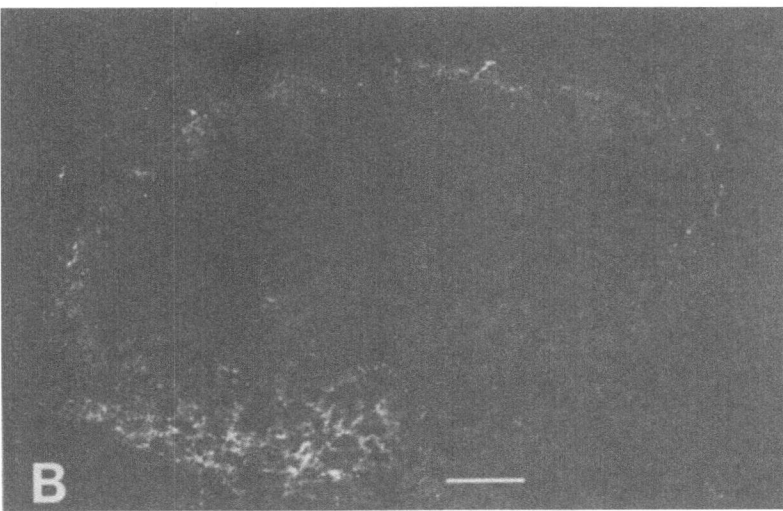

Figure 9.3 Immunofluorescence of the superficial layers of the nucleus spinalis trigemini, intact side (A) and after trigeminal deafferentation (B). Note the loss of immunoreactive sites in the denervated sensory nucleus. (Bar = 200 μm).

SP-reactive neuron cell bodies were observed in spinal root ganglia, nucleus habenula medialis, nucleus interpeduncularis, corpus striatum and globus pallidus. In agreement with the immunoassay results, no SP-containing structures were detected in the cerebellum and most regions of the cerebral cortex were devoid of reactive fibres, except for parts of the frontal and pyriform cortex.

On the basis of these results, and subsequent findings after surgical lesions (see below) it is possible to begin to construct a schematic map of the topography of

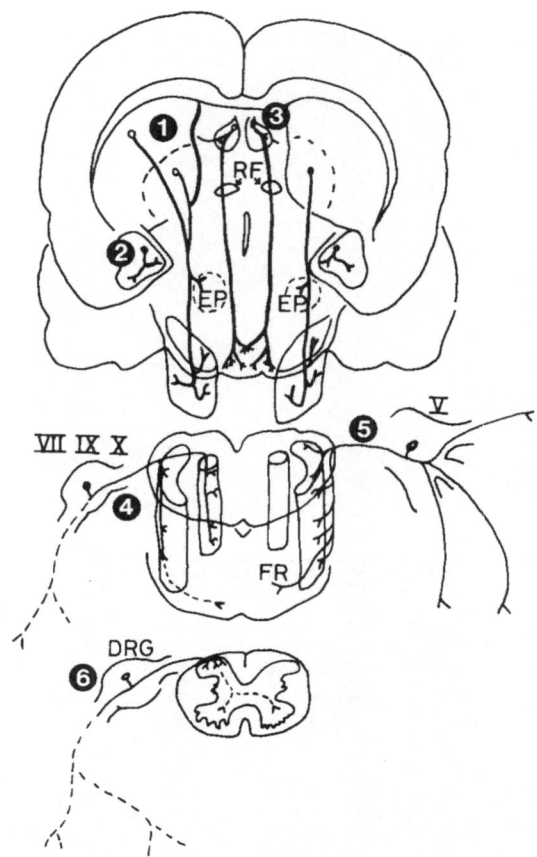

Figure 9.4 Topographical distribution of prepared SP-containing pathways in the rat CNS. The upper figure depicts: (1) the SP pathway originating in the globus pallidus and corpus striatum and terminating in the nucleus entopeduncularis and substantia nigra; (2) the intrinsic SP pathway of the amygdala; (3) the SP projection from the nucleus habenulae medialis via the fasciculus retroflexus to the area ventralis tegmenti and the nucleus interpeduncularis. The middle figure depicts the probable SP projections from: cranial nerves VII, IX and X to the nucleus tractus spinalis trigemini and nucleus tractus solitarii; (5) cranial nerve V to the nucleus tractus spinalis trigemini. The lower figure (6) depicts the afferent SP projection to the superficial layers of the dorsal horn of the spinal cord. RF, fasciculus retroflexus; EP, nucleus entopeduncularis; FR, formatis reticularis; DRG, dorsal root ganglion; (—) projected pathways.

the SP-containing neuronal pathways in rat CNS (figure 9.4). Although the peptide is widely distributed in the CNS it appears to be confined to relatively small neurons and is most highly concentrated in the terminal regions of their axons

Presence in specific neuronal pathways

Surgical lesions have been made in various parts of the rat CNS and the resulting changes in SP content examined both by radioimmunoassay and by immuno-histochemical techniques. In this way the anatomical connections of specific pathways are gradually becoming clear. Efforts have been concentrated so far on four such systems. Takahashi and Otsuka (1975) and Hökfelt *et al.* (1975a) have already demonstrated that most of the SP content of the dorsal horn disappeared after dorsal root section, suggesting that most of the SP-containing nerve terminals in this region arise from sensory neurons in dorsal root ganglia. We have shown similarly that SP largely disappears from the trigeminal nerve nucleus in rat brainstem after surgical section of the trigeminal nerve, or after lesion of the Gasserian ganglion (figure 9.3). This suggests that the abundance of SP-containing terminals in this nucleus also arises from afferent sensory fibres (Cuello *et al.*, 1977c).

We have also investigated the origin of the high density of SP nerve terminals in substantia nigra. It was found that SP largely disappeared from this brain region following large electrolytic lesions of globus pallidus or striatum (Kanazawa *et al.*, 1977a) and subsequent results, obtained after more discrete knife-cut lesions, support the conclusion that most of the SP in substantia nigra is contained in nerve terminals that originate from neurons in globus pallidus and in striatum. The fibres of the latter cells descend through the globus pallidus, internal capsule and entopeduncular nuclei to the substantia nigra. Similar findings have been reported by Brownstein *et al.* (1976), Emson *et al.* (1976) and Hong *et al.* (1977b). These lesions also destroy a descending striatonigral projection that accounts for most of the γ-aminobutyric acid (GABA) and glutamic acid decarboxylase (GAD) activity of the substantia nigra (Hattori *et al.*, 1973; Fonnum *et al.*, 1974). However, a study of the regional distribution of these two projections (GABA and SP) within the nigra as revealed by GAD and SP measurements, indicated that the GABA and SP fibres have a different topographical projection to the substantia nigra and also that the cell bodies of the SP and GABA fibres are distributed differently within the striatum. Thus vertical knife cuts made through the striatum at about the level of the posterior end of the septum produced a significantly greater depletion of SP than of GAD in the nigra, indicating that there are more SP cell bodies in the anterior striatum than GABA cell bodies. The knife cut also produced a significant depletion of SP within both zona compacta and reticulata, whereas the GAD depletion was significant only within the zona reticulata. Thus, it seems unlikely that GAD and SP co-exist in the same striato-nigral cells. The SP projection from the anterior striatum divides evenly between zona compacta and zona reticulata whereas the GABA projection from the anterior striatum is mainly to the zona reticulata (Jessell *et al.*, 1977). SP projections to the globus pallidus and entopeduncular nuclei could also be demonstrated with this type of knife cut lesion. We have also confirmed these results by kainic acid injections into the striatum (1 $\mu g/\mu l$). Kainic acid produces a selective neuronal cell loss in the striatum (Coyle and Schwarcz, 1976) resulting in loss of choline acetyltransferase, GAD and SP but no change in dopamine or tyrosine hydroxylase, indicating that it does not damage axons and terminals. Intrastriatal kainic acid injection produced a substantial depletion of SP in the substantia nigra, similar to that produced by lesions. This indicates that SP is located in neurons in the striatum projecting to the nigra and not in cortico-nigral axons that pass through the striatum. In confirmation of this result, knife cuts undercutting the cortex did not modify nigral SP levels (Jessell *et al.*, 1977).

In an earlier study using electrothermic lesions of the habenula nuclei we reported a dramatic reduction of the choline acetyltransferase and SP content of the interpeduncular nucleus/ventral tegmental area (Emson *et al.*, 1976). This study

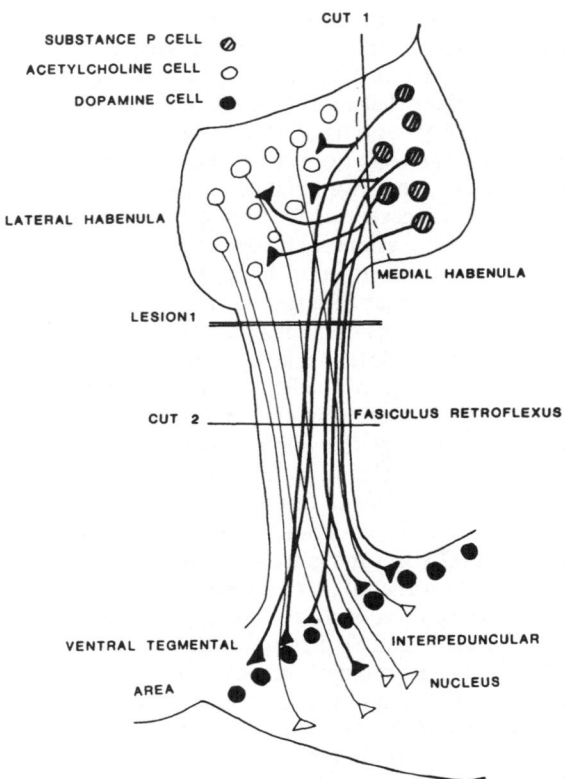

Figure 9.5 Projections of SP- and acetylcholine-containing neurons from the habenula to the ventral tegmental/interpeduncular complex. Results obtained following lesion 1 and cut 1 and 2 are summarised in table 9.1.

confirmed the findings of Mroz *et al.*, (1976) and Hong *et al.*, (1977*a*). From these results it was not possible to distinguish between the SP-containing nerves and the cholinergic projection to the ventral tegmental area. However, subsequent work (Cuello *et al.*, 1977*b*) using small knife-cut lesions has enabled us to distinguish between distinct populations of cholinergic neurons in the lateral habenula nuclei and SP-containing neurons in the medial habenula nuclei. Knife cuts which separated the medial habenula nucleus from the lateral habenula nucleus (through which the medial nucleus projects to the ventral tegmental area) resulted in a loss of SP from the lateral nucleus and from the ipsilateral ventral tegmental area (figure 9.5), table 9.1). In contrast, separation of the medial habenula nucleus from the lateral habenula nucleus by such lesions did not alter the choline acetyltransferase activity of the ventral tegmental area or of the lateral nucleus itself, but did cause a loss of this enzyme from the medial habenula. SP fibres in the ventral tegmental area are, as mentioned earlier, concentrated in the region containing the dopamine cell bodies of the A10 group. An interesting feature of these A10 neurons is that they are very strongly acetylcholinesterase-positive, like those in the substantia nigra (A9). Thus, acetylcholinesterase staining of the interpeduncular area in lesioned

Table 9.1   Choline acetyltransferase activity and SP levels in the habenula–ventral tegmental pathway after selective lesions

	Region	Choline acetyltransferase (% of control)	Substance P (% of control)
Lesion 1	Ventral tegmental area	9.5	28.6
	Medial habenula	23.1	105.2
Cut 1	Lateral habenula	107.6	57.1
	Ventral tegmental area	116.2	48.0
	Habenula	113.2	109.9
Cut 2	Fasciculus retroflexus	189.9	225.0
	Ventral tegmental area	41.0	54.7

For explanation of the site of lesions and cuts see figure 9.5. In all experiments results are expressed as the percentage of values obtained from the contralateral unlesioned region. Each value is the mean of at least four determinations.
*$P < 0.05$.

animals did not reveal any depletion, despite the fact that choline acetyltransferase was almost abolished (Emson *et al.*, 1976). However, histochemical examination of the fasciculus retroflexus after cuts that transected this pathway showed a substantial accumulation of SP and acetylcholinesterase in axons above the cut, and a disappearance of cholinesterase in the pathway ventral to the cut (Cuello *et al.*, 1977*b*). A pile-up of SP in cut axons was also seen after transection of the striatonigral pathway.

Another region under investigation is the amygdala (Emson *et al.*, 1977). Our results indicate that the SP-positive terminals in the amygdala are intrinsic to SP-containing neurons in this structure. Hemisections rostral or caudal to the amygdala failed to deplete the SP content of the amygdala, and parasagittal cuts interrupting the cortico-amygdaloid projections were also without effect. Lesions transecting the medial amygdaloid nucleus, the main SP-positive nucleus in the amygdala, did not deplete it of SP, even though a complete island of the medial amgydala was formed. This cut did, however, deplete the central and basolateral amygdaloid nuclei.

INTRACELLULAR DISTRIBUTION OF SP

The monoamine transmitters are known to be stored in high concentrations within synaptic vesicles present in the nerve terminals, axons and soma of neurons using these transmitters (De Robertis, 1964; Whittaker, 1965). There is evidence from earlier studies, using bioassay techniques, that SP may be similarly localised in intracellular storage organelles. In homogenates of dog brachial and sciatic nerves von Euler and Lishajko (1961) found that about 50 per cent of the total SP content was present in a particulate form in the microsomal fractions (probably corresponding to a synaptic vesicle localisation). In brain homogenates Lembeck and Holasek (1960) and Gaddum (1961) found that a considerable proportion of the SP content sedimented in a crude mitochondrial fraction, and subsequent work indicated a

Table 9.2 Subcellular distribution of SP in rat brain

Fraction	Ultrastructure	Protein (mg/g fresh tissue)	Substance P (pmol/g fresh tissue)	Relative specific activity
Whole homogenate		102.8 ± 4.9	179.3 ± 26.6	1
Nuclear	Cell nuclei Capillaries Myelin Debris	12.6 ± 1.1	—	—
Mitochondrial	Myelin Nerve endings Small processes Mitochondria	53.4 ± 3.2	118.7 ± 28.3	1.34 ± 0.51
Supernatant	Microsomes and soluble fraction	32.1 ± 1.3	56.3 ± 11.4	1.09 ± 0.26
Mitochondrial sub-fractions M_1	Mitochondrial fragments Microsomes Membranes Myelin	34.8 ± 4.8	18.8 ± 4.8	0.38 ± 0.17
M_2	Synaptic vesicles Membranes	2.8 ± 0.4	19.1 ± 5.5	4.35 ± 1.50
Mitochondrial M_3 supernatant	Soluble fraction	8.8 ± 0.7	6.5 ± 1.7	0.64 ± 0.13

The SP concentration in each fraction is expressed as the relative specific activity defined as: pmol SP per mg protein relative to values in whole homogenate. Each value is the mean ± s.e.m. for five determinations. The relative specific activity of fraction M_2 was significantly higher than that of the whole homogenate ($P < 0.001$). No other fraction showed any significant enrichment of SP.

synaptosomal localisation for the peptide (Ryall 1962, 1964; Inouye and Kataoka, 1962; Duffy *et al.*, 1975*a*). We have recently obtained further evidence to support a vesicular localisation for SP within nerve terminals in rat brain and spinal cord.

Application of an immunoperoxidase cytochemical staining technique showed that SP-immunoreactivity was present in some small nerve terminals and unmyelinated axons in the dorsal horn of rat spinal cord, and in the dorsal root. The most intense staining was seen over small vesicles within the axons or nerve terminals. Parallel biochemical studies were performed in which SP was estimated by radioimmunoassay in various subcellular fractions prepared from isotonic sucrose homogenates of rat forebrain. The results of these studies (table 9.2) showed that about 66 per cent of the total SP content could be recovered in a crude mitochondrial/synaptosomal fraction, and that some of the peptide was recoverable in a particulate form in synaptic vesicle fractions prepared after osomotic shock of the crude mitochondrial/synaptosomal preparations. The specific activity of SP in the synaptic vesicle fractions (pmol SP per mg protein) was 4.3 times higher than that in the original homogenate (Cuello *et al.*, 1977*a*). These results strongly support the veiw that, as with other neurotransmitters, SP is present in relatively high concentrations within synaptic vesicles in SP-containing neurons. On the other hand, Pickel *et al.*, (1977) studied the ultrastructural localisation of SP in nerve terminals in rat spinal cord, using an immunoperoxidase technique, and concluded that the peptide was particularly concentrated in the larger (60–80 nm diameter) vesicles in such structures.

The immunofluorescence findings with SP also indicate that the peptide is unevenly distributed within SP-neurons, being most highly concentrated in the nerve terminals. Indeed, positive identification of the soma of SP-containing neurons is often difficult because of their relatively weak immunofluorescence reaction.

INACTIVATION OF SP IN RAT BRAIN

If SP is released as a neurotransmitter one would expect that some mechanism should exist to terminate its actions. Although there is no evidence for any active uptake mechanism for SP by brain slices *in vitro* (Jessell *et al.*, 1976), it is known that the peptide can be degraded by neutral peptidases present in extracts of brain and other tissues (see Lembeck and Zetler, 1962 and Claybrook and Pfiffner, 1968; Benuck and Marks, 1975). The degradation of SP in rat brain has recently been investigated in this laboratory (Quik and Jessell, 1978). Slices of various regions of rat brain were incubated *in vitro* at 37 °C in Krebs–bicarbonate solution to which SP was added at a concentration of 0.7 μM; samples of the medium were removed for radioimmunoassay of SP at various intervals. It was found that SP-immunoreactivity disappeared from the incubating fluid; furthermore, the rate of disappearance was higher when slices from a SP-rich area (substantia nigra) were used than with slices from an area with low SP content (cerebellum) (figure 9.6). Various peptidase inhibitors were tested for their ability to protect SP from such degradation, and the most effective compound was found to be bacitracin (20 μM), which had a marked inhibitory effect on the inactivation process (figure 9.6). Bacitracin has also been found to have similar effects in inhibiting the degradation of other biologically active peptides, including glucagon (Desbuquois and Cuatrecasas, 1972), enkephalins (Miller *et al.*, 1977) and LHRH (McKelvy *et al.*, 1976).

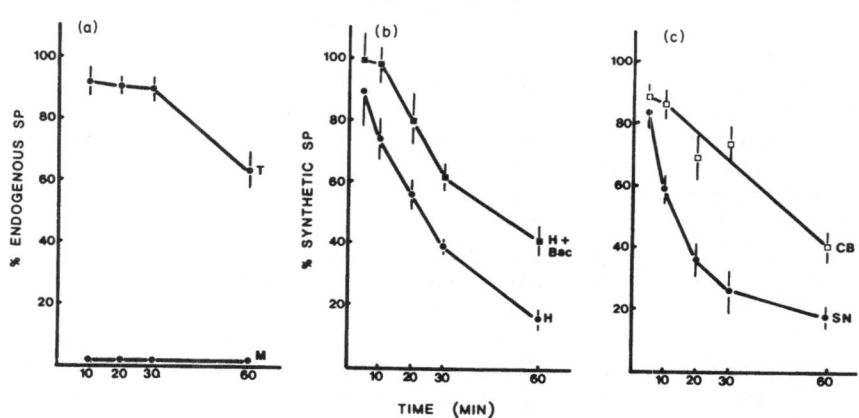

TIME (MIN)

Figure 9.6 Inactivation of endogenous and exogenous SP in rat brain slices. Portions (10–15 mg) of a suspension of rat brain slices were incubated in 0.5 ml Krebs-bicarbonate solution containing 0.5 per cent bovine serum albumin at 37 °C for various times. All values are mean ± s.e.m. (*n* = 3–6). (a) Percentage of immunoreactive substance P remaining in hypothalamic tissue slices (T) after incubation. No immunoreactive substance P was recovered in the incubation medium (M) at any time. (b) Inactivation of added synthetic SP (0.7 μM) by a suspension of rat hypothalamic slices in Krebs-bicarbonate. Values are expressed as the percentage synthetic SP remaining. H, Incubation of SP with hypothalamic slices. H + Bac, Incubation of SP with hypothalamic slices in the presence of 2×10^{-5} M bacitracin. (c) Inactivation of added synthetic SP (0.7 μM) by slices of rat substantia nigra (SN) and cerebellar cortex (CB); parallel incubations were performed using similar amounts of tissue from the two regions. Values are expressed as in (b).

RELEASE OF SP FROM RAT BRAIN AND SPINAL CORD *IN VITRO*

If SP is to be considered as a neurotransmitter in the CNS, demonstration of its release from nerve terminals in response to depolarising stimuli is an essential prerequisite. There have now been several demonstrations that such release does occur. Angelucci (1956) and Shaw and Ramwell (1968) showed the presence of SP-like activity in superfusates from spinal cord and cerebral cortex, although these studies were carried out before the advent of the specific and sensitive radioimmunoassay procedures now available. Using the latter technique, Otsuka and Konishi (1976) showed that SP-like immunoreactivity was released from isolated rat spinal cord preparations on electrical stimulation of the dorsal roots, and, furthermore, that this release was suppressed in a calcium-free medium. Schenker *et al.* (1976) also showed that the efflux of SP from rat brain synaptosome preparations could be accelerated by increasing the external potassium concentration to 60 mM, and that this was a calcium-dependent process. We have used superfused brain slice preparations for similar studies (Jessell *et al.*, 1976). In the initial studies, slices of rat hypothalamus were used, and 0.5 per cent (w/v) bovine serum albumin was added to the superfusing solution to inhibit the loss of SP by enzymic degradation or by adsorption. Radioimmunoassay of superfusate samples indicated that, during resting conditions, the rate of efflux of SP was low, corresponding to approximately 0.2–0.4 per cent of the total tissue content per minute. Brief exposure to a high-potassium medium (47 mM), however, caused more than a twofold increase in the

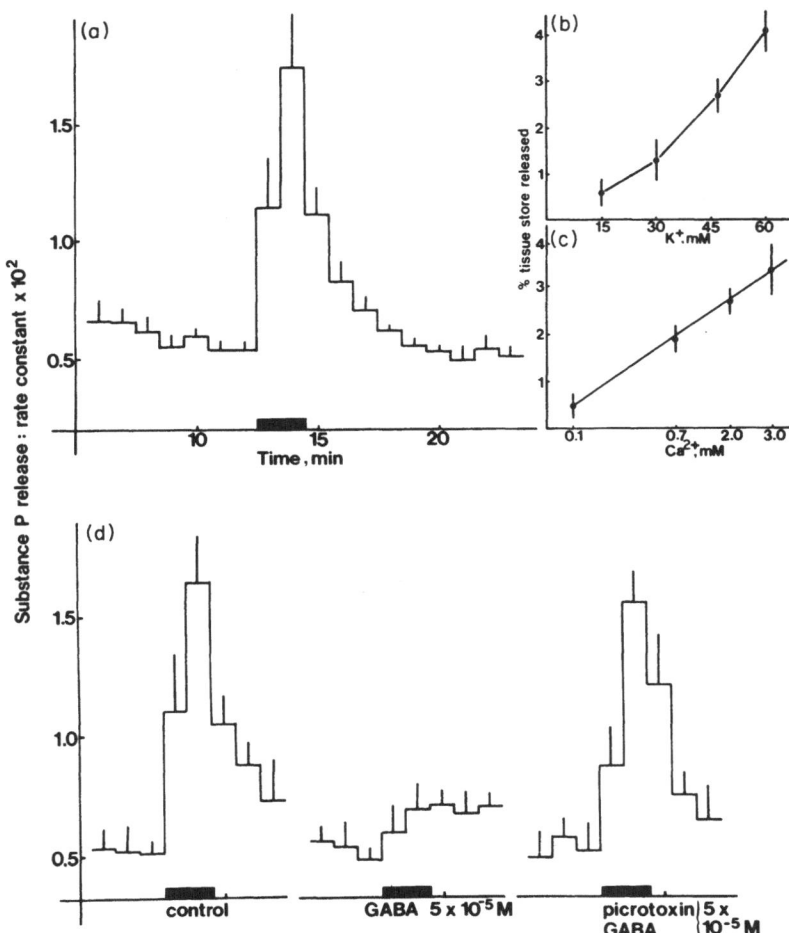

Figure 9.7   Release of SP from superfused slices of rat substantia nigra. (a) Potassium evoked-release of immunoreactive SP from slices of rat substantia nigra. (——) 47 mM K[+]. Release is calculated as the fractional efflux rate constant, and expressed as the percentage of tissue stores released per minute. Spontaneous efflux represents approximately 8 fmol/mg/min. Each point is the mean ± s.e.m. of at least eight determinations. (b) Dose dependency of potassium-evoked release of SP. Each point is the mean ± s.e.m. of 4–6 determinations. (c) Effect of Ca^{2+} concentration on release of SP evoked by 47 mM K[+]. Each point is the mean ± s.e.m. of 4–6 determinations. (d) Inhibition of the potassium-evoked release of SP by GABA, and reversal by picrotoxin. (——) 47 mM K[+]. Each point is the mean ± of six determination.

rate of SP release. Superfusion with a medium in which the calcium concentration was reduced to 0.1 mM and magnesium concentration raised to 3.5 mM had no significant effect on the spontaneous efflux of SP, but completely prevented the release normally evoked by exposure to high potassium (Jessell *et al.*, 1976).

In subsequent experiments a similar potassium-evoked release of SP-immuno-reactivity has been demonstrated *in vitro* using slices of rat substantia nigra, tri-geminal nerve nucleus and dorsal grey matter from spinal cord (Jessell, 1977*a*, *b*, *c*;

Jessell and Iversen, 1977). In all of these regions, and in the hypothalamus, SP appears to be present mainly in nerve terminals, the results indicating that SP can be released from such terminals by depolarising stimuli, and that, as with other transmitters, this process is calcium dependent.

The release of SP from slices of substantia nigra (figure 9.7) can also be increased by exposure to veratridine (50 μM), another substance which causes depolarisation. In this case the evoked release of SP did not occur if tetrodotoxin was present (1 μM); neither veratridine nor tetrodotoxin had any effect on the spontaneous release of SP from the tissue. When various concentrations of potassium were tested, a significant increase in SP release occurred with concentrations of 15 mM, and the evoked release increased linearly over the range 15–60 mM KCl.

The interactions of other transmitters and drugs with the SP release mechanism is being investigated. In slices of substantia nigra, for example, addition of GABA at a concentration of 50 μM almost completely prevented the release of SP normally evoked by exposure to 47 mM K^+, although GABA had no effect on the resting release of SP. This effect of GABA appears to be mediated by GABA receptors similar to those found elsewhere in the CNS. The effect of GABA could be mimicked by the GABA agonist muscimol (10 μM), and was largely blocked by the GABA antagonist drugs picrotoxin (50 μM) or bicuculline (50 μM) (Jessell, 1977*a*). These results suggest that GABA, acting possibly on presynaptic receptors on SP-containing nerve terminals, may be able to regulate the release of SP within the substantia nigra. Since GABA and SP may represent, respectively, inhibitory and excitatory transmitters impinging on the dopaminergic neurons in the substantia nigra (Dray and Straughan, 1976), this could represent an important controlling factor for the activity of these neurons.

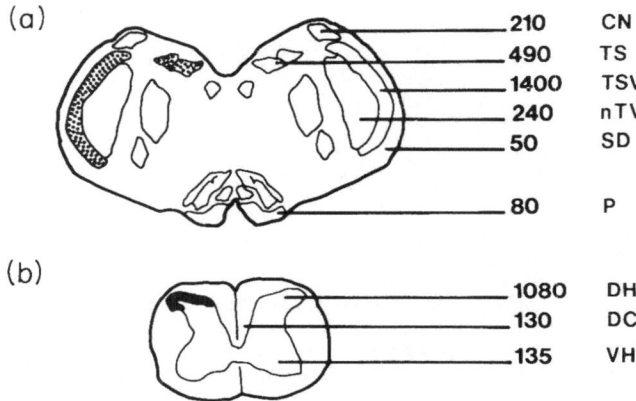

Figure 9.8　Comparative distribution of opiate receptor binding and SP levels in the rat medulla and spinal cord. Diagrammatic representation of coronal sections of (a) rat medulla, 2 mm caudal to the obex and: (b) rat cervical spinal cord. Opiate receptor binding data are adapted from Atweh and Kuhar (1977). The left hand side of each section shows areas with dense (solid) or moderate (stippled) opiate-receptor binding. The right-hand side of each section gives the SP concentrations expressed as pmol/g (wet weight). Each value is the mean of at least four determinations. CN, nucleus cuneatus; TS, tractus solitarius; TSV, tractus spinalis nervi trigemini; nTV, nucleus tractus spinalis nervi trigemini; SD, tractus spinocerebellaris dorsalis; P, pyramid; DH, dorsal horn; DC, dorsal column; VH, ventral horn.

Even more intriguing pharmacological implications arise from studies of the effects of opiate analgesics and endorphin peptides on SP release from sensory nerve terminals *in vitro* (Jessell, 1977*b*; Jessell and Iversen, 1977). It is notable that in the spinal cord and brainstem regions described by Atweh and Kuhar (1977) with high densities of opiate receptors correspond to the regions containing SP terminals (figure 9.8). In these experiments (figure 9.9a) the potassium-evoked release of SP from terminals of the trigeminal nerve was studied, using superfused slices of the trigeminal nerve nucleus. In this system the evoked but not the resting release of SP could be largely. suppressed by addition of morphine (10 μM) or levorphanol (5 μM) to the superfusion medium (figure 9.9b, d). The pharmacologically inactive enantiomer of levorphanol, dextrorphan, had no effect when added at a higher concentration (50 μM), and the inhibitory effects of morphine were mimicked by the opioid peptides β-endorphin and [D-Ala^2, Met^5]-enkephalin amide (figure 9.9f). The peptides were particularly effective, being active at concentrations as low as 0.5–1.0 μM. The inhibitory effects of [D-Ala^2, Met^5]-enkephalin amide, morphine and levorphanol were almost completely blocked by the opiate antagonist naloxone (1 μM) (figure 9.9c, e, f). The effects of opiates were not seen in slices of substantia nigra, in which addition of morphine (10 μM) or β-endorphin (0.5 μM) failed to inhibit the potassium-evoked release of SP. However, preliminary results indicate that the potassium-evoked release of SP from slices of rat spinal cord dorsal horn tissue is also suppressed by opiates.

On the basis of these results we have proposed a possible mechanism for the analgesic actions of opiates (Jessell and Iversen, 1977). It is known that opiates can cause analgesia when administered directly on to the spinal cord in the rat (Yaksh and Rudy, 1976). Furthermore, there is a considerable amount of neurophysiological evidence that in the spinal cord morphine and related opiates selectively suppress the transmission of nociceptive impulses in laminae I and V in the dorsal horn (Calvillo *et al.*, 1974; Kitahata *et al.*, 1974; Le Bars *et al.*, 1976). Duggan *et al.*, (1976) have shown that microiontophoretic application of morphine to the substantia gelatinosa region of cat spinal cord selectively suppresses the excitation of single units in lamina IV by noxious stimuli, while having no effects on the excitation of units that responded to non-noxious stimuli. The ability of opiates to suppress the stimulus-evoked release of SP from sensory nerve terminals, located predominantly in substantia gelatinosa, offers a plausible cellular mechanism to explain these phenomena. Opiate receptors are present in high density in the substantia gelatinosa, and a significant proportion disappear after lesions of the dorsal root in monkeys (Lamotte *et al.*, 1976), suggesting that opiate receptors may indeed be located presynaptically on the terminals of primary afferent fibres in this region although possible trans-neuronal effects cannot be excluded at present.

The failure of opiates to influence SP release from the rat substantia nigra indicates that such an association of opiate receptors with SP-containing nerve terminals does not occur in all regions of the CNS. Furthermore, opiates and endorphins have also been found to inhibit the release of other transmitters, notably noradrenaline (Taube *et al.*, 1976), dopamine (Loh *et al.*, 1976) and acetylcholine (Jhamandas *et al.*, 1971) in other brain regions. Nevertheless, since SP probably represents one of the major transmitters involved in the transmission of nociceptive information in primary sensory fibres (Henry, 1976; Miletić *et al.*, 1977), the interaction of opiate receptors with SP in sensory pathways seems likely to represent an important target as a primary site for opiate-induced analgesia. It is interesting to note

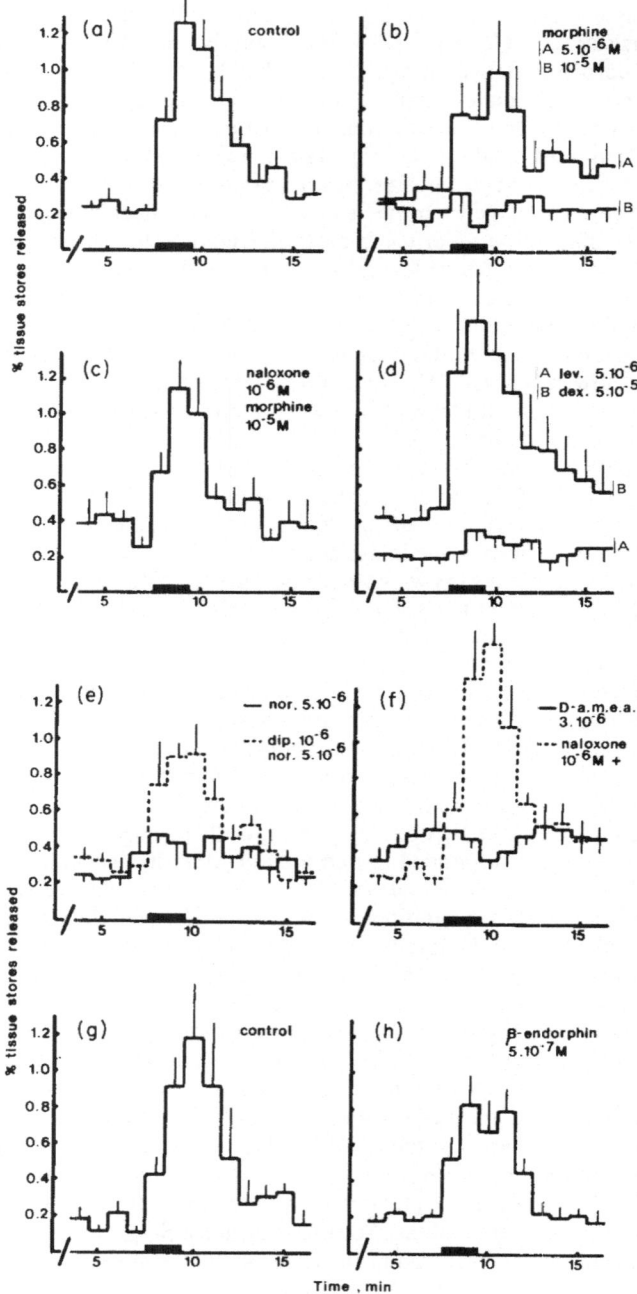

Time , min

that the concentrations of morphine and other opiates needed to suppress SP release *in vitro* are relatively high by comparison with the concentrations of the same drugs needed to displace radiolabelled ligands in receptor-binding assays (Snyder and Simantov, 1977). The concentrations needed to suppress SP release, however, are similar to those found in the brains of mice after implantation of a morphine pellet (Catlin *et al.*, 1977) or injection of levorphanol (Richter and Goldstein, 1970).

There have been previous reports that administration of SP can antagonise the analgesic actions of morphine in experimental animals (Zetler, 1956; Przic, 1961), which would be consistent with the present hypothesis. On the other hand, Stewart *et al.* (1976) found that intracerebral and even intraperitoneal injections of SP caused a morphine-like analgesia in mice. These results are at first sight difficult to reconcile with the present hypothesis. However, it is interesting to note that the analgesic effects seen after SP administration were very slow in onset by comparison with those elicited by morphine. The analgesia observed after SP administration was maximal only after 60–90 min, whereas the effects of morphine were seen much earlier. One possible explanation for the results of Stewart *et al.* (1976) might be that administration of relatively large amounts of SP could cause a progressive and persistent desensitisation of SP receptors. This would then cause a reduced responsiveness of CNS tissue to nociceptive stimuli. Desensitisation of peripheral tissues to SP is a prominent phenomenon (see, for example, Lembeck and Zetler, 1962).

SP RECEPTORS

Very little is known of the pharmacological or biochemical nature of the post-synaptic receptors for SP in the CNS. Excitation of neurons at a variety of spinal and supraspinal sites has been shown after iontophoretic or local administration of SP (Phillis and Limacher, 1974; Krnjević and Morris, 1974; Otsuka and Konishi, 1975; Davies and Dray, 1976). However, recent reports have also described inhibitory actions of SP in the vertebrate CNS (Steinacker and Highstein, 1976; Belcher and Ryall, 1977; Davies and Dray, 1977). The excitatory actions of SP on spinal cord neurons can be mimicked by related peptides such as physalaemin and eledoisin and by C-terminal fragments of the SP molecule (Konishi and Otsuka, 1974). Saito *et al.* (1975) have also suggested that baclofen (β-p-chlorophenyl GABA) may act as an antagonist of SP at spinal cord receptors.

Duffy and Powell (1975) and Duffy *et al.* (1975b) described a stimulation of adenylate cyclase activity by low concentrations of SP (0.1–1.0 μM) in homogenates of rat and human brain. Since other biologically active peptides have been found to act in this way, it seemed plausible to suppose that SP receptors in the

---

Figure 9.9 Effect of opiates on the release of SP from superfused slices of rat trigeminal nucleus. In (a)–(h) each point is the mean ± s.e.m. of at least four determinations. (a) Potassium evoked release of SP from superfused slices of rat trigeminal nucleus. (——): 47 mM K^+. Release is expressed as the percentage of tissue stores released per minute. (b) Inhibition of potassium-evoked release of SP by morphine. (c) Naloxone reversal of morphine-induced inhibition of potassium-evoked SP release. (d) Effect of levorphanol (lev) and dextrorphan (dex) on potassium-evoked SP release. (e) Inhibition of potassium-evoked SP release by normorphine and reversal by diprenorphine. (f) Inhibition of potassium-evoked SP release by [D-Ala^2, Met^5]–enkephalin amide (D–a.m.e.a.) and reversal by naloxone. (g, h) Inhibition of potassium-evoked release of SP by (porcine) β-endorphin.

CNS might be coupled to adenylate cyclase. If this were the case the adenylate cyclase.response might represent a useful model system for studying the pharmacological specificity of SP receptors.

Unfortuantely, however, we have been unable to demonstrate any effect of SP on the formation of cyclic AMP or cyclic GMP in various *in vitro* preparations from rat CNS (Quik and Iversen, in preparation). SP when added in concentrations up to 10 μM failed to cause any significant effect on adenylate cyclase activity in cell-free homogenates, nor did SP alter cyclic AMP levels in incubated slices from various regions of rat CNS, including areas rich in SP-containing nerve terminals, (striatum, hypothalamus, amygdala, spinal cord and substantia nigra). In some experiments bacitracin (30 μg/ml) was added to the incubation medium to inhibit the degradation of added SP but the results were again negative. Similarly, no effects of SP were seen on the cyclic GMP content of slices from these different brain regions.

SP IN HUMAN BRAIN

Zetler and Schlosser (1955), using a bioassay technique, reported the results of a detailed study of the distribution of SP in various regions of human brain, with values from 10 post-mortem samples. Using the radioimmunoassay technique, Duffy and Powell (1975) re-examined the regional distribution of SP in post-mortem human brains. Their results, and our own recent findings (table 9.3) (Gale *et al.*, 1977) are in general agreement with those previously reported, and in general indicate that the distribution of SP in human brain is similar to that described in animal studies. The highest concentration of SP is found in substantia nigra, with moderately high levels in other regions of basal ganglia, hypothalamus and limbic structures, and relatively low concentrations in cerebral cortex and cerebellum. Our values (table 9.3) are considerably higher than those reported by

Table 9.3   Distribution of SP in human brain

Region	Control		Huntington's chorea	Schizophrenics
Caudate	3.7 ± 0.7	(18)	3.5 ± 0.6 (19)	—
Putamen	3.3 ± 0.5	(16)	2.1 ± 0.5 (17)	—
Globus pallidus	18.1 ± 3.3	(18)	8.7 ± 1.7 (15)*	19.2 ± 4.2 (10)
Substantia nigra: compacta	47.2 ± 4.8	(13)	29.0 ± 3.7 (9)†	51.2 ± 4.2 (13)
Substantia nigra: reticulata	42.9 ± 5.0	(14)	22.1 ± 2.3 (9)*	58.7 ± 8.7 (11)
Amygdala	3.3 ± 0.5	(17)	—	4.3 ± 0.4 (17)
Hypothalamus	5.2 ± 1.1	(20)	—	—
Pineal	0.5 ± 0.2	(19)	—	—
Cortical area BA 6	2.4 ± 0.2	(9)	1.7 ± 0.3 (7)	—
Cerebellar cortex	0.2 ± 0.05	(10)	—	—
White matter	0.2 ± 0.1	(3)	—	—

Each value is expressed as pmol/mg (protein), and is the mean ± s.e.m. of the stated number of determinations (in parentheses).
*$P < 0.005$
†$P < 0.01$

Duffy and Powell (1975), and are also higher than the corresponding values from rat brain (figure 9.2).

The globus pallidus was found to contain a relatively high concentration of SP, and there was a much greater difference between globus pallidus and other areas of neostriatum than seen in rat brain. Furthermore, by comparison with rat brain, various regions of human cerebral cortex were found to contain relatively high concentrations of SP.

As reported in laboratory animals, SP-containing neurons were observed by immunocytochemistry to be present in the substantia gelatinosa of the spinal cord of man (Cuello *et al.*, 1976). Other studies in progress indicate the existence of SP-immunoreactive fibres in other areas of the human brain, including hypothalamic nuclei, the substantia nigra and the nucleus spinalis trigemini.

As reported previously (Kanazawa *et al.*, 1977*b*), the SP content of substantia nigra was significantly lower in samples from patients dying with Huntington's chorea, and a similar reduction was seen in globus pallidus but not in caudate nucleus, putamen or motor cortex (table 9.3). The reduction in SP content of the substantia nigra was less severe in the 9 choreic brains now examined than previously reported in 3 brains by Kanazawa *et al.* (1977*b*); nevertheless, it is clear that the degenerative changes in the basal ganglia which occur in Huntington's chorea do affect SP-containing neurons to some extent.

There were no significant abnormalities in the SP content of the brain regions examined in post-mortem brain samples from patients diagnosed as schizophrenic (table 9.3).

CONCLUSIONS

SP is present in a variety of discrete neuronal pathways in the CNS. It is a strong candidate as a sensory transmitter substance released at primary afferent synapses in spinal cord and in cranial nerve nuclei in the brainstem. Furthermore, SP may be particularly involved in the chemical transmission of nociceptive stimuli in the small diameter Aδ and C-fibre population. Within the brain, SP is present in various neurons in the basal ganglia, hypothalamus and limbic system in which its function is clearly not that of a primary sensory transmitter. Thus, as with other chemical transmitters, the same molecule may be used as a transmitter in functionally quite different pathways.

The pharmacological properties of SP-containing neurons and of the postsynaptic receptors with which this peptide interacts are still largely unknown. We also know little about the biosynthesis, storage, release and inactivation of this new transmitter candidate. The results obtained recently on SP release suggest that there may be important physiological and pharmacological factors influencing its release, perhaps by means of presynaptic receptors on SP-containing nerve terminals. These interactions may be particularly relevant in explaining the analgesic actions of opiate drugs.

ACKNOWLEDGEMENTS

Substance P used in the present experiments was kindly donated by Dr R. Hirschmann, Merck Sharp and Dohme, and by Dr R. O. Studer, Hoffman-La Roche. The C10 and C9 fragments of SP were kindly donated by Professor M. Otsuka.

REFERENCES

Angelucci, L. (1956). *Br.J.Pharmac.*, 11, 161–170
Atweh, S. F. and Kuhar, M. J. (1977). *Brain Res.*, 124, 53–67
Belcher, G. and Ryall, R. W. (1977). *J.Physiol., Lond.*, (in press)
Benuck, M. and Marks, N. (1975). *Biochem. biophys.Res.Commun.*, 65, 153–60
Brownstein, M. J., Mroz, E. A., Kizer, J. S., Palkovits, M. and Leeman, S. E. (1976). *Brain Res.*, 116, 299–305
Calvillo, O., Henry, J. L. and Neuman, R. S. (1974). *Can.J.Physiol.Pharmac.*, 52, 1207–11
Catlin, D. H., Schaeffer, J. C. and Loewen, M. B. (1977). *Life Sci.*, 20, 123–132
Chang, M. M. and Leeman, S. E. (1970). *J.biol.Chem.*, 245, 4784–89
Chang, M. M., Leeman, S. E. and Niall, H. D. (1971). *Nature new Biol.*, 232, 86–87
Claybrook, D. L. and Pfiffner, J. J. (1968). *Biochem.Pharmac.*, 17, 281–193
Coyle, J. T. and Schwarcz, R. (1976). *Nature*, 263, 244–46
Cuello, A. C., Polak, J. and Pearse, A. (1976). *Lancet*, ii, 1054–56
Cuello, A. C. and Kanazawa, I. (1977). *J. comp. Neurol.*, (in press)
Cuello, A. C., Jessell, T., Kanazawa, I. and Iversen, L. L. (1977a). *J. Neurochem.*, (in press)
Cuello, A. C., Emson, P. C., Jessell, T., and Paxinos, G. (1977b). *Brain Res.*, (in press)
Cuello, A. C., del Fiacco, M. and Paxinos, G. (1977c). *Brain Res.*, (in press)
De Robertis, E. (1964). *Histophysiology of Synapses and Neurosecretion.* Pergamon Press, Oxford
Desbuquois, B. and Cuatrecasas, P. (1972). *Nature new Biol.*, 236, 202–04
Davies, J. and Dray, A. (1976). *Brain Res.*, 107, 623–27
Davies, J. and Dray A. (1977). *Nature*, 268, 351–52
Dray, A. and Straughan, D. W. (1976). *J.Pharm.Pharmac.*, 28, suppl., 333–405
Duffy, M. J. and Powell, D. (1975). *Biochim.biophys.Acta.*, 385, 275–80
Duffy, M. J., Mulhall, D. and Powell, D. (1975a). *J. Neurochem.*, 25, 305–07
Duffy, M. J., Wong, J. and Powell, D. (1975b). *Neuropharmacology*, 14, 615–18
Duggan, A. W., Hall, J. G. and Headley, P. M. (1976). *Nature*, 264, 456–58
Emson, P. C., Kanazawa, I., Cuello, A. C. and Jessell, T. M. (1976). *Biochem.Soc.Trans.* 5, 187–89
Emson, P. C., Jessell, T., Paxinos, G. and Cuello, A. C. (1977). *Brain Res.*, (submitted)
Euler, U. S. von and Gaddum, J. H. (1931). *J.Physiol., Lond.*, 72, 74–87
Euler, U. S. von and Lishajko, F. (1961). *Proc.Sci.Soc.Bosnia Herzegovina*, 1, 109–12
Fonnum, F., Grofova, I., Rinvik, E., Storm-Mathisen, J. and Walberg, F. (1974). *Brain Res.*, 71, 77–92
Gaddum, J. H. (1961). *Proc.Sci.Soc.Bosnia Herzegovina*, 1, 7–13
Gale, J. S., Bird, E. D., Spokes, E., Iversen, L. L. and Jessell, T. M. (1977). *J. Neurochem.*, (in press)
Hattori, T., McGeer, P. L., Fibiger, H. C. and McGeer, E. G. (1973). *Brain Res.*, 54, 103–14
Hellauer, H. F. and Umrath, K. (1947). *J.Physiol., Lond.*, 106, 20P
Hellauer, H. F. and Umrath, K. (1948). *Pflügers Arch.ges.Physiol.*, 249, 619–30
Henry, J. L. (1976). *Brain Res.*, 114, 439–51
Hökfelt, T., Kellerth, J. O., Nilsson, G. and Pernow, B. (1975a). *Science*, 190, 889–90
Hökfelt, T., Kellerth, J. O., Nilsson, G. and Pernow, B. (1975b). *Brain Res.*, 100, 235–52
Hökfelt, T., Elde, R., Johansson, O., Luft, R., Nilsson, G. and Arimura, A. (1975c). *Neuroscience*, 1, 131–36
Hong, J. S., Costa, E. and Yang, H.-Y. T. (1977a). *Brain Res.*, 118, 523–25
Hong, J. S., Yang, H.-Y. T., Racagni, G. and Costa, E. (1977b). *Brain Res.*, 122, 541–44
Inouye, A., and Kataoka, K. (1962). *Nature*, 193, 585
Jessell, T. (1977a). *Br.J.Pharmac.*, 59, 486P
Jessell, T. (1977b). *J.Physiol., Lond.*, 270, 56–57P
Jessell, T. (1977c). *Brain Res.*, (in press)
Jessell, T. and Iversen, L. L. (1977). *Nature*, 268, 549–51
Jessell, T., Iversen, L. L. and Kanazawa, I. (1976). *Nature*, 264, 81–83
Jessell, T., Emson, P. C., Paxinos, G. and Cuello, A. C. (1977). *Brain Res*, (submitted)
Jhamandas, K., Phillis, J. W. and Pinsky, C. (1971). *Br.J.Pharmac.*, 43, 53–66
Kanazawa, I. and Jessell, T. (1976). *Brain Res.*, 117, 362–67
Kanazawa, I., Emson, P. C. and Cuello, A. C. (1977a). *Brain Res.*, 119, 447–53
Kanazawa, I., Bird, E., O'Connell, R. and Powell, D. (1977b). *Brain Res.*, 120, 387–92

Kitahata, L. M., Kosaka, Y., Taub, A., Bonikos, K. and Hoffert, M. (1974). *Anaesthesiology,* **41**, 39–48

Konishi, S. and Otsuka, M. (1974). *Brain Res.,* **65**, 397–410

Krnjević, K. and Morris, M. E. (1974). *Can.J.Physiol.Pharmac.,* **52**, 736–44

Lamotte, C., Pert, C. B. and Snyder, S. H. (1976). *Brain Res.,* **112**, 407–12

Le Bars, D., Guilbaud, G., Jurna, I. and Besson, J. M. (1976). *Brain Res.,* **115**, 518–24

Leeman, S. E. and Mroz, E. A. (1974). *Life Sci.,* **15**, 2033–44

Lembeck, F. (1953). *Naunyn-Schmiedebergs Arch.exp.Path.Pharmac.,* **219**, 197–213

Lembeck, F. and Holasek, A. (1960). *Naunyn-Schmiedebergs Arch.exp.Path.Pharmac.,* **238**, 542–45

Lembeck, F. and Zetler, G. (1962). *Int.Rev.Neurobiol.,* **4**, 159–215

Loh, H. H., Brase, D. A., Sampath-Khanna, S., Mar, J. B., Leong-Way, E. and Li, C. H. (1976). *Nature,* **264**, 567–68

McKelvy, J. F., LeBlanc, P., Laudes, C., Perrie, S., Grimm-Jorgensen, Y. and Kordon, C. (1976). *Biochem.biophys.Res.Commun.,* **73**, 507–15

Miletić, V., Kovacs, M. and Randić, M. (1977). *Fedn Proc.,* **36**, 1014

Miller, R. J., Chang, K.-J. and Cuatrecasas, P. (1977). *Biochem.biophys.Res.Commun.,* **74**, 1311–17

Mroz, E. A., Brownstein, M. J. and Leeman, S. E. (1976). *Brain Res.,* **113**, 597–99

Nilsson, G., Hökfelt, T. and Pernow, B. (1974). *Med.Biol.,* **52**, 424–27

Otsuka, M. and Konishi, S. (1975). *Cold Spring Harb. Symp. quant. Biol.,* **40**, 135–43

Otsuka, M. and Konishi, S. (1976). *Nature,* **264**, 83–84

Phillis, J. W. and Limacher, J. J. (1974). *Brain Res.,* **69**, 158–63

Pickel, V. M., Reis, D. J. and Leeman, S. E. (1977). *Brain Res.,* **122**, 534–40

Powell, D., Leeman, S. E., Tregear, G. W., Niall, H. D. and Potts, J. T. Jr (1973). *Nature new Biol.,* **241**, 252–54

Pržić, R. (1961). *Proc.Sci.Soc.Bosnia.Herzegovina,* **1.** 71

Quik, M. and Jessell, T. (1978). *J. Neurochem.,* (in press)

Richter, J. A. and Goldstein, A. (1970). *Proc.natn.Acad.Sci.U.S.A.,* **66**, 944–51

Ryall, R. W. (1962). *Nature,* **196**, 680–81

Ryall, R. W. (1964). *J.Neurochem.,* **11**, 131–45

Saito, K., Konishi, S. and Otsuka, M. (1975). *Brain Res.,* **97**, 177–80

Schenker, C., Mroz, E. A. and Leeman, S. E. (1976). *Nature,* **264**, 790–92

Shaw, J. E. and Ramwell, P. W. (1968). *Am.J.Physiol.,* **215**, 262–67

Snyder, S. H. and Simantov, R. (1977). *J.Neurochem.,* **28**, 13–20

Steinacker, A. and Highstein, S. M. (1976). *Brain Res.,* **114**, 128–33

Stewart, J. M., Getto, C. J., Neldner, K., Reeve, E. B., Krivoy, W. A. and Zimmermann, E. (1976). *Nature Lond.,* **262**, 784–85

Studer, R. O., Trzeciak, A. and Lergier, W. (1973). *Helv.chim.Acta,* **56**, *860–66*

Takahashi, T. and Otsuka, M. (1975). *Brain Res.,* **87**, 1–11

Taube, H. D., Borowski, E., Endo, T. and Starke, K. (1976). *Eur.J.Pharmac.,* **38**, 377–80

Tregear, G. W., Niall, H. D., Potts, J. T. Jr, Leeman, S. E. and Chang, M. M. (1971). *Nature new Biol.,* **232**, 86–89

Whittaker, V. P. (1965). *Progr.Biophys.molec.Biol.,* **15**, 39–96

Yaksh, T. L. and Rudy, T. A. (1976). *Science,* **192**, 1357–58

Zetler, G. (1956). *Naunyn-Schmiedebergs Arch.exp.Path.Pharmac.,* **228**, 513–38

Zetler, G. and Schlosser, L. (1955). *Naunyn-Schmiedebergs Arch.exp.Path.Pharmac.,* **224**, 159–75

Zigmond, R. and Ben-Ari, Y. (1976). *J. Neurochem.,* **26**, 1285–89

10

Endogenous opioid peptides: historical aspects

H. W. Kosterlitz (Unit for Research on Addictive Drugs, University of Aberdeen, Aberdeen, U.K.)

This short chapter serves as an introduction to the following contributions on endogenous opioid peptides and gives some of the more important historical developments in the search for an endogenous ligand for the opiate receptors. One general point has attracted our interest for some time, namely the fact that certain alkaloids found in plants have functional correlates in animal tissues. Although there are certain structural similarities between the two classes of compounds, the plant alkaloids are much more resistant to enzymic inactivation in animal tissues than the corresponding compounds of animal origin, and for this reason are often potent neurotoxic agents. Examples of such pairs of compounds are muscarine and muscimol, derived from the toadstool *Amanita muscaria*, which correspond to the neurotransmitters acetylcholine and γ-aminobutyric acid (GABA), respectively. Other pairs are nicotine and acetylcholine and also ephedrine and noradrenaline. To these may now be added morphine from *Papaver somniferum* and the class of peptides known as enkephalins and endorphins.

The analysis of the mode of action of opiates have always been difficult because of the heterogeneous nature of the central nervous system (CNS) and the rather selective effects of morphine-like drugs. There has been a long search for simpler models that goes back to 1917 when Trendelenburg showed that the peristalsis evoked in an isolated segment of the guinea pig ileum by an increase in intraluminal pressure, is inhibited by morphine at a concentration as low as that found after a single therapeutic administration. This phenomenon has been explored systematically, particularly after Paton (1957) introduced his model of the electrically stimulated preparation of the guinea pig ileum. It has been shown that the morphine receptor in the guinea pig ileum predicts well the potency of a drug as an analgesic agent in man and that these actions are competitively reversed by antagonists, for example naloxone and naltrexone (Kosterlitz and Watt, 1968; Kosterlitz and Waterfield, 1975). A second model, the mouse vas deferens, has been developed more recently (Hughes *et al.*, 1975*a*).

157

These pharmacological observations on preparations *in vitro* soon found their corollary in the CNS by the demonstration of specific recognition sites for morphine, to which tritiated ligands such as dihydromorphine, etorphine and naloxone, are bound in a saturable and stereospecific manner (Pert and Snyder, 1973; Simon *et al.*, 1973; Terenius, 1973).

It appeared unlikely that such well-defined receptors had been developed in a fortuitous manner. The view that an endogenous ligand exists was strongly supported by the observation that the antinociceptive effect caused by electrical stimulation of the periaqueductal grey of the midbrain was reversed by naloxone (Akil *et al.*, 1972). This phenomenon was best explained by the assumption that the electrical stimulation released an endogenous substance with morphine-like actions, the effect of which was reversed by the antagonist, naloxone.

When the Unit for Research on Addictive Drugs was established in October 1973, it was decided that the search for the endogenous ligand was to be one of the foremost tasks. As in any attempt of this kind, the monitoring of the progress of purification was one of the most essential considerations. In this respect, we had a considerable advantage since we were able to use two highly specific monitoring systems, namely the depression of the contractions of the guinea pig ileum and the mouse vas deferens and their reversal by specific antagonists. As antagonists, we had at our disposal not only naloxone and naltrexone but also a very rapidly acting quaternary nalorphinium compound and two optical isomers of an antagonist of the benzomorphan series. Thus, we could establish the stereospecificity of the antagonism, a most important requirement. The availability of more than one assay system proved to be very fortunate because it was shown later that the mouse vas deferens is about one order of magnitude more sensitive to the enkephalins than the guinea pig ileum.

The presence of the enkephalins in brain extracts was established within three months of the beginning of the work but the structure was not finally established until the early autumn of 1975. The main reason for this delay was the fact that it was not realised until the final stages that there were two closely related pentapeptides in the brain extracts (Hughes, 1975). The chromatographic behaviour of methionine-enkephalin and leucine-enkephalin was very similar and the two pentapeptides could not be separated by high-voltage electrophoresis. Dr Linda A. Fothergill rapidly showed that the sequence of the first four amino acids was Tyr–Gly–Gly–Phe, but found it difficult to allocate positions for the methionine and leucine which were shown to be present by amino acid analysis. We were very fortunate in obtaining the collaboration of Dr H. R. Morris of Imperial College, who established by mass spectrometry the presence of the two pentapeptides, methionine-enkephalin and leucine-enkephalin (Hughes *et al.*, 1975*b*).

One of the most surprising coincidences occurred shortly after Dr Morris had established the structure of the enkephalins. Dr Derek G. Smyth visited Imperial College and gave a talk on the prohormone β-lipotropin and showed the structure of the C-fragment or β-lipotropin$_{61-91}$ which is now more commonly known as β-endorphin. It surprised Howard Morris greatly to see that the first five amino acids of this peptide had the sequence of methionine-enkephalin. Thereafter, Derek Smyth quickly established with his colleagues at the National Institute for Medical Research the so far unknown fact that β-lipotropin$_{61-91}$ had a high affinity for the opiate receptor. We showed that it depresses the contractions of the guinea pig ileum and mouse vas deferens and that this phenomenon is reversed by naloxone.

Our paper revealing the structure of the two pentapeptides was published in the issue of *Nature* for 18 December 1975. About a fortnight before that date we had posted copies of the proof to our friends and competitors in the field. Some time later I received a letter from Dr C. H. Li of the Hormone Research Laboratory, San Francisco, in which he described some of the events which had happened. On December 11, 1975, Dr Avram Goldstein of Stanford wrote a letter to Dr Li asking him for peptide samples related to the pentapeptide derived from β-lipotropin. On December 16, Dr Li showed Avram Goldstein's letter to Dr R. Guillemin of the Salk Institute who visited Dr Li on that day. He also asked for samples of β-lipotropin and β-lipotropin$_{61-91}$.

Thus, within a few days of our publication, intensive work on the long-chain opioid peptides found in the pituitary began both in London and in San Francisco and San Diego. Needless to say, at the same time many pharmaceutical laboratories were engaged in synthesising the enkephalins and many of their analogues. A report on the results obtained by these efforts will be found in the following chapters of this volume.

REFERENCES

Akil, H., Mayer, D. J. and Liebeskind, J. C. (1972). *C.r. hebd. Seanc. Acad. Sci. Paris*, **274**, 3603–05

Hughes, J. (1975). *Brain Res.*, **88**, 295–308

Hughes, J., Kosterlitz, H. W. and Leslie, F. M. (1975*a*). *Br. J. Pharmac.*, **53**, 371–81

Hughes, J., Smith, T. W., Kosterlitz, H. W., Fothergill, L. A., Morgan, B. A. and Morris, H. R. (1975*b*). *Nature*, **258**, 577–79

Kosterlitz, H. W. and Waterfield, A. A. (1975). *A. Rev. Pharmac.*, **15**, 29–47

Kosterlitz, H. W. and Watt, A. J. (1968). *Br. J. Pharmac.*, **33**, 266–76

Paton, W. D. M. (1957). *Br. J. Pharmac.*, **12**, 119–27

Pert, C. B. and Snyder, S. H. (1973). *Science*, **179**, 1011–14

Simon, E. J., Hiller, J. M. and Edelman, I. (1973). *Proc. natn. Acad. Sci. U.S.A.*, **70**, 1947–49

Terenius, L. (1973). *Acta pharmac. tox.*, **33**, 377–84

Trendelenburg, P. (1917). *Naunyn-Schmiedebergs Arch. exp. Path. Pharmak.*, **81**, 55–129

11
Physiological and clinical relevance of endorphins

L. Terenius and A. Wahlström (Department of Medical Pharmacology, University of Uppsala, Box 573, S-751 23 Uppsala, Sweden)

INTRODUCTION

Much progress has been made in recent years towards the understanding of the action of opiates. It was demonstrated by several groups (Pert and Snyder, 1973; Simon et al., 1973; Terenius, 1973) that the first step in the interaction between opiates and nervous tissue is the reversible binding to specific receptors in the synaptic plasma membranes. Following binding to the receptor the morphine-like opiates will inhibit adenylate cyclase activity (Collier et al., 1974; Sharma et al., 1975; Traber et al., 1975). The finding of specific opiate receptors and the coupling between receptor occupation and inhibition of adenylate cyclase activity, suggested the presence of a *functional receptor unit*. Since the opiate receptor proved to be extremely specific, having practically no affinity for previously known neurotransmitters or neuroactive agents, it seemed possible that such receptors might have previously unknown natural substrates, ligands. These considerations led to the search for such ligands in the brain and evidence for their existence was obtained (Hughes, 1975; Terenius and Wahlström, 1975a). Similarly, Cox, Goldstein and co-workers (1975) identified opiate-like material in the pituitary gland. The publication of the structure of the enkephalins (Hughes et al., 1975) led to extensive chemical work and it soon became clear, that in addition to the enkephalins, several other, higher-molecular weight peptides were opiate like (Bradbury et al., 1976; Guillemin et al., 1976; Li and Chung, 1976). These peptides are now collectively called endorphins (*endo*genous mo*rphine*-like agents).

The structural relation of the endorphins to the pituitary hormone, β-lipotropin, is shown in figure 11.1. The sequence of all endorphins except leucine-enkephalin, which instead of methionine has leucine, forms part of β-lipotropin. It has therefore been suggested that β-lipotropin is a prohormone for the endorphins (Hughes et al., 1975; Bradbury et al., 1976; Lazarus et al., 1976). No corresponding precursor for leucine-enkephalin is known as yet.

Following the discovery of endogenous opiates the next important step was to define their physiological role and their potential involvement in various disease

161

Centrally Acting Peptides

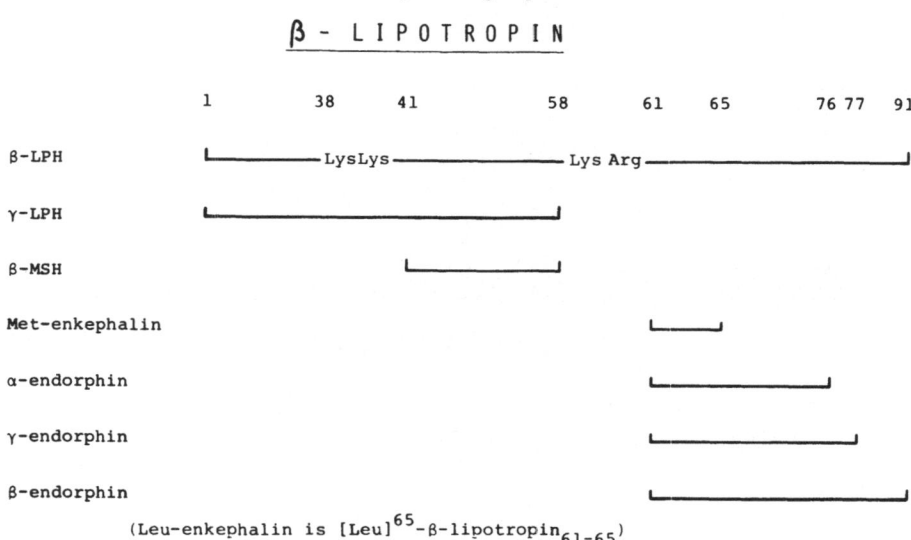

(Leu-enkephalin is $[Leu]^{65}$-β-lipotropin$_{61-65}$)

Figure 11.1 Generic relationship between the endorphins and β-lipotropin.

states. Considerable work has already been published regarding the pharmacology of the endorphins and this will only be briefly reviewed here.

Injection of endorphins into the cerebral ventricles or into the brain produces analgesia (see, for example, Belluzi *et al.*, 1976; Büscher *et al.*, 1976). In general, the longer the peptide chain, the higher the potency and the longer is the duration of action (Feldberg and Smyth, 1976). Since prolongation of the peptide chain does not markedly change opioid receptor affinity, it has been suggested that the weak activity of the enkephalins is due to metabolic degradation. This seems very likely also because synthetic analogues with replacement of glycine in position 2 by D-alanine are highly potent and long-acting analgesics (Pert *et al.*, 1976). Like opiates, met-enkephalin and β-endorphin have been found to produce dependence as judged from the naloxone-precipitated withdrawal reaction (Wei and Loh, 1976). In addition to analgesia β-endorphin is known to produce various behavioural responses in rats. Thus, β-endorphin causes catatonia at high doses (Bloom *et al.*, 1976; Jacquet and Marks, 1976), reduces the sexual peformance of male rats in moderate doses (Meyerson and Terenius, 1977) and induces excessive grooming in rats (Gispen *et al.*, 1976). All of these responses are more or less typical of morphine-like opiates.

Every important pharmacological action of the endorphins so far described is readily antagonised by narcotic antagonists such as naloxone and naltrexone. Thus, the narcotic antagonists have become extremely useful tools in the study of endorphins, particularly in defining their functional role.

SITES OF PRODUCTION OF ENDORPHINS

The two pentapeptides, leu-enkephalin and met-enkephalin, were originally isolated from brain and they have subsequently been found to be present over large areas of the central nervous system (CNS). The distribution of the enkephalins very closely

Table 11.1 Distribution of opiate receptors and enkephalin-like activity in the CNS

Author/method	Cerebral cortex	Corpus callosum	Thalamus	Hypothalamus	Globus pallidus	Amygdala	Periaqueductal grey	Cerebellum	Substantia gelatinosa
Opiate receptors									
Hiller et al. (1973) in vitro	**		**	**	*	***	**	*	
Pert et al. (1976) autoradiography	*	(*)	**	**	**	***	**	(*)	***
Kuhar et al. (1973) in vitro	*	(*)	**	**	*	***	**	(*)	
Enkephalin									
Smith et al. (1976) in vitro	*		*	***	*	***		(*)	
Elde et al. (1976) immunohisto-chemistry	*		*	**	***	**	**	(*)	***
Simantov et al. (1976) in vitro	*	*	*	***	***	**	**	(*)	

The concentrations were rated as very low (*), low *, intermediate ** and high ***.

follows the previously known distribution of opiate receptors (table 11.1). With the use of immunohistochemical methods enk-like material could be visualised in distinct areas (Elde *et al.*, 1976). The immunoreactive material was apparently localised in nerve terminals suggesting an intraneuronal synthesis of the peptides. In addition, areas which showed immunoreactive material were found in regions previously known to be rich in opiate receptors and in endorphin activity as assayed *in vitro* (Table 11.1). Recently cell bodies reacting with the antiserum have been found (Hökfelt *et al.*, 1977). The fluorescence intensity of cell bodies will increase considerably on treatment of the animal with colchicine. Colchicine is known to be an inhibitor of axonal transport and it is therefore probable that the enkephalins are being formed in the cell bodies of the neurons and then transported to the nerve terminals. Normally, the amounts in the synaptic region seem to be higher than in other parts of the neuron. To date there is no convincing evidence that the two enkephalins occupy different populations of neurons although a differential distribution of the two peptides has been described (Hughes *et al.*, chapter 12). Thus, antisera with selectivity for met-enkephalin or leu-enkephalin, respectively, give essentially equal staining patterns. If separate met-and leu-enkephalin pathways do exist, they must be running rather close to each other. Besides the CNS, enkephalins are also present in the intramural ganglia of the gastrointestinal system. However, they do not seem to be present in appreciable amounts in the pituitary gland or in peripheral nerves.

The longer endorphins have been isolated from the pituitary (Bradbury *et al.*, 1976; Li and Chung, 1976) or from combined hypophyseal–hypothalamic tissue (Guillemin *et al.*, 1976; Ling *et al.*, 1976). It seems probable that β-lipotropin serves as a prohormone and that the α-, β- and γ-endorphins represent cleavage fragments. It is not yet known which of these endorphins is the more important, nor is the exact concentration in fresh tissue known. Fluorescein-conjugated antibodies raised against α- or β-endorphin give some reaction in hypothalamic areas but mainly stain the pituitary gland (Guillemin, 1977). Hypophysectomy does not seem to change the brain endorphin content markedly (Cheung and Goldstein, 1976).

These observations lead us to the following provisional view. Enkephalins are produced intraneuronally in specific fibres in many different areas of the CNS and also in nerve plexi of the gastrointestinal tract. The longer endorphin chains are found in the pituitary but it is not yet known to what extent they enter the brain or are released into the general circulation. Any consideration of the functional role of endorphins thus will have to take into account two different systems with enkephalins being neurotransmitter candidates and the β-lipotropin-derived endorphins in the pituitary being hormonal candidates.

APPROACHES TO THE STUDY OF THE FUNCTIONAL ROLE OF ENDORPHINS

Essentially three ways have been followed:
  (1) administration of endorphins and observing their activities;
  (2) administration of the narcotic antagonists naloxone or naltrexone to inhibit the endogenous production of endorphins; and
  (3) analysis of endorphins in brain or body fluids in various physiological or disease states.
The first of these approaches is considered in detail in chapters of this volume.

The second approach is based on the assumption that narcotic antagonists will inhibit the actions of the endogenously produced endorphins. This seems reasonable, since naloxone and naltrexone have been found to antagonise almost every pharmacological action of opiates and opioid peptides. In addition, naloxone and naltrexone show practically no other effects, making them close to ideal as pharmacological tools. However, two things should be remembered. In receptor binding studies several groups have observed that opioid peptides differ from opiates in their interaction with receptors (Terenius, 1976; Simantov and Snyder, 1976*a*). Thus, naloxone and naltrexone are much less effective in inhibiting the binding of enkephalins to the receptors than in inhibiting the binding of opiates such as dihydromorphine. If this is also the case *in vivo*, more antagonist would be needed to overcome endorphin action than the action of opiates. Since it is known that 0.2–0.4 mg of naloxone injected intravenously will reverse heroin overdosing in man, several authors have considered this to be a huge dose and absence of activity at this dose level to be indicative that endorphins are not involved. The apparent safety of the drug would allow higher doses to be given and several investigators have in fact used 0.8 mg or more without producing any obvious harm to the patient. The second point relates to the short duration of naloxone action (Ngai *et al.*, 1976) and effects may be overlooked because they disappear too rapidly. An effect may also take time to develop and could require repeated dosing. This disadvantage is overcome by choosing the longer acting naltrexone as the experimental tool.

The third approach is to measure endorphins in tissue or in body fluids. Thus, several groups have measured endorphin levels in the brains of rats after various treatments (Simantov and Snyder, 1976*b*; Madden *et al.*, 1977). The endorphins are extracted in a simple procedure, processed by chromatography and measured in a radioreceptor assay (see below). The isolation procedure will probably preferentially extract enkephalins and not other endorphins although no exact information is given in the original papers. Another approach, studied extensively in our laboratory is to follow endorphins in cerebrospinal fluid (CSF) (Terenius and Wahlström, 1975*b*). The main advantage of this is that it can be used for clinical studies. It is based on the hypothesis that endorphin levels in CSF reflect CNS endorphin activity since the composition of CSF is very similar to that of the extracellular fluid of the CNS. A definite advantage is that serial samples can be drawn from the same individual. This is particularly useful in the clinical situation, where an effect of a particular treatment can be followed (see below). However, there are also interpretative difficulties in this approach. Very little is known about the actual relationship between CNS endorphin activity and the CSF levels. In clinical practice the accessible sample for analysis is the lumbar fluid and it is consequently not easy to define whether the endorphins being analysed derive from the brain, are of local spinal origin or even derive from the pituitary. Since most of the following will deal with endorphin analysis of human CSF, the characteristics of these endorphins and the method for their measurement will be considered in more detail.

Endorphins in human CSF and their measurement
The composition of CSF is chemically very close to ordinary buffer solutions and consequently samples of CSF can be introduced directly into an assay system (radioimmunoassay or radioreceptorassay). However, with the sensitivity of our assay it is an advantage to concentrate the samples. A tenfold concentration, a

10 ml CSF WITHDRAWN BY LUMBAR PUNCTURE WITH

THE PATIENT IN A SUPINE POSITION.

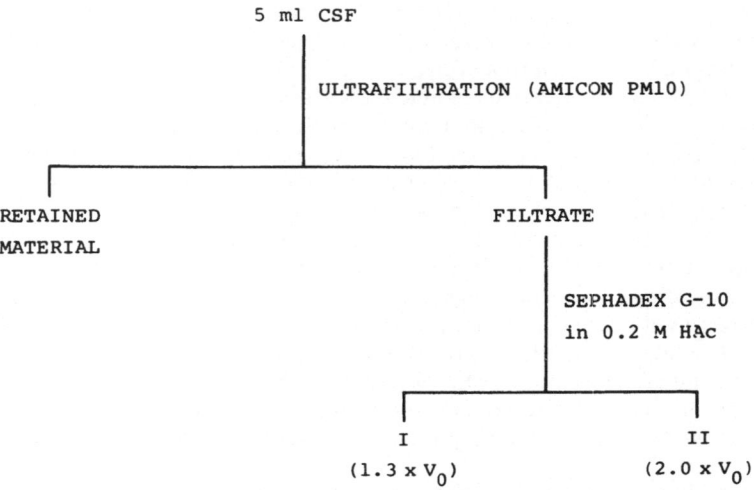

Figure 11.2   Fractionation of CSF endorphins (V_0 = void volume).

PRINCIPLES FOR TESTING OPIOID-LIKE ACTIVITY

Radioreceptorassay   Radioimmunoassay

1. Incubation   1. Incubation
       Radioindicator (RI)   Radioindicator (RI)
       Test fraction   Test fraction
       Receptor preparation   Antiserum

2. Centrifugation   2. Addition of charcoal

                                                3. Centrifugation

       Supernatant: free RI   Supernatant: antibody-bound RI
       Pellet: receptor-bound RI   Pellet (charcoal): free RI

3. Counting   4. Counting
       Pellet radioactivity   Supernatant radioactivity

Figure 11.3   Outline of and comparison between the radioreceptorassay and radioimmunoassay as used by the present authors.

chemical separation and partial desalting, was obtained by running ultrafiltered CSF samples through a Sephadex G-10 column. Ultrafiltration removed high-molecular weight components, for instance enzymes of blood origin if present. This procedure preserved endorphin activity in the samples and also protected the column from contamination. The whole separation procedure is outlined in figure 11.2. Two fractions, I and II, respectively, were isolated. These fractions are tested for opiate receptor affinity in a radioreceptorassay (figure 11.3). In a few cases, CSF samples or samples from fractions I and II were tested in radioimmuno-assay (figure 11.3), with antibodies raised against met-enkephalin or leu-enkephalin, respectively. These antibodies are specific for the enkephalins and cross-react only with molecules very similar in composition but not with, for instance, the high-molecular weight endorphins. Clearly, the two assays will measure different things: the radioreceptorassay detects receptor-active material with no other restrictions regarding the chemical composition whereas the radioimmunoassay detects only enkephalins and closely related molecules, whether biologically active or not.

The principal active molecular species in fractions I and II are being character-ised chemically. It is known that they do not represent any of the previously known endorphins (Wahlström *et al.*, 1976 and Wahlström, Fryklund, Johansson and Terenius, in preparation). Both fractions appear more basic in character than the enkephalins and probably have a lower molecular weight than α-endorphin since they differ from that compound in not being excluded from Sephadex G-10 gels. A slight cross-reaction of the fraction II endorphin with leu-enkephalin-directed antiserum may indicate a structural relationship with leu-enkephalin. A fraction with similar properties to fraction II can also be isolated from human brain (Wahlström *et al.*, in preparation) and it is likely, therefore, that it at least partly derives from the CNS. In the clinical correlations discussed later, the levels of both fraction I and II are considered.

THE PHYSIOLOGICAL IMPORTANCE OF ENDORPHINS

The existence of endogenous analgesic mechanisms have been considered in the past but the existence of endogenous analgesics of the opiate type was still largely unexpected. Despite one early and neglected report about analgesic principles in the pituitary (Murray and Miller, 1960) the concept of endogenous analgesics was born with the demonstration of the existence of specific opiate receptors with no reactivity for previously known neuroactive agents. One strong argument against the existence of such endogenous agents was the absence of subjective or objective effects of the narcotic antagonist naloxone (Jasinski *et al.*, 1967). This inactivity of naloxone *per se* showed up in several other experimental and clinical studies. The first investigation to actually suggest the possibility that naloxone might antagonise endogenous opiates was that of Jacob *et al.* (1974). They found, in a carefully con-trolled study, that when given alone naloxone slightly but significantly lowered pain thresholds in mice. This group (Jacob and Michaud, 1976) later presented similar evidence using dogs. Bell and Martin (1977) found that naloxone or nal-trexone facilitated nociceptive spinal cord reflexes. However, most studies in animals or in man largely support the concept that naloxone, or the chemically and pharmacologically related naltrexone, show practically no action of their own (Martin *et al.*, 1973; Gritz *et al.*, 1976). Provided these drugs do in fact antagonise all actions of endorphins, and available evidence would support this hypothesis, the endorphin activity must be low under normal circumstances. In teleological terms, this is not unexpected since pain is a sensory stimulus that is useful for the individual. For instance, people with congenital insensitivity to pain live a miserable life (Thrush, 1973). The system might normally be in a stand-by position ready for activation by a proper endogenous or exogenous stimulus.

Pain, analgesia and endorphins—experimental aspects

As already mentioned, naloxone given to mice or dogs will produce slight hyperalgesia (Jacob *et al.*, 1974; Jacob and Michaud, 1976). In man, a similar study revealed practically no change in pain thresholds following the administration of naloxone (El-Sobky *et al.*, 1976). It would be of interest to know whether *pain tolerance* would be affected, but such studies have not yet appeared. A study where long-term treatment of the animals with various stressful stimuli was performed resulted in fact in elevated brain endorphin levels and in a naloxone-reversible analgesia (Madden *et al.*, 1977). Although the effects were small, this study emphasises at least that the endorphin system can be activated if the experimental conditions are suitable.

In no case was hypophysectomy carried out to test whether or not the pituitary endorphins were involved. It is clear that the information summarised above gives rather little information as to the role of endorphins in physiological systems.

Pain, analgesia and endorphins—clinical aspects

Because of the problems involved in opiate therapy, alternative methods of treating patients with severe, chronic pain are constantly being sought. Besides neurosurgical intervention, various types of stimulation methods are being tried. Largely based on the Melzack–Wall gate theory, transcutaneous nerve stimulation has been developed and has occasionally been found to be of clinical value. The ancient Chinese method of acupuncture may be related in its mode of action and a method called electroacupuncture combines these two approaches. Here the electric current also causes definite muscle contractions (Andersson *et al.*, 1973; Eriksson and Sjölund, 1976). Finally, powerful analgesia can also be induced by stimulation of electrodes implanted in the brainstem of rats (Reynolds 1969; Mayer *et al.*, 1971) and of man (Adams, 1976; Boëthius *et al.*, 1976; Gybels and Cosyns, 1976; Richardson, 1976). Encouraging clinical results with electrostimulation methods have been obtained and the mechanism behind it is at present being studied. It is not unexpected that several groups have investigated whether or not endorphins are involved.

Analgesia produced by the electrostimulation of the periaqueductal grey matter of rat brain is at least partly reversible by naloxone (Akil *et al.*, 1976). Analgesia is also induced by microinjection of morphine in this region (Pert and Yaksh, 1974; Jacquet and Lajtha, 1976) and the area is rich in opiate receptors and in enkephalin pathways (Hiller *et al.*, 1973; Kuhar *et al.*, 1973; Elde *et al.*, 1976; Simantov *et al.*, 1976). Animals previously made tolerant to morphine also show tolerance to electrically induced analgesia (Mayer and Hayes, 1975). Neurophysiological studies suggest that the actual site of action of intracerebral stimulation is the spinal cord (Basbaum *et al.*, 1976; Mayer and Price, 1976). Such stimulation will selectively inhibit dorsal spinal horn neurons that respond to noxious stimuli. This region of the spinal cord is rich in enkephalin pathways (Elde *et al.*, 1976; Simantov *et al.*, 1976). Full behavioural analgesia can also be elicited by morphine injection at the level of the spinal cord of rats (Yaksh and Rudy, 1976).

Several of these observations are valid also in man. Thus, naloxone has been found to antagonise analgesia produced by brainstem stimulation (Adams, 1976; Meyerson *et al.*, 1977). A possible link between the brainstem stimulation mechanisms and various kinds of peripheral stimulation mechanisms is the observation that analgesia induced by classical Chinese acupuncture (Mayer *et al.*, 1977) or by electroacupuncture (Sjölund and Eriksson, 1976) is also reversible by naloxone.

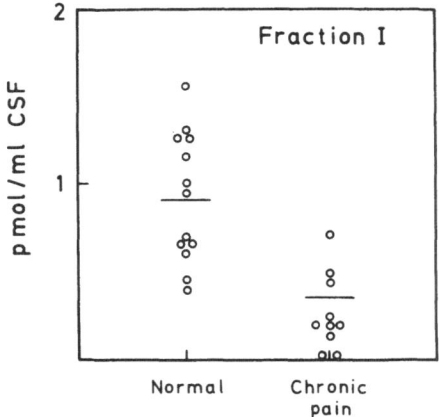

Figure 11.4 Endorphin fraction I levels in lumbar CSF in patients with chronic severe pain as compared with normal healthy volunteers. Horizontal lines represent mean values. Endorphin content expressed as pmol met-enkephalin per ml CSF.

We have been able to follow patients with chronic pain with regard to their CSF endorphin levels both before and after stimulation. The most interesting observations were made with regard to fraction I endorphin. It was noted in a pilot investigation that fraction I endorphin levels appear to be low in conditions of chronic pain (Terenius and Wahlström, 1975*b*). This observation was confirmed in the present series of patients (figure 11.4). The difference between the control and chronic pain groups is statistically significant ($P < 0.01$). When these patients with chronic pain were treated by electroacupuncture (that is, low-frequency, high-intensity electrostimulation) several responded with full analgesia (table 11.2). The endorphin levels increased in four patients in whom the stimulation involved the lower spinal cord (since electroacupuncture is applied to the same region as the painful area). The CSF sample was obtained only 30 min after ending the stimulation session and we have suggested (Sjölund *et al.*, 1977) that the CSF from higher levels may not yet have reached the lumbar region within this short interval. The validity of this conclusion is at present being investigated. These results can be

Table 11.2 Endorphin content of lumbar CSF in patients with chronic pain. Samples were taken while pain was experienced (Control) and after electroacupuncture (Stim.)

Patient no.	Pain cause	Duration	Analgesia from stimulation	Endorphin content (pmol/ml CSF) fraction I	
				Control	Stim.
1	Post-traumatic neuralgia thigh	4 years	full	<0.4	0.8
2	Mononeuritis of saphenous nerve	1 month	full	0.5	1.3
3	Herniated disc L4–L5	5 years	partial	0.9	2.0
4	Operated spinal AV malformation (L5 pain)	21 years	full	<0.4	2.6
5	Herpes zoster Th 5	8 years	full	<0.4	<0.4
6	Trigeminal neuralgia	16 years	partial	<0.4	<0.4
7	Trigeminal neuralgia	7 years	none	<0.4	<0.4
8	Trigeminal neuralgia	12 years	full	<0.4	0.5
9	Trigeminal neuralgia	5 years	full	0.5	0.6

From Sjölund et al. (1977).

Table 11.3 Endorphin content of lumbar CSF in patients with chronic pain. Samples were taken while pain was experienced (Control) and after intracerebral stimulation

Patient no.	Cause of pain	Analgesia from stimulation	Naloxone reversal	Endorphin content (pmol/ml CSF) fraction I	
				Control	Stim.
1	Breast cancer bone metastases	full	–	2.3	7.6
2	Corpus cancer	full	yes	3.1	8.4
3	Rectal pain	full	no	0.4, 0.5, < 0.4	0.5
4	Breast cancer	uncertain	–	< 0.4*	0.7
5	Bladder cancer, bone metastases	none	no	< 0.4, < 0.4	0.7, 0.8
6	Bladder cancer	none	–	< 0.4*	< 0.4*

From Meyerson et al. (1977).
*Treated with Ketobemidone

interpreted as favouring a stimulated release of endorphins at the segmental level (see above).

Besides the clinical interest in these methods, the observation that acupuncture or electroacupuncture stimulates endorphin activity is important because it may give us a lead to an afferent physiological activation process. The fibres activated by these procedures are not known and may not necessarily be the pain-conducting fibres. The stimulation of muscular activity seems to be of importance and it may be speculated that deep sensory fibres trigger the response. Definition of the sensory pathways activating endorphin activity in the spinal cord is an important neurophysiological problem.

In another series of patients, intracerebral stimulation of mesencephalic centres was carried out (Meyerson *et al.*, 1977). Here the patient donated one CSF sample after being without stimulation for 24 hours and another sample was obtained 24 hours later. During these last 24 hours the patient was allowed to use the stimulator at intervals of 4 hours or more frequently. Some patients also participated in naloxone trials during a period of stimulation. From table 11.3 it can be seen that two patients reported full analgesia which in one tested case was naloxone reversible. Another case reported pain relief but showed no reaction on naloxone injection and no elevation of CSF endorphins. In three patients no analgesia was recorded, one was naloxone negative and one showed a marginal elevation of CSF endorphins. Although the series is small, there is good agreement between the clinical response and results of CSF analysis. It also shows that focal electrostimulation *per se* does not increase CSF endorphins.

It is well known that individuals differ in their pain tolerance and that psychological states will influence reaction to pain. For instance, pain may not be perceived by severely wounded patients. It is possible that endorphin activity in supraspinal centres accounts for differences of this kind. Other high CNS mechanisms may also contribute since analgesia induced by hypnosis does not appear to be influenced by naloxone (Goldstein and Hilgard, 1975). However, hypnosis is a complex phenomenon and interpretation of these results is not easy. This brings us to another area of research into endorphin activities.

PSYCHE AND ENDORPHINS—EXPERIMENTAL ASPECTS

The actions of morphine on the CNS have been characterised as being 'analgesia, drowsiness, changes in mood and mental clouding'. The analgesic effect is on the reaction to a painful stimulus rather than on the pain threshold (Jaffe, 1970). It therefore seems reasonable to investigate whether endorphins might influence the expression of various behavioural responses. Indeed, several studies point to such a relationship. Thus, Gispen *et al.* (1976) found that β-endorphin elicits stereotyped grooming behaviour in rats, an effect also seen with morphine. Meyerson and Terenius (1977) found that β-endorphin, introduced into the lateral ventricles, reduced the sexual performance of male rats at doses that did not influence general motor activity or the approach to a castrated male. The finding that high doses of β-endorphin like morphine, produce a cataleptic state (Bloom *et al.*, 1976; Jacquet and Marks, 1976) has been taken as evidence that changes in endorphin homeostasis may be causally related to the development of mental illness, notably schizophrenia (Guillemin, 1977).

The main importance of these findings is that they emphasise the need for further investigations into other endorphin actions than analgesia. Much work will be needed to define their role in higher CNS activities. Their relationship to the action of morphine-like opiates also needs to be investigated. Several of the responses described above, for instance catalepsy, are typical also of opiates.

PSYCHE AND ENDORPHINS–CLINICAL ASPECTS

Since morphine-like substance have such wide effects on different CNS functions we have been investigating the possible role of endorphins in disease states unrelated to pain, particularly different psychiatric disorders. A study on endorphin levels in CSF was initiated and naloxone was administered to patients in conditions where endorphin hyperactivity seemed probable.

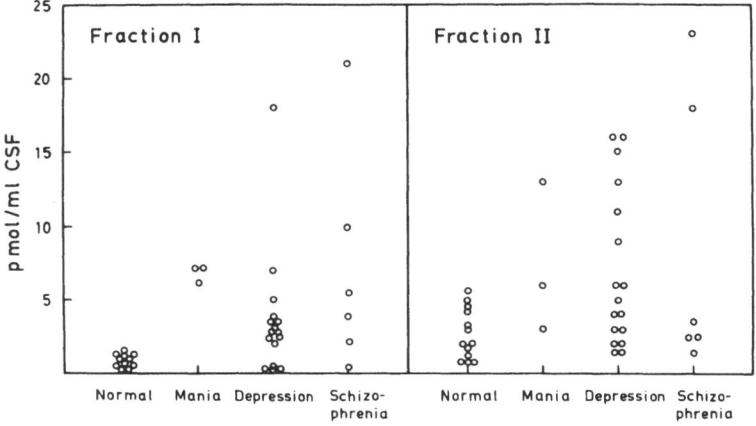

Figure 11.5 Endorphin levels in lumbar CSF of patients with various psychiatric diagnoses compared with normal healthy volunteers.

The results on CSF analysis of endorphins are summarised in figure 11.5. It is immediately clear that while control values from healthy volunteers fall within rather narrow limits, there is a large spread of values in psychotic patients. Since there is no evidence that the values follow a normal distribution, the results have also been tabulated as shown in table 11.4. The high levels in depression of both fractions I and II were striking. This observation prompted us to investigate the effect of naloxone in these patients. Five cases of severe endogenous depression,

Table 11.4 Distribution of patients with respect to their CSF levels of endorphin fractions I and II

Patient category	Below normal range I	II	Within normal range I	II	Above normal range I	II	No. of patients tested
Chronic pain	7	3	4	8	0	0	11
Mania	0	0	0	1	3	2	3
Depression	3	0	2	9	12	8	17
Schizophrenia (drug free)	0	0	1	4	5	2	6
Schizophrenia (treated)	1	0	2	2	4	5	7

Table 11.5   Effect of naloxone on severity of depression and on CSF endorphins

Patient	Day	Treatment	Length of trial (days)	Severity of depression		Endorphins (pmol/ml CSF) fraction I
				Zung	CPRS	
1	1	Naloxone 0.4 mg × 3 s.c.	7	50	8.0	6.7
	7			36		5.0
	lapse of 2 months					
	1	Naloxone 0.8 mg × 3 s.c.	12	66	10.0	2.6
	12			57	9.0	1.9
2	1	Naloxone 0.4 mg × 3 s.c.	8	71	10.5	4.2
	4	Naloxone 0.8 mg × 3 s.c.				
	8			75	11.5	1.7
3	1	Naloxone 0.4 mg × 3 s.c.	10	63	9.5	3.6
	10			63	10.5	2.5
4	1	Naloxone 0.4 mg × 3 s.c.	10	58	5.0	3.6
	5			44	1.5	3.1
	10			44	2.5	–
5	1	Saline × 3 i.m.	4	70	12.0	20
	4	Naloxone 0.4 mg × 3 i.m.	6	75	12.5	15
	10			76	9.0	7

From Terenius *et al.* (1977)
s.c., subcutaneously; i.m., intramuscularly

resistant to conventional therapy participated, one patient participated in two trials. Naloxone was given over a period of at least one week. CSF samples were withdrawn both before and following the trial. In every treatment session, endorphin fraction I values dropped, but the effect was not generally large enough to restore 'normal' levels (table 11.5). The clinical course was largely unfavourable, two patients reporting a deterioration of their mood levels within 1 to 2 days after the treatment ended.

In patients with the cyclic manic-depression syndrome high endorphin levels were also observed. In a few cases a longitudinal study was possible and some indication of a rise in endorphin values in the manic phase was evident (Terenius *et al.*, 1976). No tests with naloxone have so far been carried out on this category of patients.

Schizophrenia and endorphins

Figure 11.5 also illustrates the elevated levels of endorphins in patients with chronic schizophrenia. The levels were particularly high during a period when these patients had been off antipsychotic medication for several weeks and a successful trial with clozapine reduced the CSF levels (Terenius *et al.*, 1976). The elevated levels of CSF endorphins led to the endorphin hypothesis of schizophrenia (Gunne *et al.* 1977). It has been known for many years that the classical partial morphine antagonist nalorphine produces dysphoric reactions including delusions and hallucinations. The hallucinations are mainly auditory. These effects are also produced by other partial antagonist analgesics, particularly cyclazocine (Lasagna *et al.*, 1964). In many respects these side-effects are similar to the symptoms seen in schizophrenics. In particular the induced auditory hallucinations are very similar to those experienced by schizophrenics. A logical step was to test naloxone in chronic schizophrenics who reported hallucinations. In four of six tested cases, an acute dose of naloxone induced an improvement of the hallucinations within minutes after the injection. The effect was reversible and not shown by placebo injections. The experiment favours a direct involvement of endorphin in the production of certain symptoms and we therefore advanced the hypothesis that the formation of cyclazocine-like endorphin is involved in the aetiology of schizophrenia (Gunne *et al.*, 1977).

Several other groups have been testing the endorphin hypothesis and conflicting results have appeared (personal communications to the authors). The selection of patients, the criteria for response and the dosage regimen are variables which may have been different from our study and could account for the conflicting results. Further research will be necessary to evaluate a potential therapeutic role for narcotic antagonists in psychosis.

Clinical use of endorphins

To date, no reports on the actions of endorphins in man have been published. This section can therefore be only speculative. It can be foreseen that the pharmaceutical industry will produce peptide analogues that are more suitable than the natural ones for clinical use. Great expectations were originally attributed to the endorphins as potential non-addictive strong analgesics. In animal experiments, however, both β-endorphin and met-enkephalin seem addictive as judged from the withdrawal reaction in chronically exposed animals (Wei and Loh, 1976). The relevance of this observation to the clinical situation is not yet known and will require extensive trials.

CONCLUDING REMARKS

The discovery of the endorphins has opened up a new field in physiology. The developments have been rapid since the pharmacological actions of morphine-like agents are well documented and narcotic antagonists are available as pharmacological tools to define the functional role of the endorphins.

Most investigators agree that there is no significant tonic release of endorphins. Artificial stimulation of the system (for example, by acupuncture) has to be intense and of rather long duration. It could be that under more natural conditions the endorphin systems are being activated by injury or severe stress. The rapid development of tolerance to all opioids could be a mechanism that protects the individual from prolonged analgesia with the consequent inability to experience painful stimuli.

It is possible that more or less general changes in endorphin homeostasis may be causally related to the development of various psychic disorders. Another area of interest might be movement disorders as judged from the distribution of enkephalin fibres.

Finally, our familiarity with the effects of opiates should not automatically be transposed to the endorphin field. The contribution of endorphin fibres in the neuronal circuitry remains largely unknown.

ACKNOWLEDGEMENTS

This work is the result of stimulating co-operation with many clinical colleagues and the staff of our group. It is supported by the Swedish Medical Research Council.

REFERENCES

Adams, J. E. (1976). *Pain,* 2, 161–66
Akil, H., Mayer, D. J. and Liebeskind, J. C. (1976). *Science,* 191, 961–62
Andersson, S. S., Ericsson, T., Holmgren, E. and Lindqvist, G. (1973). *Brain Res.,* 63, 393–96
Basbaum, A. I., Clauton, C. H. and Fields, H. L. (1976). *Proc. natn. Acad. Sci. U.S.A.,* 73, 4685–88
Bell, J. A. and Martin, W. R. (1977). *Eur.J.Pharmac.,* 42, 147–54
Belluzzi, J. D., Grant, N., Garsky, V., Sarantakis, D., Wise, C. D. and Stein, L. (1976). *Nature,* 260, 625–26
Bloom, F., Segal, D., Ling, N. and Guillemin, R. (1976). *Science,* 194, 630–33
Boëthius, J., Lindblom, U., Meyerson, B. A. and Widen, L. (1976). In *Sensory Functions of the Skin* (ed. Y. Zotterman), Pergamon Press, Oxford and New York, 531–48
Bradbury, A. F., Smyth, D. G. and Snell, C. R. (1976). *Biochem.biophys.Res. Commun.,* 69, 950–56
Büscher, H. H., Hill, R. C., Römer, D., Cardinaux, F., Closse, A., Hauser, D. and Pless, J. (1976). *Nature,* 261, 423–25
Cheung, A. L. and Goldstein, A. (1976). *Life Sci.,* 19, 1005–08
Collier, H. O. J. and Roy, A. C. (1974). *Nature,* 248, 24–27
Cox, B. M., Opheim, K. E., Teschemacher, H. and Goldstein, A. (1975). *Life Sci.,* 16, 1777–82
Elde, R., Hökfelt, T., Johansson, O. and Terenius, L. (1976). *Neuroscience,* 1, 349–51
El-Sobky, A., Dostrovsky, J. O. and Wall, P. D. (1976). *Nature,* 263, 783–84
Eriksson, M. and Sjölund, B. (1976). In *Sensory Functions of the Skin* (ed. Y. Zotterman), Pergamon Press, Oxford and New York, 575–81.
Feldberg, W. and Smyth, D. G. (1976). *J. Physiol. Lond.,* 260, 30–31P
Gispen, W. H., Wiegant, V. M., Bradbury, A. F., Hulme, E. C., Smyth, D. G., Snell, C. R. and de Wied, D. (1976). *Nature,* 264, 794–95

Goldstein, A. and Hilgard, E. R. (1975). *Proc. natn. Acad. Sci. U.S.A.*, 72, 2041–43
Gritz, E. R., Shiffman, S. M., Jarvik, M. E., Schlesinger, J. and Charuvastra, V. C. (1976). *Clin.Pharmac.Ther.*, 19, 773–76
Guillemin, R. (1977). *New Engl. J. Med.*, 1, 226–28
Guillemin, R., Ling, N. and Burgus, R. (1976). *C.r.hebd.Séanc.Acad.Sci. Paris, C*, 282, 783–85
Gunne, L.-M., Lindstrom, L. and Terenius, L. (1977). *J. Neural Transm.*, 40, 13–19
Gybels, J. and Cosyns, P. (1976). In *Sensory Functions of the Skin* (ed. Y. Zotterman), Pergamon Press, Oxford and New York, 521–30
Hiller, J. M., Pearson, J. and Simon, E. J. (1973). *Res.Commun.Chem.Path.Pharmac.*, 6, 1052–61
Hökfelt, T., Elde, R., Johansson, O., Terenius, L. and Stein, L. (1977). *Neurosci. Lett.*, (in press)
Hughes, J. (1975). *Brain Res.*, 88, 295–308
Hughes, J., Smith, T. W., Kosterlitz, H. W., Fothergill, L. A., Morgan, B. A. and Morris, H.B. (1975). *Nature*, 258, 577–79
Jacob, J. J. and Michaud, G. M. (1976). *Archs.int.Pharmacodyn.Ther.*, 222, 332–40
Jacob, J. J., Tremblay, E. C. and Colombel, M.-C. (1974). *Psychopharmacologia* (Berl.), 37, 217–23
Jacquet, Y. F. and Lajtha, A. (1976). *Brain Res.*, 103, 501–13
Jacquet, Y. F. and Marks, N. (1976). *Science*, 194, 632–35
Jaffe, J. H. (1970). In *The Pharmacological Basis of Therapeutics*, 4th edn. (ed. L. S. Goodman and A. Gilman), Macmillan, London, p. 239
Jasinski, D. R., Martin, W. R. and Haertzen, C. A. (1967). *J.Pharmac.exp.Ther.*, 157, 420–26
Kuhar, M. J., Pert, C. B. and Snyder, S. H. (1973). *Nature*, 245, 447–50
Lasagna, L., De Kornfeld, T. J. and Pearson, J. W. (1964). *J. Pharmac.*, 144, 12–16
Lazarus, L. H., Ling, N. and Guillemin, R. (1976). *Proc. natn. Acad. Sci. U.S.A.*, 73, 2156–59
Li, C. H. and Chung, D. (1976). *Proc. natn. Acad. Sci. U.S.A.*, 73, 1145–48
Ling, N., Burgus, R. and Guillemin, R. (1976). *Proc. natn. Acad. Sci. U.S.A.*, 73, 3942–46
Madden, J., Akil, H., Patrick, R. L. and Barchas, J. D. (1977). *Nature*, 265, 358–60
Martin, W. R., Jasinski, D. R. and Mansky, P. A. (1973). *Archs. gen. Psychiatry*, 28, 784–91
Mayer, D. J. and Hayes, R. L. (1975). *Science*, 188, 941–43
Mayer, D. and Price, D. (1976). *Pain*, 2, 379–404
Mayer, D. J., Price, D. D. and Rafii, A. (1977). *Brain Res.*, 121, 368–72
Mayer, D. J., Wolfe, T. L., Akil, H., Carder, B. and Liebeskind, J. C. (1971). *Science*, 174, 1351–54
Meyerson, B. J. and Terenius, L. (1977). *Eur.J.Pharmac.*, 42, 191–92
Meyerson, B. A., Terenius, L. and Wahlström, A. (1977). To be published
Murray, W. J. and Miller, J. W. (1960). *J.Pharmac.exp.Ther.*, 128, 380–83
Ngai, S. H., Berkowitz, B. A., Yang, J. C., Hempstead, J. and Spector, S. (1976). *Anesthesiology*, 44, 398–401
Pert, C. B. and Snyder, S. H. (1973). *Science*, 179, 1011–14
Pert, A. and Yaksh, T. (1974). *Brain Res.*, 80, 135–40
Pert, C. B., Pert, A., Chang, J. K. and Fong, B. T. W. (1976). *Science*, 194, 330–32
Reynolds, D. V. (1969). *Science*, 164, 444–45
Richardson, D. E. (1976). *I.E.E.E. Trans. Biomed. Eng.*, BME-23, 304–06
Sharma, S. K., Nirenberg, M. and Klee, W. A. (1975). *Proc. natn. Acad. Sci. U.S.A.*, 72, 590–94
Simantov, R., Kuhar, M. J., Pasternak, G. W. and Snyder, S. H. (1976). *Brain Res.*, 106, 189–97
Simantov, R. and Snyder, S. H. (1976a). In *Opiates and Endogenous Opioid Peptides*, North-Holland, Amsterdam–New York–Oxford, p. 41–48
Simantov, R. and Snyder, S. H. (1976b). *Nature*, 262, 505–07
Simon, E. J., Hiller, J. M. and Edelman, I. (1973). *Proc. natn. Acad. Sci. U.S.A.*, 70, 1947–49
Sjölund, B. and Eriksson, M. (1976). *Lancet*, ii, 1085
Sjölund, B., Terenius, L. and Eriksson, M. (1977). *Acta physiol. scand.*, 100, 382–84
Terenius, L. (1973). *Acta Pharmac.Tox.*, 32, 317–20
Terenius, L. (1977). *Psychoneuroendocrinology*, 2, 53–58
Terenius, L. and Wahlström, A. (1975a). *Acta physiol.scand.*, 94, 74–81
Terenius, L. and Wahlström, A. (1975b). *Life Sci.*, 16, 1759–64
Terenius, L., Wahlström, A. and Agren, H. (1977). *Psychopharmacologica*, 54, 31–33

Terenius, L., Wahlström, A., Lindström, L. and Widerlöv, E. (1976). *Neurosci.Lett.*, 3, 157–62
Thrush, D. C. (1973). *Brain*, 96, 369–86
Traber, J., Fischer, K , Latzin, S. and Hamprecht, B. (1975). *Nature*, 253, 120–22
Wahlström, A., Johansson, L. and Terenius, L. (1976). In *Opiates and Endogenous Opioid Peptides*, Elsevier/North-Holland Biomedical Press, Amsterdam, The Netherlands, 49–56
Wei, E. and Loh, H. (1976). *Science*, 193, 1262–63
Yaksh, T. L. and Rudy, T. A. (1976). *Science*, 192, 1357–58

12
Pharmacological and biochemical aspects of the enkephalins

J. Hughes, H. W. Kosterlitz, A. T. McKnight, R. P. Sosa, J. A. H. Lord and
A. A. Waterfield (Unit for Research on Addictive Drugs,
University of Aberdeen, Marischal College, Aberdeen, U.K.)

INTRODUCTION

Kosterlitz and Hughes (1975) originally suggested that the enkephalins may be
inhibitory neurotransmitters or neuromodulators. Recent work in this field and
particularly in the immunofluorescent localisation of the enkephalins has
supported this original concept. However, the concept of an endogenous ligand
for the opiate receptor did not take into account the possibility of multiple ligands.
When the structures of methionine- and leucine-enkephalin were reported it was
pointed out (Hughes et al., 1975) that the sequence of met-enkephalin was included
as residues 61–65 of β-lipotropin (β-LPH; Li and Chung, 1976a). Following this
observation several different fragments of β-LPH were isolated from the pituitary
gland and were shown to be potent agonists at opiate receptor sites (Feldberg and
Smyth, 1976; Guillemin et al., 1976; Lazarus et al., 1976; Li and Chung, 1976b).
This picture is further confused by the reports that other opioid peptides exist in
the pituitary (Gentleman et al., 1976), brain (Ross et al., 1976; Smith et al., 1976)
and cerebrospinal fluid (Terenius, chapter 11). These unidentified peptides are not
identical with any known fragment of β-LPH and some at least may be derived
from the putative precursor of leu-enkephalin.

It is apparent, therefore, that considerable pitfalls may await investigators in this
field both in the measurement of opioid peptides where there is the possibility of
cross-reactivity in assay systems with unidentified peptides, and in the interpretation of drug effects which may involve more than one peptide system. A full evalua-
tion of the physiological significance of the opioid peptides requires a fuller understanding of their distribution, biogenesis, release and receptor interactions. A number of these points are discussed in the following sections.

ISOLATION AND DISTRIBUTION

Our studies have concentrated mainly on the two enkephalins, using only the electrically stimulated mouse vas deferens as assay tissue (Henderson et al., 1972). It is

179

Centrally Acting Peptides

possible to assay 1 ng of enkephalin with this system and although bioassay is not as sensitive or rapid as assays using antibody or opiate receptor binding it does have the advantage of a built-in specificity—it only responds to biologically active peptides. Further, each sample may be checked individually for naloxone antagonism to ensure receptor specificity. However, methods of separation have to be used to distinguish individual peptides. Amberlite XAD-2 resin specifically separates met-enkephalin, leu-enkephalin and α-endorphin from longer fragments of β-LPH (Smith *et al.*, 1976) and using this method it was possible to show that the enkephalins could not possibly be derived from the breakdown of longer LPH fragments during the isolation or chromatography procedures. For example, addition of 5–10 μg β-LPH or of β LPH$_{61-91}$ to tissue homogenates did not result in an additional recovery of enkephalins when compared with the recovery from untreated tissues containing 0.4–0.8 μg of endogenous enkephalin.

Separation of the enkephalins is most easily achieved by thin-layer chromatography on silica gel (Smith *et al.*, 1976) or, alternatively, the peptides may be differentially assayed after incubation with cyanogen bromide which selectively destroys met-enkephalin activity. It should be noted that oxidation of the terminal methionine in met-enkephalin may cause losses of this peptide. This problem can be overcome by including glutathione or ascorbic acid (1 μg/ml) in any solution containing the peptide; this is particularly important before thin-layer chromatography of the peptides.

The enkephalins are distributed widely throughout the brain and gastrointestinal tract. The immunofluorescent localisation of enkephalin containing nerves is described by Elde, Snyder and Miller in chapters 2, 6 and 13, respectively. However, the separate distribution of met- and leu-enkephalin has yet to be shown by immunofluorescence due to problems of cross-reactivity with this technique. In contrast we were able to show that the ratio of these two peptides varies considerably from one brain region to another (table 12.1). Thus we find almost equal concentrations of the two peptides in the cerebral cortex whereas there is proportionately much more met-enkephalin in mesencephalic and hypothalamic areas. However, the highest concentrations of both peptides are found in the striatum and

Table 12.1 Distribution of enkephalins in guinea pig brain

	Cyanogen bromide assay			t.l.c. assay ratio
	met-enk	leu-enk (pmol/g)	Ratio	
Cortex	67 ± 9	51 ± 5	1.3	1.5
Hippocampus	150 ± 11	70 ± 9	2.1	1.7
Striatum	750 ± 41	180 ± 25	4.2	4.0
Pons + Medulla	139 ± 12	20 ± 5	6.9	5.8
Thalamus	157 ± 12	21 ± 3	7.4	6.3
Hypothalamus	453 ± 52	54 ± 18	8.4	—
Cerebellum	< 10	< 10		

The peptides were differentially assayed by determining opioid activity before and after treatment with cyanogen bromide. The means ± s.e. of four experiments are shown. In two separate experiments the ratio of the two peptides was determined after separation by t.l.c.

hypothalamus. This may indicate important roles for the enkephalins in extra-pyramidal and endocrine control mechanisms.

Our results differ from those of Yang *et al.* (1977) who found considerably higher concentrations of met-enkephalin in the rat brain with a corresponding higher met-enkephalin/leu-enkephalin ratio. These authors used microwave irradiation to kill their animals and concluded that this method avoided proteolytic losses associated with the more usual method of decapitating animals used by us and by Simantov and Snyder (1976). If this is correct then this is an extremely important observation. However, these results also indicated a greater lability of met-enkephalin in brain compared to leu-enkephalin; this has not been observed in studies *in vitro*. Further, fixation of brain tissue *in vivo* by formaldehyde perfusion through the carotid artery does not significantly alter enkephalin concentrations (Hughes *et al.*, 1977; table 12.2). It will be important to determine whether microwave irradiation causes formation of met-enkephalin from precursor protein as this apparently occurs with Substance P (Iversen, chapter 9). Preliminary results in our laboratory indicate that the met-enkephalin/leu-enkephalin ratio is not altered by microwave irradiation and that the increased recovery of peptides following irradia-

Table 12.2   Enkephalin content of rat brain under different conditions

Conditions	Enkephalin content (pmol/g)
(A) 3 min after exsanguination	189 ± 19
(B) 1 hour after exsanguination	82 ± 11*
(C) Pentobarbitone (60 mg/kg)	207 ± 16
(D) Pentobarbitone + formaldehyde	235 ± 23

Total enkephalin activity is expressed as met-enkephalin, results are the mean ± s.e. of four experiments. In groups (A) and (B) the rats were stunned and the head cut off; the brains were dissected out and the cerebellum removed 3 min and 60 min, respectively, after death. Groups (C) and (D) were anaesthetised with pentobarbitone (60 mg/kg), after 30 min the head was removed and the brain extracted (C) or 10 ml of 10 per cent formaldehyde in 0.9 per cent saline was perfused via a 20 ml syringe through one carotid artery and then the brain was removed (D). *Significantly different from A, C, and D ($P < 0.01$).

Table 12.3   Enkephalin distribution in guinea pig and rabbit ileum

	Cyanogen bromide assay			t.l.c. ratio
	met-enk	leu-enk (pmol/g)	Ratio	
Guinea pig myenteric plexus	410 ± 50	151 ± 16	2.7	3.8
Rabbit myenteric plexus	270 ± 36	113 ± 14	2.4	2.2
Guinea pig circular muscle	49 ± 4	11 ± 1	4.7	4.2
Rabbit circular muscle	25 ± 2	21 ± 6	1.5	0.9

Met- and leu-enkephalin concentrations were determined by assaying opioid activity before and after cyanogen bromide treatment. The mean ± s.e. of four experiments are shown. The ratio of met-enkephalin to leu-enkephalin was also determined after t.l.c., the means of two separate experiments are shown.

tion may be due to factors other than the rapid inactivation of brain enzymes.

It is unlikely that proteolysis is a source of loss of enkephalins in the gut since this tissue can be maintained in a viable state *in vitro* for many hours without significant losses or changes in the met-enkephalin/leu-enkephalin ratio. Both peptides have been detected in the gastrointestinal tracts of the rat, mouse, guinea pig and rabbit. The concentrations of the peptides in the small intestine of the latter two species are similar to those found in the brain (Smith *et al.*, 1976; table 12.3).

Recent experiments have also established that small amounts of enkephalin-like material are present in peripheral tissues other than the gut (Hughes *et al.*, 1977). Amounts ranging from 1 to 20 pmol/g have been extracted from rat vagus nerve and atria, rabbit sympathetic ganglia, guinea pig kidney and lung. Surprisingly, perhaps, no such activity has been detected in extracts of the mouse vas deferens.

Endorphins

At present the large opiate peptides α-, β- and γ-endorphin have only been identified definitively in the pituitary. Immunofluorescence microscopy has revealed the presence of α- and β-endorphin-containing cells in the pars intermedia and to a lesser extent in the pars distalis or adenohypophysis (Bloom *et al.*, 1977). We have used gel filtration and thin-layer chromatography (t.l.c.) to examine extracts of rabbit pituitary, rabbit striatum and guinea pig myenteric plexus for enkephalin and endorphin activity. The pituitary was found to contain up to 34 nmol/g of endorphin (assayed as β-endorphin) and negligible amounts of enkephalin. The striatum contained 0.82 nmol/g of enkephalin and 0.3 nmol/g of endorphin activity which was tentatively identified on t.l.c. as β-endorphin. In contrast, the myenteric plexus contained as much enkephalin as the striatum but had no detectable endorphin activity.

The results from various studies on the distribution of the enkephalins and endorphins appear broadly to agree. The enkephalins are widely distributed in the central nervous system (CNS) and peripheral tissues whereas the endorphins are present in high concentration in the pituitary although an extra-pituitary distribution also seems likely. It remains to be determined whether β-endorphin is invariably formed as an intermediate if LPH is the precursor of met-enkephalin. However, the virtual absence of enkephalin in the pituitary and apparent lack of endorphin in the myenteric plexus suggests that there is no clear association between these peptides.

BIOGENESIS OF ENKEPHALINS

It has been assumed but not proven that LPH is the precursor of met-enkephalin and by analogy it is thought that a similar precursor might exist for leu-enkephalin. The possible steps involved in the production of enkephalin are shown in figure 12.1. Certain predictions can be made from this model. If the precursor is stored or transported before conversion to enkephalin then an appreciable lag would be expected between the introduction of labelled amino acids into the system and the appearance of labelled enkephalin. This may not be the case if the precursor is directly converted to enkephalin. Enkephalin is stored in a stable form in the crude synaptosomal, P_2, fraction of brain homogenates and it is likely that this involves transport of either precursor or enkephalin into some form of granular store. Proteolytic cleavage is also likely to involve more than one enzyme and we would predict that

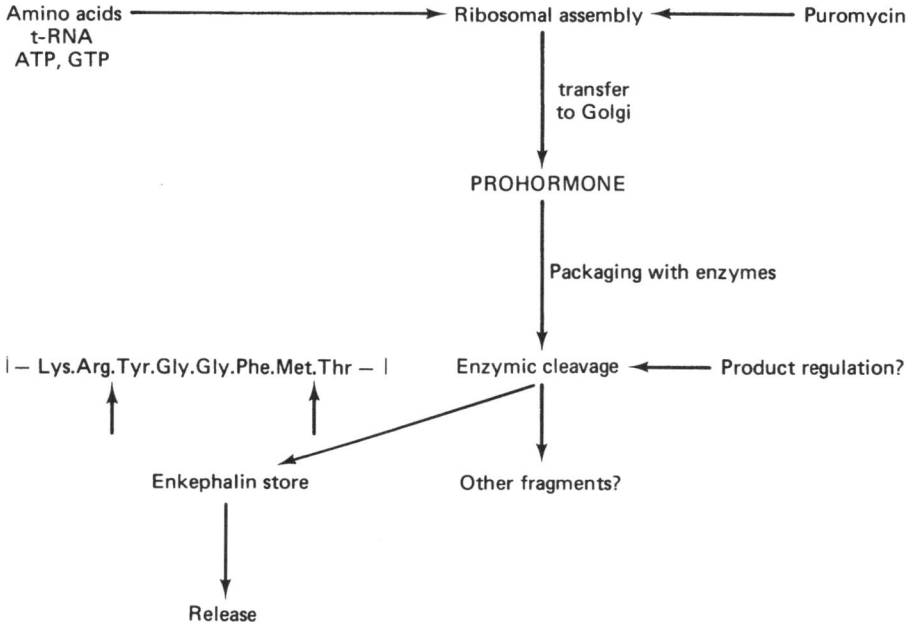

Figure 12.1 Biogenesis of enkephalins. Ribosomal assembly of precursor protein and packaging of precursor occurs in the perikaryon. By analogy with neurosecretory neurons it is likely that the precursor and cleavage enzymes are co-packaged and that conversion to enkephalin occurs during transport of the granules to the nerve terminals. This would explain the difficulty of visualising cell bodies by immunofluorescence techniques. Colchicine by blocking axonal transport would lead to an increase in enkephalin content of the cell body. Regulation of enkephalin production may be dependent on the activity of the cleavage enzymes rather than on prohormone synthesis which may proceed at a constant rate determined only by energetic and amino acid requirements. Thus cleavage activity could be dependent on the enkephalin concentration.

cleavage of the methionine—threonine bond in LPH would involve a highly specific endopeptidase.

There is an underlying beauty to this apparently complex system for the production of a putative neurotransmitter. The same basic system is capable of producing varied products from the same precursor by altering the specificity or number of cleavage enzymes present in the system. Identification of such enzymes will be of great importance to a further understanding of these systems.

Incorporation of amino acids into enkephalins

Clouet and Ratner (1976) reported the incorporation of [³H] glycine into brain enkephalin after intracisternal injection. They concluded that there is a very rapid incorporation (15–30 min) and turnover (2–3 hours) of the peptides. However, there must be some doubt regarding these results since the authors did not use carrier enkephalin as a marker to follow the purification of their extracts and it is unlikely that their procedure yielded radiochemically pure enkephalin. Further, the amounts of endogenous enkephalin recovered by Clouet and Ratner, as measured by the rather non-specific fluorescamine assay, are at least two orders of magnitude greater than that reported by workers using specific bioassay or immunoassay.

We have studied the incorporation of high specific activity tyrosine into rat brain and guinea pig myenteric plexus. The enkephalins were extracted with 0.1 M HCl and 40 μg of carrier peptide added. The peptides were then isolated on Amberlite XAD-2 resin (Smith *et al.*, 1976) and purified by ion exchange and t.l.c. (table 12.4). Electrophoresis of the enkephalin fractions from the final t.l.c. showed that the radioactive t.l.c. spot migrated as one homogeneous peak which was identical to the marker enkephalin. Further, we have also shown that the specific activity of the enkephalins is constant in the final two chromatography steps; this is an important indication of radiochemical purity.

Table 12.4   Isolation of isotopically labelled enkephalins

---

(1)   Homogenise tissues in 0.1 M HCl with addition of carrier enkephalins.

(2)   Absorption and elution from Amberlite XAD–2 resin.

(3)   h.p.l.c. on 100 × 0.4 cm column of SCX resin. Initial condition 0.2 M acetic acid, final elution with 0.35 M ammonium acetate (pH 8.35) in 30 per cent methanol.

(4)   h.p.l.c. on 100 × 0.4 cm column of AE–SAX resin. Initial condition 0.08 M ammonium acetate (pH 9.5), final elution with 0.04 M acetic acid in 30 per cent methanol.

(5)   t.l.c. on silica gel with ethyl acetate: pyridine: water: acetic acid (100: 42: 25: 11). Complete resolution of met- and leu-enkephalin.

(6)   Electrophoresis at 20 V/cm on glass fibre strips impregnated with silica gel.

Our preliminary results with intracisternal injection of [^{3}H] tyrosine show that labelled enkephalin accumulates slowly in the brain reaching a peak within 4–6 hours and remaining constant for up to 12 hours. The levels of radioactivity slowly decline over the 4 days following a single injection. These results are quite similar to the findings of Jones and Pickering (1972) who observed a similar time course for incorporation of [^{3}H] tyrosine into vasopressin and oxytocin. Our results, although preliminary, support the concept of incorporation of amino acids into a precursor which is then slowly converted to enkephalin. More direct evidence for this view was obtained from the isolated myenteric plexus.

Incorporation of amino acids into enkephalin in the guinea pig myenteric plexus also proceeds slowly. We have found that incubation with label for 3 hours followed by a further 3 hours' incubation with cold amino acids results in an appreciable labelling of the met- and leu-enkephalin stores. The incorporation of [^{3}H] tyrosine was markedly inhibited by exposure of the tissue to either cyclohexamide or puromycin (table 12.5) at concentrations that caused an inhibition of similar magnitude of protein synthesis. These results are the first direct support for the precursor hypothesis and indeed confirm that leu-enkephalin may be formed from precursor protein. It is also of interest that the ratio of incorporation of [^{3}H] tyrosine into

Table 12.5 Incorporation of [^{3}H] tyrosine into guinea pig myenteric
plexus enkephalin stores

	Met-enkephalin	Leu-enkephalin	Protein
	(dpm/g tissue)		(dpm/mg)
Untreated	3756 ± 255	1426 ± 62	24 561 ± 2422
Puromycin	317 ± 57	295 ± 127	1529 ± 335

The isolated plexus was incubated with 3.3 μCi/ml of L-[2,3,5,6-^{3}H]-tyrosine (specific activity
= 80 Ci/mmol) for 4 hours and then with cold amino acids (1 μg/ml) for a further 3 hours.
Puromycin (0.1 mM) was included in the bathing fluid for 15 min prior to incubation with
label and for the rest of the experiment. Incorporation of label into met- and leu-enkephalin
was inhibited by 92 and 80 per cent, respectively, in the presence of puromycin. There was a
similar inhibition of protein synthesis. Means of three experiments.

the two peptides was very similar to the endogenous met-enkephalin/leu-enkephalin
ratios in brain and gut.

There is obviously much to be determined about the biogenesis and utilisation
of the enkephalins. There are numerous points at which the production of these
peptides may be physiologically controlled and which may in turn lead to a
pharmacological exploitation. It also seems likely that measurement of the turnover
of the enkephalins will be more likely to yield information on the role of these peptides than measurement of steady-state tissue concentrations. This point is discussed
further in a later section.

METABOLISM AND INACTIVATION

Hughes (1975) originally described the rapid loss of biological activity of the enkephalins on incubation with various tissue extracts. It has now been demonstrated by
a number of workers that a major difference between β-endorphin and the enkephalins is the rapid loss of activity of the latter and the stability of the former on
incubation with nerve membranes or brain extracts (Pert *et al.*, 1976; Hughes and
Kosterlitz, 1977; Meek *et al.*, 1977; Miller *et al.*, 1977). These studies indicate that
the half-life of enkephalin *in vitro* or *in vivo* is of the order of 1 min or less whereas
α- and β-endorphin retain their activity for several hours. It is perhaps important to
note that we have been unable to detect the formation of enkephalin activity from
α- or β-endorphin even on prolonged incubation with brain slices or synaptosomal
preparations.

Although the enkephalins are rapidly metabolised by carboxypeptidase-A, it
appears that N-terminal hydrolysis by aminopeptidase is the prime means of inactivation, at least *in vitro* (Hambrook *et al.*, 1976; Meek *et al.*, 1977). N-terminal
hydrolysis leads to the liberation of free tyrosine and the inactive tetrapeptides
Gly–Gly–Phe–Met and Gly–Gly–Phe–Leu. These peptides may be more stable than
the enkephalins and may perhaps provide a further means of studying enkephalin
turnover.

The importance of N-terminal hydrolysis in limiting the biological action of the
enkephalins was illustrated by the synthesis of the stable analogue Tyr–D-Ala–Gly–
Phe–Met (Pert *et al.*, 1976; Miller *et al.*, 1977). This compound is resistant to peptidase action and, unlike the enkephalins, is a potent analgesic when administered

intraventricularly. Conversely, amidation of the C-terminal in enkephalin has little effect on its biological potency. At present no specific inhibitors of enkephalin metabolism or inactivation have been described. Bacitracin (100 μg/ml), aprotinin (200 units/ml), o-phenanthroline and p-chloromercuriphenyl sulphonic acid have all been shown to possess some enzyme-inhibiting activity. However, no really suitable agent for use in physiological or binding studies has yet been found. This problem can be circumvented in receptor binding studies by working at 0 °C but this is of no use in release studies.

RELEASE OF ENKEPHALINS

Confirmation of a neurotransmitter role for the enkephalins demands that these substances be released from nerve endings under physiological conditions. This has yet to be conclusively shown although evidence is beginning to accumulate.

Indirect evidence

Waterfield and Kosterlitz (1975) showed that the narcotic antagonists naloxone and (−)-2-allyl-β-9-methyl-5-phenyl-2-hydroxy-6, 7-benzomorphan (GPA 1843) increase the spontaneous and electrically evoked release of acetylcholine from the guinea pig myenteric plexus–longitudinal muscle preparation. This was stereospecific effect since the (+)-isomer of GPA 1843 had no such action. Similarly, Puig *et al.* (1977) showed that the same antagonists were able to partially prevent the depression of single twitches that had been caused by high frequency electrical stimulation of the myenteric plexus. We have confirmed this effect (figure 12.2) and were also able to show that carboxypeptidase when added to fluid bathing the ileum was able to reverse the inhibitory effect of added enkephalin but not that of high frequency stimulation (figure 12.2). The inhibition by high frequency stimulation of subsequent single stimuli is thus unlikely to be due to an accumulation of enkephalin in the bathing fluid. It is more likely that tetanic stimulation initiates a period of continuous firing of enkephalinergic neurons that outlasts the initial stimulus.

Evidence for enkephalin release under more physiological conditions was obtained by Van Neuten *et al.* (1976). They studied the peristaltic reflex and observed that when the preparation became fatigued and the reflex response erratic then addition of naloxone would restore the regularity of the reflex. It seems, therefore, that endogenously released enkephalin may act to restrict acetylcholine release in the guinea pig ileum and it is likely that this tissue will prove of great value in studying the possible neurotransmitter roles of enkephalin.

Direct evidence

The enkephalins have been found to be localised in the P_2 or synaptosomal fraction of brain homogenates (Simantov *et al.*, 1976; Smith *et al.*, 1976). We have studied the effect of raised levels of potassium ions on enkephalin release from the P_2 fraction of rabbit striatum (table 12.5). A low resting release of enkephalin was seen with this preparation but this release was markedly increased on increasing K^+ concentration to 50 mM. The K^+ induced release was found to vary directly with the Ca^{2+} concentration (table 12.6). Similar results were obtained with striatal slices from the rabbit and guinea pig. Unfortunately, experiments using electrical

stimulation of superfused brain slices have given inconsistent results in our laboratory. We suspect that the massive instantaneous release engendered by raised the
K^+ concentration allows at least some of the enkephalin to escape enzymic destruction, whereas the release of enkephalin by electrical stimulation is probably a
slower process and does not swamp the inactivating enzymes.

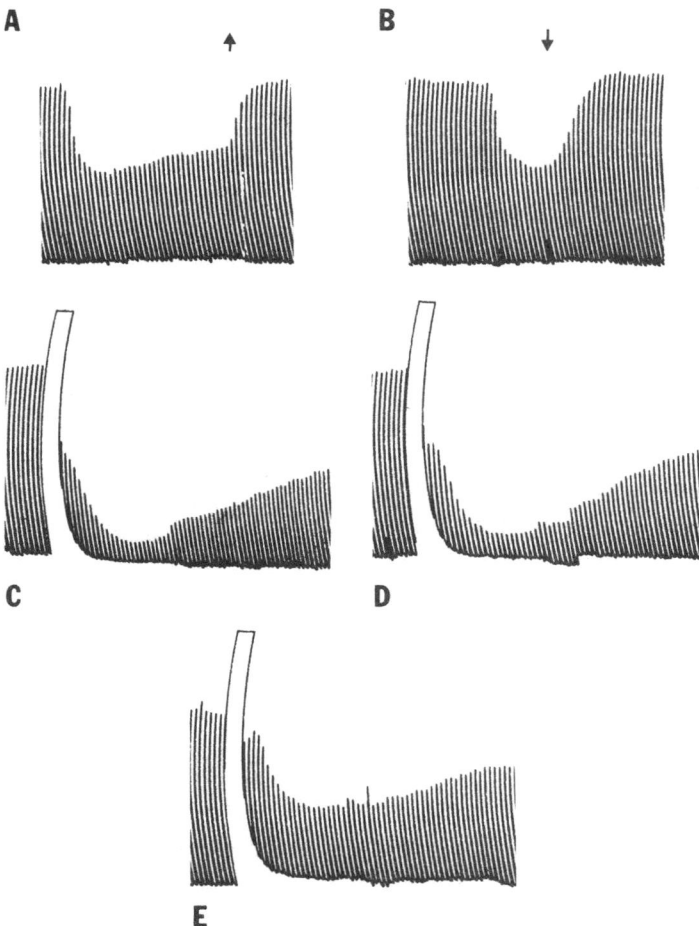

Figure 12.2 Mobilisation of endogenous opiate in the isolated longitudinal muscle–myenteric
plexus preparation of the guinea pig. The preparation was electrically stimulated at 0.1 Hz
throughout the experiment. Panel A shows the effect of 100 nM met-enkephalin, the inhibition
of the twitch slowly declined during the 6 min exposure to the peptide; the inhibition was
rapidly terminated on washing out the enkephalin (↑). In B, carboxypeptidase-A (5 U/ml) was
added to the 3 ml bath (↓); this caused a rapid recovery of the twitch without washing out the
enkephalin (100 nM). In C, D and E the tissue was subjected to a 30 s period of stimulation at
10 Hz, in the absence of further treatment (C) this caused a long lasting inhibition of the single
twitches ($t_{1/2}$ for recovery = 400 s). In D the bath contained 20 U/ml of carboxypeptidase-A;
this did not influence the subsequent recovery of the single twitches ($t_{1/2}$ = 390 s) following the
stimulation at 10 Hz. In E the bath contained naloxone (100 nM); in this case the inhibition of
the single twitches was much reduced and the recovery period was almost halved ($t_{1/2}$ = 210 s).

Table 12.6 Enkephalin release from rabbit brain synaptosomes

[Ca^{2+}] (mM)	[K^+]	n	Fractional release
2.54	1.19	11	0.4 ± 0.2
2.54	50	4	1.9 ± 0.3
5.08	50	4	4.2 ± 0.2
0	50	3	0.8 ± 0.1

A crude synaptosomal fraction (P_2) of rabbit striatum was prepared by homogenising 0.6–0.8 g of tissue in 0.32 M sucrose, the nuclear fraction (10 000 g-min) was discarded and the P_2 fraction was prepared by centrifuging the supernatant at 17 000 g for 20 min. The P_2 fraction was resuspended in Krebs solution and layered on a 1 cm diameter filter superfused with Krebs at 36 °C and gassed with 95 per cent O_2/5 per cent CO_2. The fractional release refers to the total enkephalin found in the superfusate over a 15 min period expressed as a percentage of the total tissue content of enkephalin. In each experiment basal release was determined twice, before and after one 15 min exposure to 50 mM K^+.

The release of enkephalins from the guinea pig myenteric plexus by prolonged electrical stimulation has been described recently (Schulz *et al.*, 1977). These authors were also able to measure a resting release in the absence of stimulation. We have been unable to detect any consistent release from this preparation under a variety of conditions even using a system for the continuous collection of the bathing fluid. The reason for this discrepancy is unclear, particularly since Schulz *et al.* found extremely large amounts of material in their bathing fluids. We have similarly been unable to detect the release of intact labelled enkephalin from the myenteric plexus after incubation with [^{3}H] tyrosine. This was despite the clear incorporation of [^{3}H] tyrosine into endogenous enkephalin stores; we calculate that under these conditions we should be able to detect the release of as little as 10–100 pg of labelled enkephalin.

It appears that one of the prime requisites for a neurotransmitter, namely that it should be shown to be released from nerve endings, has been partially fulfilled for the enkephalins. However, judgement should perhaps be reserved until an unequivocal release can be shown with discrete stimuli.

Enkephalin release and turnover *in vivo*
The question has arisen as to whether the level of enkephalin in the brain might reflect changes in enkephalinergic nerve activity. In one series of experiments (Akil *et al.*, 1976) it was reported that rats subjected to stress had a longer latency to a nociceptive heat stimulus than control animals. This prolonged latency of the nociceptive response was accompanied by an apparent increase in the level of opioid peptides in the brain as measured by receptor binding assay. Simantov and Snyder (1976) used a combination of gel filtration and receptor assay and found elevated levels of enkephalin-like peptides in the brains of rats made dependent on morphine. This latter effect was very rapidly reversed on administration of naloxone.

We have carried out similar experiments on mice treated acutely and chronically with morphine and naloxone (table 12.7). Our results showed no significant changes in enkephalin concentration as measured by bioassay after Amberlite XAD-2 chromatography and chloroform extraction. In a large series of experiments we have found that brain and gut concentrations of the enkephalins are remarkably

Table 12.7   Effect of morphine and naloxone on enkephalin
content of mouse brain

Treatment	Total enkephalin content (ng/g)
None	204 ± 17
Chronic morphine*	212 ± 10
None	210 ± 23
Saline	220 ± 16
Acute naloxone†	202 ± 23
Acute morphine‡	193 ± 18

The total enkephalin content was assayed against met-enkephalin after decapitation, 0.1 M HCl extraction and chromatography on Amberlite XAD-2. All the samples were adjusted to pH 10 and extracted three times with chloroform to remove alkaloids from the samples. Addition of 100 μg morphine or naloxone to control samples showed that the alkaloids were completely extracted by chloroform.
*75 mg pellet implanted subcutaneously for 5 days
†3 mg/kg s.c. each hour for 3 hours
‡15 mg/kg single dose i.p., animals killed 1 hour later
There was no significant difference between the groups.

constant and unaffected by chronic dosing with drugs such as reserpine, 6-hydroxy-dopamine and physostigmine.

It is possible that the discrepancy between our results and those of Simantov and Snyder (1976) are due to differences in the methodology particularly in the preparation and assay of tissue samples. We consider it unlikely that changes in endogenous enkephalin levels will provide a satisfactory index of central enkephalinergic activity unless it can be shown that either (a) increased activity leads to an increase in free enkephalin in the interstitial fluid or to a decrease in stored enkephalin, or (b) decreased activity leads to an increase in stored enkephalin. It is likely that changes of this magnitude would only be seen in extreme conditions as with other central transmitters such as noradrenaline. Indeed, changes in the concentration of brain amines only appear to provide an index of central activity under special conditions, for instance, when synthesis is inhibited. Our experiments on the incorporation of labelled tyrosine into brain enkephalin stores suggest that under normal conditions this is a relatively slow process. This situation is of course quite distinct from that of putative biogenic amine neurotransmitters. It remains to be seen whether feedback mechanisms operate to control enkephalin biogenesis (figure 12.1). In the absence of such information and without detailed knowledge of the size of the prototype-peptide pools, assessment of enkephalin release or turnover based on endogenous enkephalin levels is likely to prove hazardous.

In view of the rapid destruction of enkephalin it seems unlikely that enkephalin release would be accompanied by an increase in extracellular brain peptide levels. However, opioid-like material has been detected in human cerebrospinal fluid. (CSF) by receptor assay (Wahlström *et al.*, 1976). A fraction of this material appears to co-chromatograph with met-enkephalin while a further fraction has yet to be identified. We have obtained similar results using bioassay and, in addition, have found that there are increased levels of opioid activity in the CSF of patients receiving intracerebral brain stimulation for the relief of chronic pain. We have yet to characterise this opioid material fully but it does appear that at least a portion of the activity is due to met-enkephalin. It may be that under suitable conditions enkephalin is either released close to the ventricular spaces or in sufficient concentration that it overflows into the CSF. However, there is general agreement between our results and those of Wahlström *et al.* (1976) that under normal conditions the CSF concentration of enkephalin is very low.

MULTIPLE OPIATE RECEPTORS

The concept of multiple opiate receptors (Lord *et al.*, 1976, 1977) is perhaps central to the problem of why there are multiple endogenous ligands for the opiate receptor. Evidence for the presence of more than one species of opiate receptor is of considerable importance for an understanding of the respective roles of the short- and long-chain opioid peptides. Martin and his colleagues (Martin *et al.*, 1976; Gilbert and Martin, 1976) have postulated the existence of at least three different types of opiate receptors; for classical morphine-like drugs (μ-receptor), for certain benzomorphans such as ketocyclazocine and ethylketocyclazocine which do not substitute for morphine in dependent monkeys (κ-receptor) and for drugs like the prototype *N*-allylnorcyclazocine (σ-receptor).

We have used the principle of parallel assay to investigate this problem (Lord *et al.*, 1977). This involves the biological assessment of agonist and antagonist potencies in the guinea pig ileum and mouse vas deferens and the inhibition of binding of [^{3}H] naloxone and of [^{3}H] leu-enkephalin in homogenates of guinea pig brain. A comparison of the biological potencies of the opioid peptides and morphine (figure 12.3) indicates interesting differences between the guinea pig ileum and mouse vas deferens. Thus met-enkephalin is somewhat less potent than β-endorphin in the guinea pig ileum while α-endorphin and leu-enkephalin are much

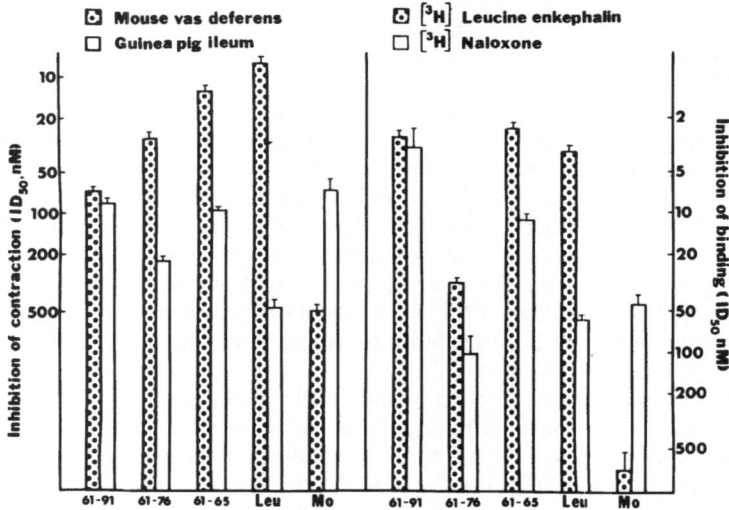

Figure 12.3 The agonist activities of various compounds in the mouse vas deferens and guinea pig ileum and their potencies to inhibit [^{3}H] leu-enkephalin and [^{3}H] naloxone binding in homogenates of guinea pig brain (pH 7.4, 0 °C, no sodium, 150 min incubation). The numbers on the abscissa indicate the amino acid sequence of β-lipotropin. Leu, leu-enkephalin; Mo, morphine.

less active. Compared with the guinea pig ileum the absolute potency of β-endorphin is the same in the mouse vas deferens whereas there is a progressive increase in potency for α-endorphin, met-enkephalin and leu-enkephalin and a marked drop in the potency of morphine (figure 12.3). In the binding studies, the pattern of inhibition of [^{3}H] leu-enkephalin and of [^{3}H] naloxone binding is markedly different (figure 12.3). Against [^{3}H] leu-enkephalin, met-enkephalin has 64 per cent, leu-enkephalin 43 per cent and morphine 0.2 per cent of the potency of β-endorphin; against naloxone, the corresponding values are 19, 4 and 6 per cent.

It would be expected from the principle of parallel assay that if tissues have identical receptors then the rank order of potency of a series of agonists should vary in parallel. This is clearly not the case in our experiments and the evidence suggests that the receptor populations in the mouse vas deferens and guinea pig ileum are not identical and that the receptor population in the guinea pig brain is heterogeneous. The pharmacological responses of the guinea pig ileum to the peptides and morphine appear to parallel their abilities to inhibit [^{3}H] naloxone binding rather than [^{3}H] leu-enkephalin binding. The pharmacological responses in the mouse vas deferens are more related to the inhibition of [^{3}H] leu-enkephalin binding.

We suggest that the predominant receptor type in the guinea pig ileum is that of the μ-receptor responding to morphine-like compounds (Martin *et al.*, 1976), whereas the mouse vas deferens contains predominantly another type of receptor we have defined as the σ-receptor (Lord *et al.*, 1976). Strong support for this view comes from a study of the effectiveness of antagonists in the two bioassay tissues. In the guinea pig ileum, morphine and the opioid peptides are readily antagonised by naloxone with a K_e of approximately 2 nM. A similar K_e is found for naloxone against morphine in the mouse vas deferens but this value is increased tenfold when naloxone is tested against the enkephalins, α- and β-endorphin and [D-Ala^2, Met^5]-enkephalin in this tissue (Lord *et al.*, 1977). This reduced potency of naloxone as an antagonist of the opioid peptides strongly supports our concept for different receptor populations in the mouse vas deferens and guinea pig ileum.

CONCLUSIONS

There is insufficient evidence at present to allocate a specific peptide or physiological function to a specific receptor type. Gross differences are apparent between the various endorphins and enkephalins on the basis of their potencies in different test systems but in the absence of specific antagonists no clear distinctions can be made. Perhaps the most interesting finding is that β-endorphin is equally effective in displacing [^{3}H] leu-enkephalin and [^{3}H] naloxone in binding assays, and is also equipotent in the guinea pig ileum and mouse vas deferens. Thus β-endorphin may be viewed as the prototype opioid peptide interacting equally well with several or perhaps all opiate receptors. The shorter peptides on the other hand may have been evolved for more specific functions. The shorter peptides not only acquired a different receptor specificity but also became sensitive to rapid proteolytic inactivation. This 'specialisation' of β-endorphin thus leads to the production of peptides with characteristics that fit them for a neurotransmitter role of a rapid and transient nature. The beauty of this hypothetical model is that by merely modifying existing biosynthetic mechanisms the body is able to produce and store compounds with a postulated neuroendocrine function (β-endorphin) or neurotransmitter function (enkephalins). It is perhaps relevant that one other established system also displays these characteristics, thus synthesis of dopamine → noradrenaline → adrenaline successively produces neurotransmitters of different characteristics and finally a blood-borne hormone.

It remains to be established whether the other unidentified opioid peptides are of physiological significance. However, it is now fairly certain that there must be a pro-peptide or precursor for leu-enkephalin and elucidation of this structure may greatly aid our understanding of this complex system.

ACKNOWLEDGEMENTS

This study was supported by grants from the Medical Research Council and from the U.S. National Institute on Drug Abuse (DA 00662) to H.W.K.

REFERENCES

Akil, H., Madden, J., Patrick, R. L. and Barchas, J. D. (1976). In *Opiates and Endogenous Opioid Peptides* (ed. H. W. Kosterlitz), North-Holland, Amsterdam, pp. 63–70
Bloom, F., Battenberg, E., Rossier, J., Ling, N., Leppaluoto, J., Vargo, T. M. and Guillemin, R. (1977). *Life Sci.*, 20, 43–48
Clouet, D. H. and Ratner, M. (1976). In *Opiates and Endogenous Opioid Peptides* (ed. H. W. Kosterlitz), North-Holland, Amsterdam, pp. 71–78
Feldberg, W. and Smyth, D.G. (1976). *J. Physiol., Lond.*, 266, 30–31P
Gentleman, S., Ross, M., Lowney, L. I., Cox, B. M. and Goldstein, A. (1976). In *Opiates and Endogenous Opioid Peptides* (ed. H. W. Kosterlitz), North-Holland, Amsterdam, pp. 27–34
Gilbert, P. E. and Martin, W. R. (1976). *J. Pharmac. exp. Ther.*, 198, 66–82
Guillemin, R., Ling, N. and Burgus, R. (1976). *C.r. hebt. Séanc. Acad. Sci., Paris, D*, 282, 783–85
Hambrook, J. M., Morgan, B. A., Rance, M. J. and Smith, C. F. C. (1976). *Nature*, 262, 782–83
Henderson, G., Hughes, J. and Kosterlitz, H. W. (1972). *Br. J. Pharmac.*, 46, 764–66
Hughes, J. (1975). *Brain Res.*, 88, 295–308
Hughes, J. and Kosterlitz, H. W. (1977). *Br. med. Bull.*, 33, 157–61
Hughes, J., Smith, T. W. Kosterlitz, H. W., Fothergill, L. A., Morgan, B. A. and Morris, H. R. (1975). *Nature*, 258, 577–79
Hughes, J., Kosterlitz, H. W. and Smith, T. W. (1977). *Br. J. Pharmac.*, 61, 639–48
Jones, C. W. and Pickering, B. T. (1972). *J. Physiol., Lond.*, 227, 553–64
Kosterlitz, H. W. and Hughes, J. (1975). *Life Sci.*, 17, 91–96
Lazarus, L. H., Ling, N. and Guillemin, R. (1976). *Proc. natn. Acad. Sci. U.S.A.*, 73, 2156–59
Li, C. H. and Chung, D. (1976a). *Nature*, 260, 622–24
Li, C. H. and Chung, D. (1976b). *Proc. natn. Acad. Sci. U.S.A.*, 73, 1145–48
Lord, J. A. H., Waterfield, A. A., Hughes, J. and Kosterlitz, H. W. (1976). In *Opiates and Endogenous Opioid Peptides* (ed. H. W. Kosterlitz), North-Holland, Amsterdam, pp. 275–80
Lord, J. A. H., Waterfield, A. A., Hughes, J. and Kosterlitz, H. W. (1977). *Nature*, 267, 495–99
Martin, W. R., Eades, C. G., Thompson, J. A., Huppler, R. E. and Gilbert, P. E. (1976). *J. Pharmac. exp. Ther.*, 197, 517–32
Meek, J. L., Yang, H.-Y. T. and Costa, E. (1977). *Neuropharmacology*, 16, 151–54
Miller, R. J., Chang, K. J., Cuatrecasas, P. and Wilkinson, S. (1977). *Biochem. biophys. Res. Commun.*, 74, 1311–17
Pert, C. B., Bowie, D. L., Fong, B. T. W. and Chang, J.-K. (1976). In *Opiates and Endogenous Opioid Peptides* (ed. H. W. Kosterlitz), North-Holland, Amsterdam, pp. 79–86
Puig, M. M., Gascón, P., Craviso, G. L. and Musacchio, J. M. (1977). *Science*, 195, 419–20
Ross, M., Tsung-ping, S., Cox, B. M. and Goldstein, A. (1976). In *Opiates and Endogenous Opioid Peptides* (ed. H. W. Kosterlitz), North-Holland, Amsterdam, pp. 35–40
Schulz, R., Wüster, M., Simantov, R., Snyder, S. and Herz, A. (1977). *Eur. J. Pharmac.*, 41, 347–48
Simantov, R. and Snyder, S. H. (1976). *Nature*, 262, 505
Simantov, R., Snowman, A. M. and Snyder, S. (1976). *Brain Res.*, 107, 650–62

Smith, T. W., Hughes, J., Kosterlitz, H. W. and Sosa, R. P. (1976). In *Opiates and Endogenous Opioid Peptides* (ed. H. W. Kosterlitz), North-Holland, Amsterdam, pp. 57–62
Van Neuten, J. M., Janssen, P. A. J. and Fontaine, J. (1976). *Life Sci.,* 18, 803
Waterfield, A. A. and Kosterlitz, H. W. (1975). *Life Sci.,* 16, 1787–92
Wahlstrom, A., Johansson, L. and Terenius, L. (1976). In *Opiates and Endogenous Opioid Peptides* (ed. H. W. Kosterlitz), North-Holland, Amsterdam, pp. 49–56
Yang, H.-Y., Hong, J. S. and Costa, E. (1977). *Neuropharmacology,* 16, 303–07

13

Distribution and pharmacology of the enkephalins and related opiate peptides

Richard J. Miller*, Kwen-Jen Chang and Pedro Cuatrecasas
(Wellcome Research Laboratories, Research Triangle Park,
North Carolina 27709 U.S.A.) and
S. Wilkinson, L. Lowe, C. Beddell and R. Follenfant
(Wellcome Research Laboratories, Beckenham, Kent, U.K.)

Hughes and his colleagues were the first to describe the structure of the enkephalins, two naturally occurring pentapeptides which have high affinity for the opiate receptor (Hughes *et al.*, 1975). It has been pointed out that the structure of these peptides is related to that of β-lipotropin, a 91 amino acid hormone found in the pituitary gland. Moreover further peptides have now been described which have opiate activity and represent longer sequences of the C-terminal portion of β-lipotropin. These longer peptides are generally known as endorphins. The largest of these is β-endorphin which represents residues 61-91 of β-lipotropin. [Met^5]-enkephalin actually represents residues 61-65 of β-lipotropin (Simantov and Snyder, 1976*a*; Lazarus *et al.*, 1976; Cox *et al.*, 1976; Bradbury *et al.*, 1976*a*).

The discovery of these opiate peptides has aroused considerable interest, and information concerning their biochemistry, pharmacology and function has accumulated rapidly. This chapter is concerned mainly with the structural pharmacology of the opiate pentapeptides. Some data concerning the distribution and biochemistry of the enkephalins are also presented.

DISTRIBUTION

The enkephalins and endorphins have been detected both by direct chemical means and by radioimmunoassay. Sensitive radioimmunoassays for [Leu^5]- and [Met^5]-enkephalins and for α and β-endorphins have now been developed. The antisera developed are useful not only in assaying such peptides in tissues and biological

*Present address: Department of Pharmacological and Physiological Science, University of Chicago.

Table 13.1   Regional distribution of [Met^5]- and [Leu^5]-enkephalins in rat brain

Brain area	[Met^5]-enkephalin (pmol/mg protein)	[Leu^5]-enkephalin (pmol/mg protein)	$\dfrac{[Met^5]}{[Leu^5]}$
Whole brain (3)	0.88 ± 0.1	0.21 ± 0.06	4.2
Striatum (6)	5.9 ± 0.7	0.62 ± 0.03	9.5
Hypothalamus (6)	5.42 ± 0.4	0.40 ± 0.07	13.6
Olfactory tubercle/ nucleus accumbens (6)	6.1 ± 0.9	0.66 ± 0.07	9.2
Midbrain (6)	1.90 ± 0.1	0.21 ± 0.01	9.0
Hippocampus (6)	0.91 ± 0.1	0.06 ± 0.01	15.2
Cerebellum (6)	< 0.01	< 0.01	–
Pituitary (6)	0.03 ± 0.009	0.08 ± 0.008	0.4
Cortex (6)	0.62 ± 0.04	0.44 ± 0.02	1.4
Globus pallidus (3)	6.7 ± 0.8	0.77 ± 0.09	8.7
Thalamus (6)	1.04 ± 0.08	0.34 ± 0.01	3.1

Rat brains were dissected and homogenised in 0.1 N HCl at 4 °C. After centrifugation the pellet was rehomogenised in 0.1 N HCl and centrifuged. The supernatant was added to the first supernatant. The extract was lyophilised and then taken up in buffer (0.1 M sodium phosphate, pH 7.4) and aliquots assayed for [Met^5]- and [Leu^5]-enkephalin content by radioimmunoassay.

fluids but also for mapping enkephalin-containing cells by immunohistochemical techniques. Table 13.1 shows the distribution of [Leu^5]- and [Met^5]-enkephalins in several gross brain areas as measured by radioimmunoassay. (Miller *et al.*, 1978).

The antibodies used have been developed at the Wellcome Research Laboratories and are directed against the C-terminal protion of the enkephalin molecule. They have very high specificity with a cross-reactivity of < 1per cent. Reports of antisera directed against the N-terminal of the molecule seem to show less specificity as might be predicted *a priori* (Weissman *et al.*, 1976; Simantov *et al.*, 1977). It is clear that the relative distribution of the enkephalins in rat brain is similar to that already reported for total opiate activity. It is also clear that [Met^5]-enkephalin/[Leu^5]-enkephalin ratio differs considerably in different parts of the brain. The significance of this observation is hard to assess at this time. Several endorphins have also been detected in brain and pituitary. For example radioimmunoassay has been used to measure concentrations of α and β-endorphins in rat brain and pituitary. It is hard to say at this time precisely what proportion of the total opiate activity found in the brain is due to the enkephalins and what proportion consists of the larger endorphins. However, it seems reasonable to conclude that although the enkephalins contribute very substantially to brain opiate activity, this does not seem to be true of the pituitary, where the endorphins predominate (Snyder and Simantov, 1977). It should be noted that the concentration of total opiate activity that can be extracted from the pituitary is far higher than that from any other tissue (unpublished observations). It may also be that the pituitary contains another opiate substance named anodynin (Pert *et al.*, 1976).

Antisera to the enkephalins and endorphins have also been used to localise these peptides by fluoresence immunohistochemistry (Elde *et al.*, 1976; Bloom *et al.*, 1977; Simantov *et al.*, 1977). The distribution of enkephalins mapped in this way is similar to that found by radioimmunoassay. The antibodies used in our radioimmunoassay give very intense immunofluorescent staining in some brain areas. (F. Bloom, M. Kuhar, personal communication). Antibodies to α and β-endorphin have also been used to map neurons in the brain that contain the endorphins (F. Bloom, personal communication). In addition, extremely intense staining is obtained with these antisera in the intermediate lobe of the pituitary. In collabora-

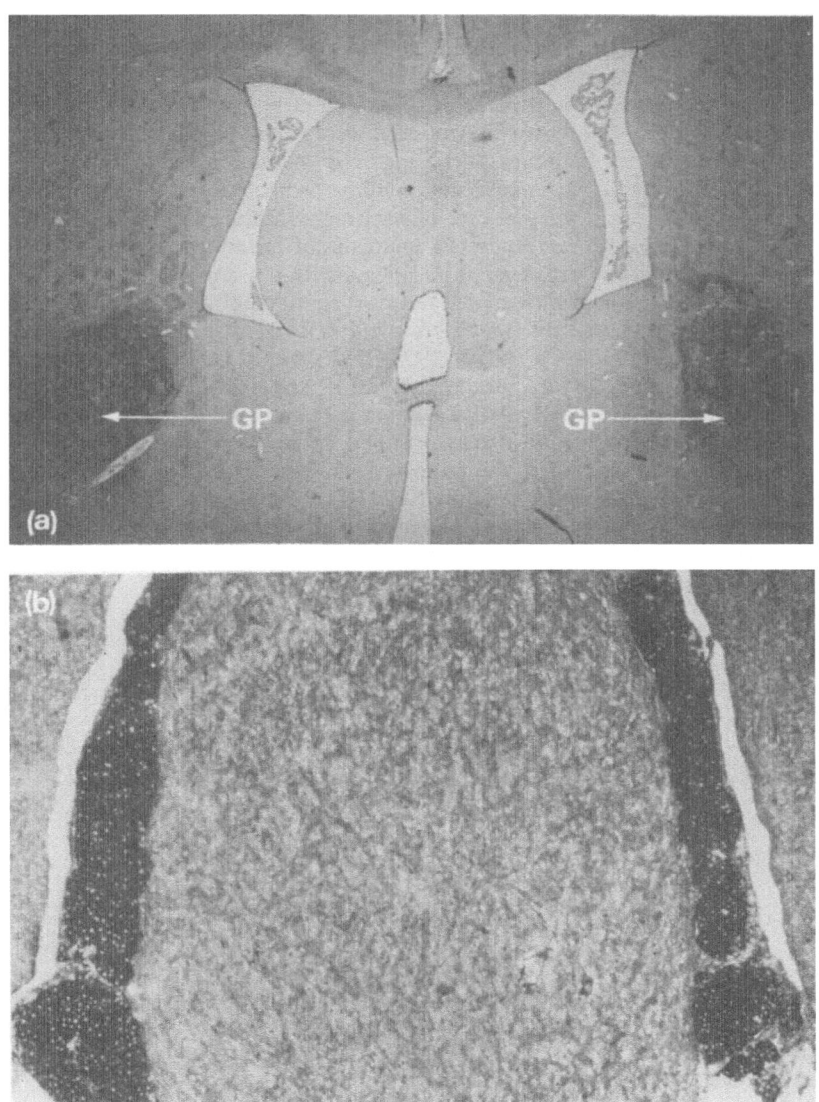

Figure 13.1 (a) Staining of enkephalin-like immunoreactivity in brain and pituitary (a coronal section of rat brain showing staining in the globus pallidus (G.P.). (b) Section through rat pituitary showing heavy staining in the immediate lobe. (M. Sar, W. Stumpf, R.J.M., K.J.C., and P.C., unpublished observations).

tion with M. Sar and W. Stumpf we have performed immunoperoxidase staining experiments, an example of the resulting staining patterns is shown in figure 13.1.

STRUCTURE/ACTIVITY RELATIONSHIPS

As the enkephalins are naturally occurring molecules it would not be surprising if specific enzymes existed for their metabolism. Indeed early experiments performed with the enkephalins in several pharmacological assay systems, such as the guinea pig ileum, showed their effects to be transient (Hughes, 1975). Such transient effects were also observed in their analgesic action after intracerebral injection (Belluzzi *et al.*, 1976). Moreover the peptides were found to be rapidly metabolised in biochemical receptor binding assays using brain membranes, although it was subsequently found that the enkephalin could be protected by running such assays at low temperatures or in the presence of bacitracin (Miller *et al.*, 1977a).

We have synthesised several hundred analogues of the enkephalins. One early observation was that the replacement of glycine with a D-amino acid at the [2] position produced analogues that were resistant to metabolism (Pert *et al.*, 1976; Miller *et al.*, 1977). If the substituent is D-alanine then the high affinity for the opiate receptor is retained (Pert *et al.*, 1976; Beddell *et al.*, 1977). The two main factors that determine the biological activity of an enkephalin analogue are probably its receptor affinity and relative resistance to metabolism. In addition, however, other factors such as ease of transport and lipophilic character are likely to be important in specific instances. The increased stability of analogues containing the [D-Ala^2] residue is illustrated in figure 13.2. [Leu^5]-enkephalin is effective in inhibiting the contractions of the guinea pig ileum but its effect is transient; however, in the case of the [D-Ala^2, D-Leu^5] analogue, the inhibition is constant until the peptide is washed out of the organ bath.

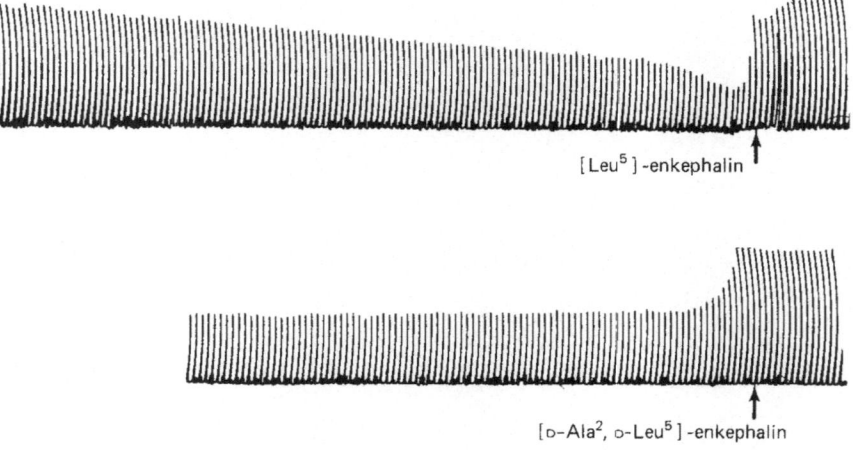

[Leu^5]-enkephalin

[D-Ala^2, D-Leu^5]-enkephalin

Figure 13.2   Effect of enkephalins on contraction of the electrically stimulated guinea pig ileum. Upper trace show the effect of [Leu^5]-enkephalin. Lower trace shows the effect of [D-Ala^2, D-Leu^5]-enkephalin. Drugs added as indicated by arrows.

It is useful to consider the enkephalins to begin with as typical opiate agonists and then to see how precisely they fall into this category. No antagonist or partial agonist activity has yet been demonstrated for any analogue. The following areas will be considered.

(1) Opiate receptor affinity.
(2) Activity in the guinea pig ileum.
(3) Activity in the mouse vas deferens.
(4) Analgesic activity.
(5) Anti-diarrhoeal activity.
(6) Anti-tussive activity.

To date, enkephalin analogues have been synthesised that show activity in all six of these categories. However, the activity of these analogues is more completely documented in the first four test systems at this time. The structural requirements for enkephalin analogues will be illustrated by reference to selected examples.

D-Amino acid analogues

One particularly useful series of analogues is obtained by the sequential substitution of D-amino acids in each position in the enkephalin molecule. The effect of such substitutions on the activity of the molecule in the mouse vas deferens and opiate receptor binding assays is illustrated in table 13.2. It is clear that in order to retain full biological activity in these two systems D-amino acid substitutions are only permissible at certain positions. For example [D-Tyr^1], [D-Ala^3], and [D-Phe^4] analogues show a considerable reduction in activity. However, substitution of a D-amino acid at the 2 or 5 position, or at both positions simultaneously, does not necessarily lead to deleterious effects. In fact some substitutions seem to improve activity, presumably because they serve to protect the molecule from metabolism as well as producing a small intrinsic increase in receptor affinity in some cases.

Table 13.3 lists some of the most potent analogues found so far in each of the biological assay systems. It can be seen for example that activity in the mouse vas deferens and opiate receptor affinity are highly correlated. A survey of the analogues in table 13.3 together with others listed elsewhere (Beddell *et al.*, 1977) allows the following general observations to be made with respect to the mouse vas deferens, guinea pig ileum and opiate receptor interactions.

(1) As the length of the side chain of the amino acid becomes less hydrophobic, there is a decrease in potency in the mouse vas deferens and in opiate receptor affinity. It may be important in relation to the preparation and storage of [Met^5]-enkephalin solutions that oxidation to the sulphoxide analogue markedly reduces activity.

(2) The low potency of [Tyr^4]-enkephalin and also of the analogue [Tyr-D-Ala-Gly-Leu-Leu] suggests that for high activity the 4-position must be occupied by a non-phenolic aromatic hydrophobic amino acid residue. However, a parachlorophenylalanine residue is permitted in this position. Indeed analogues with this substitution are among the most potent that have been tested.

(3) The glycine at position [3] appears to be essential for high biological activity. It is difficult to explain this in terms of peptide conformations alone, as no con-

Centrally Acting Peptides

Table 13.2 Activity of enkephalin analogues *in vitro* on mouse vas deferens and inhibition of binding of [^{3}H] naloxone to rat brain opiate receptors, both estimated as $10^{-7}/IC_{50}M$. The first compound is [Leu^5]-enkephalin

	Structural formula				Activity on vas deferens $10^{-7}/IC_{50}M$	Receptor binding $10^{-7}/IC_{50}M$
1	2	3	4	5		
Tyr	Gly	Gly	Phe	Leu	8.17	9.09
D–Tyr	Gly	Gly	Phe	Leu	0.034	–
Tyr	D–Ala	Gly	Phe	Leu	57.3	31.3
D–Tyr	D–Ala	Gly	Phe	Leu	0.15	0.051
Tyr	D–Ala	Gly	Phe	D–Leu	263.0	38.5
D -Tyr	D–Ala	Gly	Phe	D–Leu	0.064	<0.010
Tyr	Gly	Gly	Phe	Leu	8.17	9.09
Tyr	Ala	Gly	Phe	Leu	0.006	0.020
Tyr	D–Ala	Gly	Phe	Leu	57.3	31.3
Tyr	Gly	D–Ala	Phe	Leu	0.021	0.002
Tyr	Ala	D–Ala	Phe	Leu	0.009	<0.010
Tyr	D–Ala	D–Ala	Phe	Leu	0.075	0.026
Tyr	Gly	Ala	Phe	Leu	0.018	0.011
Tyr	Ala	Ala	Phe	Leu	0.009	<0.010
Tyr	D–Ala	Ala	Phe	Leu	2.58	1.26
Tyr	Gly	Gly	Phe	Leu	8.17	9.09
Tyr	Gly	Ala	Phe	Leu	0.18	0.011
Tyr	Gly	D–Ala	Phe	Leu	0.021	0.002
Tyr	D–Ala	Gly	Phe	Leu	57.3	31.3
Tyr	D–Ala	Ala	Phe	Leu	2.50	1.26
Tyr	D–Ala	D–Ala	Phe	Leu	0.075	0.026
Tyr	Ala	Gly	Phe	Leu	0.006	0.02
Tyr	Ala	Ala	Phe	Leu	0.009	<0.010
Tyr	Ala	D–Ala	Phe	Leu	0.009	<0.010
Tyr	Gly	Gly	Phe	Leu	8.17	9.09
Tyr	Gly	Gly	D–Phe	Leu	0.002	<0.011
Tyr	D–Ala	Gly	Phe	Leu	57.3	31.3
Tyr	D–Ala	Gly	D–Phe	Leu	0.043	0.011
Tyr	Gly	Gly	Phe	Leu	8.17	9.09
Tyr	Gly	Gly	Phe	D–Leu	9.66	4.00
Tyr	D–Ala	Gly	Phe	Leu	57.3	31.3
Tyr	D–Ala	Gly	Phe	D–Leu	263.0	38.5
Morphine					0.215	28.6

Table 13.3 Comparison of activity of enkephalins and their analogues
in several pharmacological assay systems

Guinea pig ileum. Potency relative to morphine (morphine = 1)

Analogue							Relative potency
	Tyr	D–Ala	Gly	Phe	D–LeuOMe		6.04
	Tyr	D–Ala	Gly	Phe	D–Leu		1.95
	Tyr	D–Ala	Gly	Phe	D–MetOMe		1.76
	Tyr	D–Ala	Gly	Phe	MetOMe		1.38
	Tyr	D–Ala	Asn	Phe	Leu		1.28
	Tyr	D–Ala	Gly	pClPhe	D–Leu		1.0
NMe	Tyr	Gly	Gly	Phe	Leu		0.93
	Tyr	D–Ala	Gly	Phe	D–Met		0.92
	Tyr	D–Ala	Gly	Phe	Leu	Thr	0.87
NMe	Tyr	Gly	Gly	Phe	LeuOMe		0.75
	Tyr	D–Ala	Gly	Phe	D–Leu		0.66
	Tyr	D–Ala	Gly	Phe	Met	Thr	0.63
NMe$_2$	Tyr	D–Ala	Gly	Phe	LeuOMe		0.65

Mouse vas deferens. Potency relative to morphine (morphine = 1)

Analogue							Relative potency
	Tyr	D–Ala	Gly	pClPhe	D–Leu		1695
	Tyr	D–Ala	Gly	Phe	D–Leu		1227
	Tyr	D–Ala	Gly	Phe	Leu	Thr	461
	Tyr	D–Ala	Gly	Phe	Leu		267
	Tyr	D–Ala	Gly	Phe	Met		229
	Tyr	D–Ala	Gly	Phe	Met	Thr	229
NMe	Tyr	D–Ala	Gly	Phe	D–Leu		229
	Tyr	D–Ala	Gly	Phe	D–Met		228
	Tyr	D–Ala	Gly	pClPhe	D–LeuOMe		226
	Tyr	D–Ala	Gly	Phe	Nle		209
	Tyr	D–Ala	Gly	Phe	D–LeuOpClPh		177

Anti-diarrhoeal activity. ED$_{50}$ mg/kg *subcutaneously*

Analogue						ED$_{50}$ mg/kg
NMe	Tyr	D–Ala	Gly	Phe	D–LeuNH$_2$	0.3
	Tyr	D–Ala	Gly	PheOMe	–	0.7
NMe	Tyr	D–Ala	Gly	Phe	D–MetNHEt	0.7
NMe	Tyr	D–Ala	Gly	Phe	D–MetNH$_2$	0.8
	Tyr	D–Ala	Gly	pClPhe	D–LeuNHEt	1.0
NMe	Tyr	D–Ala	Gly	Phe	D–Met	2.0
NMe	Tyr	D–Ala	Gly	Phe	D–MetOMe	2.0
	Tyr	D–Ala	Gly	Phe	D–Met	2.0
	Tyr	D–Ala	Gly	Phe	D–Leu	3.0
	Tyr	D–Ala	Asn	Phe	Leu	3.0
	Tyr	D–Ala	Gly	Phe	Thr	3.0
	Tyr	Ile	Asn	Met	Leu	8.0

Table 13.3 Continued

Opiate receptor affinity relative to morphine (morphine = 1)

Analogue						Relative affinity
Tyr	D–Ala	Gly	pClPhe	D–Leu		3.5
Tyr	D–Ala	Gly	pClPhe	D–LeuOMe		1.75
Tyr	D–Ala	Gly	Phe	D–Leu		1.34
Tyr	D–Ala	Gly	Phe	Leu		1.09
Tyr	D–Ala	Gly	Phe	Met		0.97
Tyr	D–Ala	Gly	Phe	MetOMe		0.7
Tyr	D–Ala	Gly	Phe	Met	Thr	0.62
Tyr	D–Ala	Gly	Phe	D–Met		0.44
Tyr	D–Ala	Gly	Phe	Leu	Thr	0.44
Tyr	D–Ala	Gly	Phe	Nle		0.4
Tyr	D–Ala	Gly	Phe	LeuOMe		0.4
Tyr	D–Ala	Gly	Phe	D–LeuOMe		0.4

Analgesic activity (hot plate) ED_{50} *for mice (μg/mouse injected intraventricularly)*

Analogue								ED_{50} μg/mouse
NMe	Tyr	D–Ala	Gly	Phe	D–MetNH$_2$			0.005
	Tyr	D–Ala	Gly	Phe	D–LeuNHEt			0.01
NMe	Tyr	D–Ala	Gly	Phe	D–Met			0.007
NMe	Tyr	D–Ala	Gly	Phe	D–MetOMe			0.02
NMe	Tyr	D–Ala	Gly	Phe	D–LeuNH$_2$			0.05
	Tyr	D–Ala	Gly	Phe	D–LeuOpClPh			0.05
NMe	Tyr	D–Ala	Gly	Phe	D–LeuOMe			0.05
NMe	Tyr	D–Ala	Gly	Phe	D–Leu			0.07
	Tyr	D–Ala	Gly	pClPhe	D–Leu			0.07
NMe	Tyr	D–Ala	Gly	Phe	D–MetNHEt			0.07
	Tyr	D–Ala	Gly	Phe	D–LeuOMe			0.07
	Tyr	D–Ala	Gly	Phe	D–Leu	D–Thr		0.1
	Tyr	D–Ala	Gly	Phe	D–Leu	Phe	Gly	0.1
	Tyr	D–Ala	Gly	Phe	D–Leu	Thr		0.1
	Tyr	D–Ala	Gly	PheOMe	–			0.2
	Tyr	D–Ala	Gly	Phe	Met	Thr		0.24
	Tyr	D–Ala	Gly	Phe	D–Leu			0.5
	Tyr	D–Ala	Gly	Phe	D–Met			0.5
	Tyr	D–Ala	Gly	Phe	Thr			0.5
	Tyr	D–Ala	Gly	Phe	D–LeuOMe			0.5
	Tyr	D–Ala	Gly	Phe	–			0.7

formation is known for a small peptide which shows absolute stringency for a glycine. It is also unlikely that a glycine-containing peptide would be much more stable metabolically than a peptide containing alanine at position [3] unless special degradative mechanisms were involved. Thus it is suggested that [Gly^3] is involved in a close and specific interaction with the receptor, and provides a degree of peptide chain flexibility by virtue of the absence of a sterically hindered side chain. In keeping with this hypothesis are the decreases in biological activity and receptor affinity seen with [Ala^3], [Ser^3] and [Pro^3] analogues. There are some exceptions, however, one analogue [Tyr-D-Ala-Asn-Phe-Leu] has proved to have a relatively high selective activity in the guinea pig ileum and anti-diarrhoeal tests. The basis of this effect is not clear at this time, but could represent some fundamental differences in tissue or receptor specificity.

(4) The substitution of [Gly^2] by a natural amino acid lowers biological activity and decreases opiate receptor affinity. It is of interest that substitution by Ser, Pro or Ala in either the [2] or [3] positions are relatively and respectively the same.

(5) The modification of [Tyr^1] by the addition of a single carbon unit, either attached to the phenolic function as in *O*-methyltyrosine, or inserted in the main peptide chain as with β-homotyrosine, reduced activity and binding. The importance of the free group in the aromatic ring is emphasised by the decrease of receptor affinity found with the [Phe^1]-analogue. Activity is retained if the phenolic function is derivatised with a labile moiety such as acetyl. It should be noted that [Tyr^1] and [Phe^4] are not interchangeable.

(6) Increasing the length of the peptides has rather unpredictable results. Extending the chain at the N-terminus with an arginine decreases biological activity and receptor affinity. Activity is retained on extending the C-terminal by addition of Thr but only when an L-amino acid is in position [5].

(7) Esterification of the peptide tends to reduce receptor affinity and activity in the mouse vas deferens, although several esterified analogues have an increased potency in the guinea pig ileum assay.

(8) There is a progressive decrease in opiate receptor affinity and activity in the mouse vas deferens on passing from analogues bearing a primary amino terminal residue through a secondary to a tertiary amine. Again in the guinea pig ileum the relative potencies of these analogues are not quite so predictable.

(9) It is clear that the tetrapeptides, although they are less potent than the corresponding pentapeptides, do retain full intrinsic activity in preparations such as the guinea pig ileum. Analogues such as [Tyr-D-Ala-Gly-Phe], which are protected from metabolism, have an appreciable analgesic activity when injected intracerebrally. Certain tetrapeptides are also very potent in the anti-diarrhoeal test.

Pharmacological aspects

There are some interesting differences with respect to the relative potencies of pentapeptides in the two bioassay systems. It is a consistent observation that the enkephalins and their analogues are considerably more potent relative to morphine on the mouse vas deferens than in the guinea pig ileum (table 13.4; Hughes *et al.*, 1975). The basis of this effect is not clear. It has been previously shown that some narcotics are more potent in the guinea pig ileum than in the mouse vas deferens (Hutchinson *et al.*, 1975). As a result of this and other observations some authors have postulated the existence of various subclasses of opiate receptors.

Table 13.4   Comparison of the relative effects of enkephalins and certain analogues on the guinea pig ileum and mouse vas deferens

Peptide	$IC_{50}M$ (morphine/IC_{50} (test compound))		
	Vas deferens	Receptor binding (Na^+ absent)	Guinea pig ileum
[Leu^5]-enkephalin	38	0.318	0.13
[Met^5]-enkephalin	32	0.175	0.36
[D–Ala^2, Leu^5]-enkephalin	267	1.094	0.66
[D–Ala^2, Met^5]-enkephalin	229	0.921	0.41
[D–Ala^2, D–Leu^5]-enkephalin	1227	1.346	1.95
[D–Ala^2, D–Met^5]-enkephalin	228	0.438	0.92

With respect to the analgesic and anti-diarrhoeal effects of the enkephalins other factors in addition to receptor affinity are clearly important. Many of the most potent analogues in these two assay systems contain modifications of the N- and C-termini and are esters, secondary amines or amides. It is probable, therefore, that increased hydrophobicity is an important factor in producing these effects. In the case of the anti-diarrhoeal action, peptides with structures considerably removed from the original pentapeptide have proved effective as is the case with [Tyr–Ile–Asn–Met–Leu] (table 13.3). In assays for anti-tussive activity a significant number of analogues have not yet been screened in order for structure/activity relationships to become clear, however, anti-tussive effects have been clearly demonstrated with several compounds given subcutaneously.

Apart from measuring the affinity of various drugs for the opiate receptor, the [^{3}H] naloxone binding assay has also been used to predict whether a drug will behave as an agonist, partial agonist or antagonist (Pert and Snyder, 1974). This may be achieved by examining the effects of sodium ions at physiological concentrations on the interaction of the drug with the opiate receptor. Thus sodium ions decrease the binding of agonists and increase the binding of antagonists to the receptor. The degree to which the sodium shift occurs, defined as the ratio of the $IC_{50}(M)$ values in the presence and absence, respectively, of sodium ions, correlates well with the 'purity' of agonist activity. Thus the sodium shift for drugs such as morphine is larger than for partial agonists such as cyclazocine and it is unity for naloxone. It has been found in general that analogues of [Leu^5]-enkephalin have sodium shifts > 10 and should therefore behave as full agonists which they seem to do. [Met^5]-enkephalin has a sodium shift < 10 and should therefore behave as a partial agonist. Similar findings have been reported for β-endorphin. Indeed β-endorphin and anodynin have sodium shifts of unity (Bradbury *et al.* 1976a, and unpublished observations). It does not appear that β-endorphin has antagonist activity and the significance of the sodium shift with respect to opiate peptides is somewhat doubtful.

BINDING OF ENKEPHALIN TO THE OPIATE RECEPTOR

The interaction of the enkephalins with opiate receptors may also be investigated by performing binding assays with radiolabelled enkephalins rather than radiolabelled opiates. Tritiated enkephalins are available commercially and binding assays with these ligands have been reported (Lord *et al.*, 1976; Simantov and Snyder, 1976*b*). We have developed receptor binding assays utilising radioiodonated derivatives of [D-Ala^2]-enkephalins. The advantage of these derivatives is their extremely high specific activity and resistance to metabolic degradation. The enkephalins are iodinated using the conventional chloramine-T method and then purified on Biogel P2 and DEAE–Sephadex columns. This yields a mixture of mono- and diiodinated derivatives. It can be shown from experiments with unlabelled iodinated derivatives that monoiodinated compounds retain high affinity for the opiate receptor but that the affinity of the diiodinated derivatives is reduced (Miller *et al.* 1977*b*).

Both [^{125}I] [D-Ala^2, Leu^5] and [^{125}I] [D-Ala^2, D-Leu^5]-enkephalins bind stereospecifically to receptors in both brain and certain cell lines in culture (figure 13.3). The neuroblastoma × glioma hybrid line NG108-15 and the neuroblastoma line N4TG1 have a particularly high concentration of receptors. Figure 13.3 shows

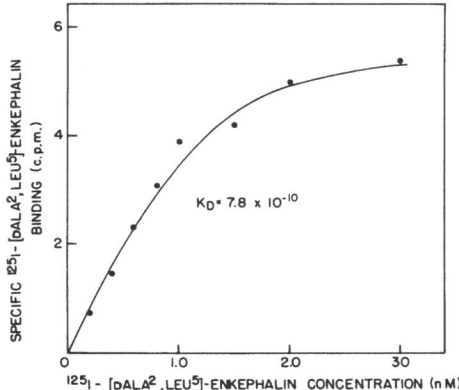

Figure 13.3 Stereospecific binding of [^{125}I]-labelled [D-Ala^2, Leu^5]-enkephalin to opiate receptors in rat brain membrane preparation.

that the binding of [^{125}I] [D-Ala^2, Leu^5]-enkephalin to brain membranes is saturable and of high affinity. At higher concentrations in experiments using [^{3}H] [D-Ala^2, Leu^5]-enkephalin a second binding site of lower affinity is also apparent. The existence of two binding sites has been previously described both for opiate agonists and for antagonists. Several lines of evidence suggest that the enkephalin binding site is the opiate receptor. First, the binding is stereospecific with respect to levorphanol and dextrorphan, the former compound being much more potent than the latter in displacing bound enkephalin. Secondly, the binding is altered by sodium and manganese ions in the same manner as the binding of a conventional opiate agonist. Thus, physiological concentrations of sodium ions decrease enkephalin binding whereas manganese ions enhance it.

Some interesting differences have been detected in the ability of drugs to displace bound enkephalin relative to bound [^{3}H] naloxone. A comparison of these relative effects is shown in table 13.5. It is clear that some opiates are considerably less potent in displacing bound enkephalin than bound [^{3}H] naloxone. Morphine, fentanyl and pethidine for example would fall into this category. However, this does not apply to all drugs. Etorphine and levallorphan, for example, are equally

Table 13.5 Comparison of the effects of various narcotic drugs and opiate peptides in inhibition of stereospecific binding of [³H] naloxone or [¹²⁵I] [D–Ala², Leu⁵]-enkephalin binding to opiate receptors in rat brain or of [¹²⁵I] [D–Ala², D–Leu⁵]-enkephalin to opiate receptors in N4TG1 (neuroblastoma) cells

Drug	[³H] naloxone binding rat brain	IC_{50}(M)	
		[¹²⁵I] [D–Ala², Leu⁵]-enkephalin binding rat brain	[¹²⁵I] [D–Ala², D–Leu⁵]-enkephalin binding N4TG1 cells
Naloxone	4×10^{-9}	2×10^{-9}	9×10^{-9}
Nalorphine	3×10^{-9}	7×10^{-9}	4×10^{-8}
Levorphanol	1.5×10^{-9}	3.6×10^{-9}	1.4×10^{-8}
Etorphine	8×10^{-10}	5×10^{-10}	8×10^{-10}
Morphine	3×10^{-9}	4×10^{-8}	7×10^{-8}
Levallorphan	2×10^{-9}	9×10^{-10}	6×10^{-9}
Pentazocine	9×10^{-8}	2×10^{-8}	1×10^{-7}
Phenazocine	4×10^{-9}	7×10^{-9}	8×10^{-9}
Phentanyl	1×10^{-9}	5×10^{-8}	4×10^{-7}
Pethidine	1×10^{-6}	2×10^{-5}	9×10^{-5}
Dextrorphan	3×10^{-5}	$> 10^{-6}$	7×10^{-5}
Methadone	1×10^{-8}	8×10^{-8}	9×10^{-8}
Etonitazene	8×10^{-9}	5×10^{-7}	2×10^{-6}
Codeine	7×10^{-5}	8.5×10^{-6}	5×10^{-5}
[Met⁵]-enkephalin	2.0×10^{-8}	6×10^{-9}	9.8×10^{-10}
[Leu⁵]-enkephalin	1.1×10^{-8}	9.6×10^{-9}	2.2×10^{-9}
[D–Ala², D–Leu⁵]-enkephalin	2.6×10^{-9}	1×10^{-9}	7×10^{-10}
[D–Ala², Leu⁵]-enkephalin	3.2×10^{-9}	9×10^{-10}	1×10^{-9}
β-endorphin	8×10^{-9}	2×10^{-9}	8×10^{-10}
β-lipotropin	$> 10^{-6}$	$> 10^{-6}$	$> 10^{-6}$

potent in displacing enkephalin or naloxone. In contrast to the above it can also be seen that opiate peptides are relatively more potent in displacing bound enkephalin than in displacing bound naloxone. It therefore seems as though different opiates and opiate peptides may interact with slightly different portions of the opiate receptor. In this respect the function of the two aromatic rings in enkephalin may be important in interacting with sites postulated to exist in the opiate receptor for the binding of drugs such as etorphine. Such a topography of the opiate receptor has been recently discussed (Feinberg *et al.*, 1976).

CONFORMATION OF ENKEPHALIN

From the pharmacological data discussed above it is interesting to try and gain some insight into the conformation of the pentapeptide. In this respect the stereo-specificity exhibited by the molecule at each position provides particularly useful information. Table 13.2 illustrates the effect of substitution of a D-amino acid on the potency of enkephalin in the mouse vas deferens assay and also its effect on opiate receptor affinity.

The various conformations of a peptide can be described in terms of torsion angles of each bond in the peptide chain. Assuming that the peptide group is 'trans' and planar and considering the allowed contact distance between atoms, then a two dimensional plot of ϕ and ψ, the torsion angles about the other bands in the peptide chain may be constructed (figures 13.4 and 13.5). The regions of accessible conformational space for a glycine and an L-alanine residue are shown in figure 13.5(a) and (b) respectively. Consider the conformations shown in table 13.6. The

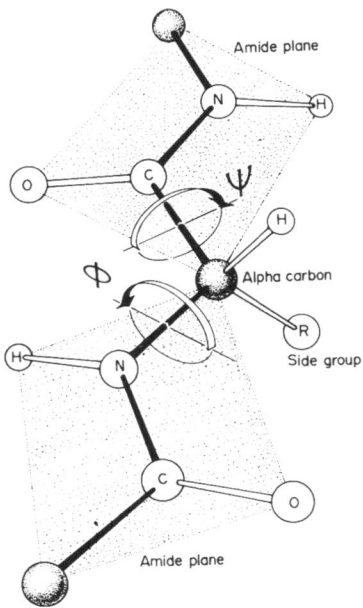

Figure 13.4 Representation of fragment of a peptide chain illustrating the dihedral angles (ϕ) and (ψ).

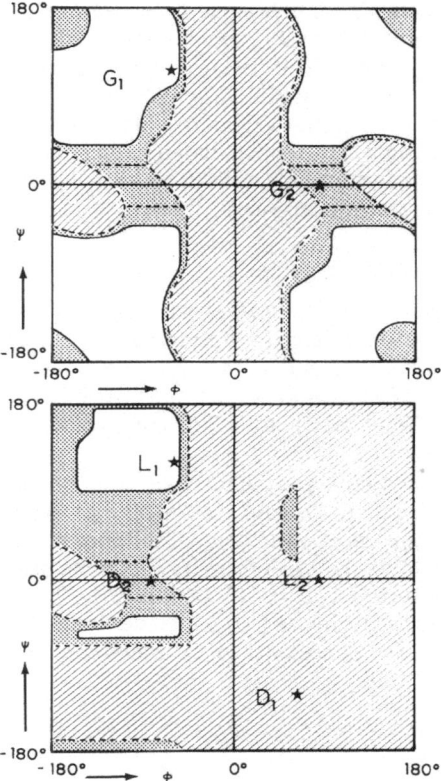

Figure 13.5 Conformational maps for glycine (top) and L-alanine. Disallowed combinations of ϕ and ψ are hatched. For these, atoms would be too close. Allowed areas are clear. Stippled areas (allowed) correspond to marginally close approaches (e.g. hydrogen bonds). The two residues (1, 2) in a β11-bend conformation are plotted according to each being glycine (G), an L-amino acid (L) or a D- amino acid (D).

configurations allowed at each residue position in these conformations may be estimated by plotting the ϕ and ψ values. Each residue is plotted according to its being glycine, an L-amino acid or a D-amino acid. A D-residue is plotted on the map for an L-residue by inverting the ϕ and ψ angles before plotting. It can be seen that a D-residue at position [1] and an L-residue at position [2] are disallowed. Table 13.7 summarises the results of this study for the conformations in table 13.6 and mirror related conformations denoted by a prime. The top line of table 13.8 summarises the activity and binding data from table 13.3. The configurations at each position in the peptide that favour activity and binding are shown. For example, at position [1], L-tyrosine favours activity and binding and D-tyrosine does not. At position [2] on the other hand the reverse is the case. Activity and binding are low when L-alanine is present, but high when D-alanine or glycine are present, provided that substitutions elsewhere in the molecule do not inhibit activity and binding. Apart from the optical configuration allowed at each position in the

Table 13.6 Torsion angles for some hydrogen-bonded peptide conformations. The ϕ and ψ angles for the minimum number of residues needed to describe each conformation are shown, successive residues being indicated by subscript

Conformation	ϕ_1	ψ_1	ϕ_2	ψ_2	ϕ_3	ψ_3	Reference
γ-turn	172	128	68	-61	-131	162	Nemethy and Printz, 1972
Inverse γ-turn	70	-170	-86	57	-155	-60	Matthews, 1972
V-turn	-80	80	80	-80			Lewis et al., 1973
β1-bend	-60	-30	-90	0			Lewis et al., 1973
β11-bend	-60	120	80	0			Lewis et al., 1973
β111-bend	-60	-30	-60	-30			Lewis et al., 1973
2_7-helix	-75	70					Ramakrishnan and Ramachandran, 1965
2.2_7^R-helix	-78	59					Ramakrishnan and Ramachandran, 1965
3_{10}^R-helix	-49	-26					Ramakrishnan and Ramachandran, 1965
α^R-helix	-48	-57					Ramakrishnan and Ramachandran, 1965
γ^R-helix	84	78					Ramakrishnan and Ramachandran, 1965

Table 13.7 Stereoselectivity of various hydrogen bonded peptide conformations estimated from Ramachandran plots. Conformations at the boundary between allowed and disallowed regions are presumed to be allowed

Conformation	Position 1	Position 2	Position 3
γ-turn	G(L)	D	L
γ'-turn	G(D)	L	D
Inverse γ-turn	D(L)	L	L
Inverse γ'-turn	L(D)	D	D
V-turn	L	D	
V'-turn	D	L	
β1-bend	A	L	
β1'-bend	A	D	
β11-bend	L	D	
β11'-bend	D	L	
β111-bend	A	A	
β111'-bend	A	A	
2_7, 2.2_7^R-helices	L		
$2'_7$, 2.2_7^L-helices	D		
3_{10}^R, α^R-helices	A		
3_{10}^L, α^L-helices	A		
γ^R helix, γ^L-helix	G		

G, glycine only
L, L-residue or glycine only
D, D-residue or glycine only
A, any residue configuration (that is L, D, G)
( ), within 15° of allowed region, may be possible

peptide, glycine is regarded as permitted in all cases, since there is no area on the conformational map accessible to an L- or D-residue that excludes glycine. We assume that, if loss of activity results from a replacement of an L- or D-residue by glycine, this is due to a loss of interaction with the receptor rather than destabilisation of the peptide conformation. Clearly we are neglecting possible stabilising interactions between the residue side chains within the peptide. In the remainder of table 13.7 we consider each conformation in turn in each successive possible alignment in the peptide chain and show the configurational specificity appropriate to the conformation. In the right hand column of table 13.7 is indicated agreement or contradiction between configurational specificity of a conformation in a given alignment with the peptide residues and that deduced for enkephalin from activity and binding data. Violation occurs when the activity and binding data allow an L- or D- configuration for a residue and this is disallowed in the conformation. Table 13.8 summarises the resultant possible conformations for enkephalin, the last two, however, explain very little. The inverse γ-turn, the V'-turn and the β11'-bend can explain the preference for a D-residue in position [2]. These and the β1-bend could also account for the preference for an L-residue in position [4]. However, the inverse γ-turn can simultaneously explain both. Finally, no conformation can explain the preference for glycine in position [3].

Although the inverse γ-turn explains the data most completely it would be premature to conclude that this was the conformation of enkephalin at the receptor.

Table 13.8 Comparison between the observed configurational selectivity of enkephalin (top) and that predicted if enkephalin were to adopt any of the hydrogen bonded conformations described in table 13.7

Structure	Position 1	2	3	4	5	Validity*
Enkephalin	L†	D†	G	L	A†	
γ-turn	*G(L)*	D	L	L	L	x(?)
		G(L)	D	L		x
			G(L)	*D*	L	x
γ'-turn	*G(D)*	*L*	D			x
		G(D)	L	*D*		x
			G(D)	L	*D*	x
Inverse γ-turn		D(L)	L	L		√
Inverse γ'-turn		*L(D)*	D	*D*		x
V-turn		*L*	D			x
			L	*D*		x
V'-turn		D	L			√
			D	L		√
β1-bend		A	L			√
			A	D		√
β1'-bend		A	D			√
			A	D		x
β11-bend		*L*	D			√
			L	*D*		√
β11'-bend		D	L			√
			D	L		√
β111 or 111'-bend		D	L			√
			A	A		√
2_7, 2.2_7^R-helices	L	*L*	L	L	L	x
$2'_7$, 2.2_7^L-helices	*D*	D	D	*D*	D	x
γ^R helix, γ^L-helix	G	G	G	*G*	G	x
$3_{10}^{R,L}$ or $\alpha^{R,L}$-helices	A	A	A	A	A	√

*√ means allowed, x means disallowed.
†L or D includes G. A is L, D or G.
Letters in *italics* indicate disagreement with enkephalin.

Table 13.9 A summary of the conformations for enkephalin indicated as valid in table 13.8

Structure	Position				
	1	2	3	4	5
Enkephalin	L†	D†	G	L	A*
Inverse γ-turn		D(L)	L	L	
V'-turn		D	L		
			D	L	
β1-bend		A	L		
			A	L	
β1'-bend		D	L		
			D	L	
β11'-bend		D	L		
			D	L	
β111 or 111'-bend		A	A		
			A	A	
$3_{10}^{R,L}$ or $\alpha^{R,L}$-helices	A	A	A	A	A

*L or D includes G.
A is L, D or G.

For example, certain assumptions have been made here which may not be valid, such as the planar nature of the peptide group. Bradbury *et al.* (1976c) have advocated a β-bend at positions [2] and [3]. Jones *et al.* (1976) and Roques *et al.* have advocated a β1-turn in positions [3] and [4] from proton magnetic resonance data from enkephalin solutions. Our studies would not necessarily exclude either of these conformations.

REFERENCES

Beddell, C. R., Clark, R. B., Hardy, G. W., Lowe, L. A., Ubatuba, F. B., Vane, J. R., Wilkinson, S., Miller, R. J., Chang, K–J. and Cuatrecasas, P. (1977). *Proc. R. Soc. B*, **198**, 249–65
Belluzzi, J. D., Grant, N., Garsky, V., Sarantakis, D., Wise, C. D. and Stein, L. (1976). *Nature*, **260**, 625–26
Bloom, F., Battenberg, E., Rossier, J., Ling, N., Leppaluoto, J., Vargo, T. M. and Guillemin, R. (1977). *Life Sci.*, **20**, 43–48
Bradbury, A. F., Feldberg, W. F., Smyth, D. G. and Snell, C. K. (1976a) In *Opiates and Endogenous Opiate Peptides* (ed. H. Kosterlitz), North-Holland, Amsterdam, pp. 9–18
Bradbury, A. F., Smyth, D. G., Snell, C. K., Birdsall, N. J. M., and Hulme, E. C. (1976b). *Nature*, **260**, 794–95
Bradbury, A. F., Smyth, D. G. and Snell, C. R. (1976c). *Nature*, **260**, 165–66
Cox, B. M., Goldstein, A. and Li, C. H. (1976). *Proc. natn. Acad. Sci. U.S.A.*, **73**, 1821–23
Elde, R., Hökfelt, T., Johansson O. and Terenius L. (1976). *Neurosci. Lett.*, **1**, 349–51
Feinberg, A., Creese, I. and Snyder S. H. (1976). *Proc. natn. Acad. Sci. U.S.A.*, **73**, 4215–19
Hughes, J. (1975). *Brain Res.*, **88**, 295–308

Hughes, J., Smith, T. W., Kosterlitz, H. W., Forthergill, L. A., Morgan, B. A. and Morris, H. R. (1975). *Nature*, **258**, 577–79

Hutchinson, M., Kostelitz, H. W., Frances, C. M., Terenius, L. and Waterfield, A. M. (1975). *Br. J. Pharmac.*, **55**, 541–42

Jones, C. R., Gibbons, W. A., and Garsky, V. (1976). *Nature*, **262**, 779–82

Lazarus, L. H., Ling, N. and Guillemin, R. (1976). *Proc. natn. Acad. Sci. U.S.A.*, **73**, 2156–59

Lewis, P. N., Momany, F. A. and Scheraga, H. A (1973). *Biochim. biophys. Acta*, **303**, 211–29

Lord, J. A. H., Waterfield, A. M., Hughes, J. and Kosterlitz, H. (1976). In *Opiates and Endogenous Opiate Peptides* (ed. H. Kosterlitz) North-Holland, Amsterdam, pp. 275–80

Matthews, B. W. (1972). *Macromolecules*, **5**, 818–19

Miller, R. J., Chang, K-J., Cuatrecasas, P. and Wilkinson, S. (1977*a*). *Biochem. biophys. Res. Commun.*, **74**, 1311–18

Miller, R. J., Chang, K-J., Leighton, G. and Cuatrecasas, P. (1977*b*). *Life Sci.*, (in press)

Miller, R. J., Chang, K-J., Cooper, B. and Cuatrecasas, P. (1978). *J. biol. Chem.*, (in press)

Nemethy, G. and Printz, M. P. (1972). *Macromolecules*, **5**, 755–58

Pert, C. B. and Snyder, S. H. (1974). *Molec. Pharmac.*, **10**, 868–78

Pert, C. B., Bowie, D. L., Fong, B. T. S. and Chang, J-K. (1976). In *Opiates and Endogenous Opiate Peptides* (ed. H. Kosterlitz), North-Holland, Amsterdam, pp. 79–86

Pert, C. B., Pert, A. and Tallman, J. (1976). *Proc. natn. Acad. Sci. U.S.A.*, **73**, 2226–30

Ramakrishnan, C. and Ramachandran, G. N. (1965). *Biophys. J.*, **5**, 909–33

Roques, B. P., Garbay-Jallreguiberry, C., Oberlin, R., Anteunis, M. and Lala, A. K. (1976). *Nature*, **267**, 778–9

Simantov, R. and Snyder, S. H. (1976*a*) *Proc. natn. Acad. Sci. U.S.A.*, **73**, 2515–19

Simantov, R. and Snyder, S. H. (1976*b*). In *Opiates and Endogenous Opiate Peptides* (ed. H. Kosterlitz), North-Holland, Amsterdam, pp. 41–49

Simantov, R., Kuhar, M. J. and Snyder, S. H. (1977). *Proc. natn. Acad. Sci. U.S.A.*, (in press).

Snyder, S. H. and Simantov, R. (1977). *J. Neurochem.*, **28**, 1320

Weissman, B. A., Gershon, H. and Pert, C. B. (1976). *FEBS Lett.*, **70**, 245–48

14
Electrophysiological effects
of opiates and opioid peptides

P. B. Bradley, R. J. Gayton and Lynn A. Lambert
(Department of Pharmacology, The Medical School, Birmingham B15 2TJ, U.K.)

INTRODUCTION

Any discussion of the electrophysiological effects of opiate drugs in the central nervous system (CNS) should include effects on the electroencephalogram (EEG) and evoked potentials, as well as effects on single neurons. However, studies of 'gross' electrophysiological changes induced by opiates (Bradley, 1968; Navarro and Elliott, 1971) have not contributed significantly to our understanding of their mode of action, nor have any such studies been undertaken with the opioid peptides. Thus, this chapter will be restricted to a discussion of the effects of opiate drugs and opioid peptides on single neurons in the CNS.

Such studies can be carried out in two ways; first, the activity of a single neuron can be recorded with a microelectrode and the drug injected intravenously. In two studies of this kind, the responses of neurons in (a) the hypothalamus (Eidelberg and Bond, 1972) and (b) the locus coeruleus (Korf *et al.*, 1974) to intravenous injections of morphine were examined. Both groups reported depression of neuronal firing and the reversal or blocking of this effect by the morphine antagonist, naloxone. This method suffers from a number of defects, the principal one being that the effect of the drug may be produced at a site distant from the neuron that is being recorded.

A second method, which is to be preferred, is that of microiontophoresis. Since this technique has already been referred to in this symposium (Renaud, chapter 5) and also described in numerous places in the literature, it will not be described in detail here. Microiontophoresis has been used extensively over the past five years or so for studying the effects of opiate drugs on single neurons and, more recently, for examining the actions of opiate peptides. There is a wealth of data from these studies which have been carried out with neurons in many parts of the brain and spinal cord, and we have summarised these data in a series of tables.

Table 14.1 summarises the data obtained from studies of the actions of morphine and related drugs on single neurons in various regions of the brain. The drugs studied

Table 14.1 Effects of morphine and related opiate drugs on single neurons in the brain

Region	Species	Predominant effect			Naloxone reversible	Charge (μC)
		Morphine	Levorphanol	Dextrorphan		
Cortex	Rat (1)	→			Yes	6
	Rat (2)	↓*	→	0	Yes	3
	Rat (8)	↑			Yes	2
Thalamus	Rat (2)	↓*			Yes	3
	Cat (9)	→				18
Caudate nucleus	Rat (2)	↓*			Yes	3
	Rat (3)	→	→	0	Yes	1
	Rat (11)	→			?	?
Nucleus accumbens	Rat (10)				Yes	3
Hippocampus	Rat (2)	↑*			Yes	3
	Rat (5)	0			Yes	24
L.C.	Rat (4)	→	→	0	Yes	1
Raphe	Cat (7)	→			No	1
P.A.G.	Rat (11)	→			Yes	12
	Cat (7)	→			No	1
Brainstem	Rat (2)	↓*	→	0	Yes	3
	Rat (6)	↓†			Yes	3
	Rat (8)	↑			Yes	2
	Cat (7)	→			No	1

↓ = depression; ↑ = excitation; 0 = no effect; *results with normorphine; †results with morphine and etorphine.
(1) Satoh et al., 1976; (2) Nicoll et al., 1977; (3) Gayton and Bradley, 1976; (4) Bird and Kuhar, 1977; (5) Segal, 1977; (6) Bradley and Dray, 1974; Bramwell and Bradley, 1974; Bradley and Bramwell, 1977; (7) Gent and Wolstencroft, 1976a; (8) Davies and Dray, 1976a; (9) Duggan and Hall, 1977; (10) McCarthy et al., 1977; (11) Frederickson and Norris, 1976; Frederickson et al., 1976.

were morphine, normorphine, etorphine, levorphanol and the relatively inactive isomer of levorphanol, dextrorphan. The table also shows the effects of the morphine antagonist, naloxone, on the responses to opiate drugs. Most regions of the brain have been covered, and in many cases by more than one laboratory. The cerebellum has been omitted here, although there are some preliminary data for this region (see Nicholl *et al.*, 1977). Table 14.1 shows that the predominant effect of morphine was depression of neuronal activity, and this effect was found in all regions except the hippocampus where excitation was the only consistent effect. It should be emphasised, however, that even where the predominant effect was depression, excitation of some neurons was also observed. For example, Davies and Dray (1976*a*) found a high proportion of neurons in the cortex and brainstem of the rat to be excited by morphine and this finding is at variance with those of most other laboratories (see below for discussion). In the rat, the depressant action of morphine (and related drugs) was antagonised by or reversed by naloxone, whereas in the cat, naloxone was ineffective against inhibition produced by morphine (Gent and Wolstencroft, 1976*a*). Where levorphanol and dextrorphan were tested, levorphanol mimicked morphine depression whereas dextrorphan did not. Also shown in table 14.1 (final column) is an estimate of the mean charge used for iontophoresis to produce the effects observed. This is expressed in microcoulombs and since we know the transport number for morphine and that, within certain limits, the amount released from micropipettes is proportional to the charge (Bradley and Dray, 1974) these figures represent an approximate measure of the amount of drug applied. What is striking is that, with certain exceptions, there is very little variation in the charge used in different studies and the differences which do exist cannot account for the different findings.

AN ANALYSIS OF OPIATE ACTION ON SINGLE NEURONS

The effects of opiate drugs on neuronal activity are illustrated in more detail by examples from studies in this laboratory. The experiments were carried out on urethane-anaesthetised rats and all the brainstem neurons studied were spontaneously active. One of the first observations that was made (Bradley and Dray, 1974) was that morphine excited some neurons and depressed others. When the transport number for morphine was measured using radiolabelled material, it became apparent that the drug was very potent in producing these effects since only about 1.5 pmol were released from the micropipettes. At this time, the excitatory response to morphine appeared to be the more interesting one since it often showed desentisation or 'tachyphylaxis'. However, subsequent studies (Bramwell and Bradley, 1974; Bradley and Bramwell, 1977) revealed that only the depression of neuronal activity by morphine was reversed or blocked by naloxone and in our experience, morphine-induced excitation has never been affected by naloxone (figure 14.1). Levorphanol was found to mimic both excitation and depression produced by morphine but again only depression by levorphanol was antagonised by naloxone. Dextrorphan sometimes mimicked the effects of morphine but in most cases it either had no effect or opposite effects; none of the effects of dextrorphan was antagonised by naloxone. It is interesting to note that etorphine, a more potent analgesic drug, produced only depression (Bradley *et al.*, 1976; Bradley, Gayton and Lambert, unpublished). In a recent study in the caudate nucleus (Gayton and Bradley, 1976), similar effects of morphine and levorphanol were found though excitation was

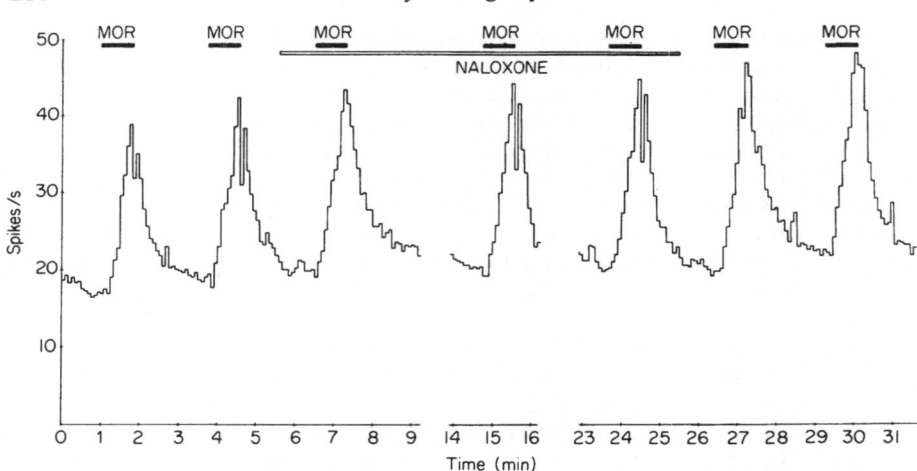

Figure 14.1 Excitatory effect of microiontophoretically applied morphine on the firing rate of a spontaneously active brainstem neuron. The firing rate, in spikes/s in successive 5 s epochs, is plotted against time (min). Iontophoretic applications are indicated by the horizontal bars. As can be seen, reproducible excitations were produced by regular applications of morphine (MOR 50 nA for 50 s) at 2 min intervals. An iontophoretic application of naloxone (25 nA for 20 min) failed to antagonise the morphine excitation (from Bradley and Bramwell, 1977).

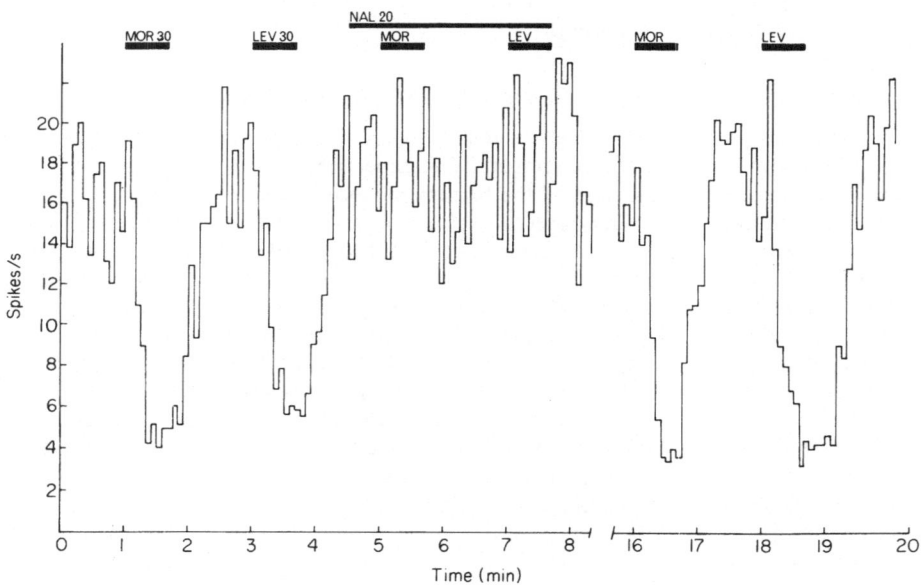

Figure 14.2 Effects of microiontophoretically applied morphine (MOR), levorphanol (LEV) and naloxone (NAL) on a neuron in the caudate nucleus, driven by slow release of D, L-homocysteic acid (0.2 nA). Both morphine and levorphanol at currents of 30 nA produced depression of firing and these effects were completely antagonised by an iontophoretic application of naloxone (20 nA), followed by recovery.

comparatively rare, and again the depression of neuronal activity was antagonised by naloxone (figure 14.2).

Thus, we concluded that the depression of neuronal activity by morphine was the more important response since it showed stereospecificity and was antagonised by naloxone, whereas morphine-induced excitation showed neither of these features. Which of the many pharmacological actions of morphine may be related to depression of neuronal activity is difficult to say, but in the brainstem we believe that this effect may be involved in the analgesic actions of the drug, first because the cells we were studying were in the region where neurons responding to nociceptive stimuli have been identified (Benjamin, 1970) and secondly because microinjection of analgesic drugs in this region of the brain produces analgesia (Takagi *et al.*, 1976).

EFFECTS OF OPIOID PEPTIDES

The discovery of endogenous opioid peptides in brain (Hughes, 1975), together with their identification as methionine- and leucine-enkephalin and the demonstration of morphine- like actions peripherally (Hughes *et al.*, 1975) resulted in these substances being studied extensively as soon as they became available. Table 14.2 shows the results obtained from a survey of the effects of the two enkephalins, applied microiontophoretically, to single neurons in various brain regions. It is surprising that there are fewer reports of the effects of leu- than of met-enkephalin but this may simply reflect either differences in availability or early reports that met- enkephalin was more active (Hughes *et al.*, 1975).

It can be seen from table 14.2 that the predominant effect observed was again depression of neuronal activity and this was found in all areas that have been studied, with the exception of the hippocampus, where excitation was the main effect (Nicoll *et al.*, 1977; Hill and Pepper, personal communication). Davies and Dray (1976a) who had observed excitation to be the predominant effect of morphine in the cortex and brainstem, found similar effects with the enkephalins. All the responses reported in the rat appeared to be naloxone sensitive, including excitation, whereas those found in the cat were not (Gent and Wolstencroft, 1976a, b).

In this laboratory, studies have been carried out with four opioid peptides. These are: met- and leu-enkephalin, together with two analogues, [D-Ala^2, Met^5]-enkephalin (Tyr-D-Ala-Gly-Phe-Met-NH$_2$) and a Burroughs Wellcome compound, BW 180C (Tyr-D-Ala-Gly-Phe-D-Leu-HCl). The [D-Ala^2, Met^5]-enkephalin analogue was chosen for study because it is reported to be active at central opiate receptors and, at the same time, to be more resistant to enzymatic degradation *in vitro* than either met- or leu-enkephalin (Pert *et al.*, 1976). Furthermore, this compound has been found to cause analgesia when injected intracerebrally (Pert, 1976). While [D-Ala^2, Met^5]-enkephalin represents a stable analogue of met-enkephalin, the BW 180C compound is an analogue of leu-enkephalin and has been reported to possess antinociceptive actions (Baxter *et al.*, 1977).

We have already reported the effects of met-enkephalin applied iontophoretically to spontaneously active bramstem neurons (Bradley *et al.*, 1976). Figure 14.3A illustrates the effect of an application of met-enkephalin on the discharge rate of a neuron in the brainstem and this is typical of the responses we have seen with the two enkephalins. Figure 14.3B is a continuous record of the firing rate of the same neuron and compares the responses of met-enkephalin with those of etorphine. The

Centrally Acting Peptides

Table 14.2 Effects of methionine- and leucine-enkephalin on single neurons in the brain

Region	Species	Met-enk	Leu-enk	Naloxone reversible
Cortex	Rat (1)	↓		Yes
	Rat (2)	↓		Yes
	Rat (8)	↑	↑	Yes
	Rat (12)	↓		Yes
Thalamus	Rat (2)	↓		Yes
	Rat (12)	↓		Yes
Caudate nucleus	Rat (2)	↓		Yes
	Rat (11)	↓		Yes
Nucleus accumbens	Rat (10)	↓	↓	Yes
Hippocampus	Rat (2)	↑		Yes
	Rat (12)	↑		Yes
L.C.	Rat (4)	↓		Yes
Raphe	Cat (7)	↓		No
P.A.G.	Rat (11)	↓		Yes
	Cat (7)	↓	↓	No
Brainstem	Rat (2)	↓		Yes
	Rat (6)	↓	↓	Yes
	Rat (8)	↑	↑	Yes
	Rat (12)	↓		Yes
	Cat (7)	↓	↓	No

↓ = depression; ↑ = excitation; 0 = no effect
(1) Zieglgänsberger and Fry, 1976; Zieglgänsberger *et al.*, 1976; (2) Nicoll *et al.*, 1977; (4) Young, Bird and Kuhar, 1977; (6) Bradley *et al.*, 1976; (7) Gent and Wolstencroft, 1976a; (8) Davies and Dray, 1976a; (10) McCarthy *et al.*, 1977; (11) Frederickson and Norris, 1976; Frederickson *et al.*, 1976; (12) Hill, Pepper and Mitchell, 1976a, b; Hill and Pepper, personal communication.

latter characteristically produced depression with a slow onset and recovery. In Figure 14.3B iontophoretically applied naloxone blocked the responses to both etorphine and enkephalin with subsequent recovery. During the naloxone application there was an increase in the firing rate of the neuron and this fell again after the naloxone application was terminated. The cause of this effect is unclear but it may not be due to opiate receptor blockade.

A comparison of the effects of met- and leu-enkephalin applied to the same neuron showed that in some cases both were equally, or nearly equally effective (figure 14.4A) whereas on other neurons, met-enkephalin was effective but leu-enkephalin was not (figure 14.4B). This variation in activity might be due to a number of factors, for example variation between different samples of enkephalins. However, although we tested enkephalins from four different sources and there

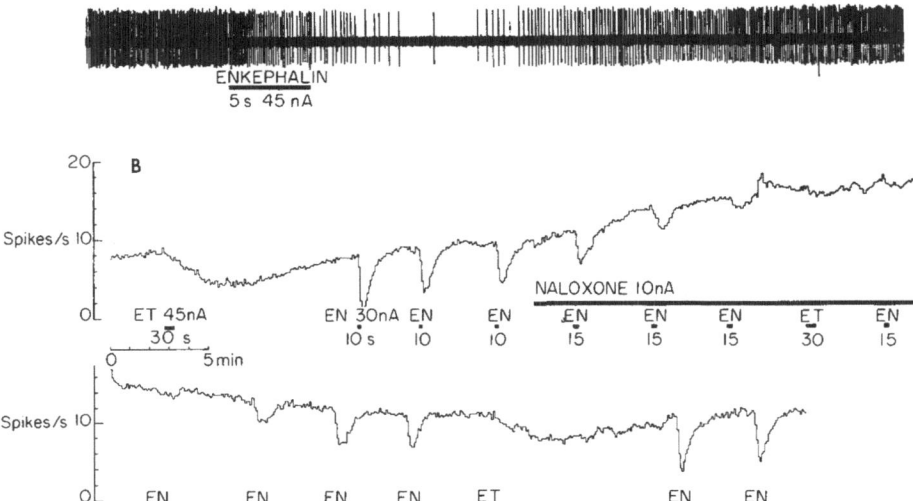

Figure 14.3   Effects of methionine-enkephalin, etorphine and naloxone on a spontaneously active brainstem neuron. (A) the inhibitory effect of met-enkephalin (45 nA for 5 s) on neuronal firing with recovery. (B) Continuous plot of the firing rate of the same neuron showing the effects of repeated applications of etorphine (ET, 45 nA for 30 s) and met-enkephalin (EN, 30 nA for 10 s). During an iontophoretic application of naloxone at 10 nA, the inhibitory responses to both etorphine and met-enkephalin were blocked, but recovered after the termination of the naloxone application (from Bradley *et al.*, 1976).

was some variation in solubility, this did not account for the variation in activity. A second possibility to be considered is variation in the amount released from the micropipettes. However, measurement of the transport number for leu-enkephalin (Candy and Lambert, unpublished) showed that although there could be variation between different pipettes, even those with low transport numbers were effective *in vivo*. Furthermore, the electrodes tested showed very linear release characteristics. Therefore we do not think that this factor can account for the variation in effectiveness of leu-enkephalin.

Desensitisation or 'tachyphylaxis' to the enkephalins has been reported from studies of neurons in the cortex and striatum (Zieglgänsberger and Fry, 1976). This has not been observed in the medulla and figure 14.5 shows the responses of a medullary neuron to repeated applications of leu-enkephalin.

[D-Ala^2, Met^5]-enkephalin, applied to the same neuron, produced a depression that was much longer lasting than that seen with met- or leu-enkephalin (figure 14.4A and B, figure 14.6) and in many cases the response to this compound appeared to be of greater amplitude. The depressant effects of [D-Ala^2, Met^5]-enkephalin were antagonised by naloxone (figure 14.6). The BW 180C compound also produced prolonged depression, very similar to that seen with [D-Ala^2, Met^5]-enkephalin, and the effects of this substance were also antagonised by naloxone.

A comparison of the effects of these four opioid peptides with those of mor-

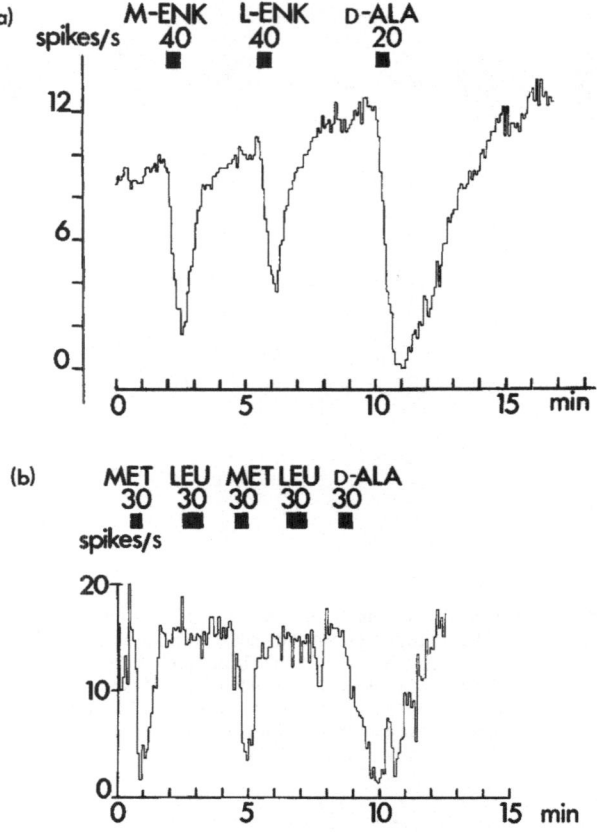

Figure 14.4 Effects of met-enkephalin (M-ENK), leu-enkephalin (L-ENK) and D-Ala^2]-enkephalin (D-ALA) on two different brainstem neurons. In (A) met- and leu-enkephalin (each at 40 nA) were approximately equally effective in depressing the firing rate, while [D-Ala^2]-enkephalin (20 nA) produced a greater and longer-lasting effect. In (B) leu-enkephalin (LEU) was ineffective compared with met-enkephalin (MET). [D-Ala^2]-enkephalin again produced a long-lasting response.

phine showed that the peptides mimicked the actions of morphine where the latter was depressant and they often produced depression when morphine was excitatory. Some excitations have been seen with met-enkephalin (Bradley *et al.*, 1976) but not with the other three peptides.

The results of our studies with the four opioid peptides and etorphine are summarised in Table 14.3. It can be seen that the percentage of cells inhibited is very high, almost 90 per cent for etorphine and [D-Ala^2, Met^5] whereas for morphine it is lower, of the order of 65 per cent. A number of the cells studied were marked with Pontamine Sky Blue to identify them histologically and the positions of a representative sample are illustrated in figure 14.7, together with an indication of the compounds to which they responded. Many of the peptide-sensitive neurons were in the region of the nucleus reticularis gigantocellularis.

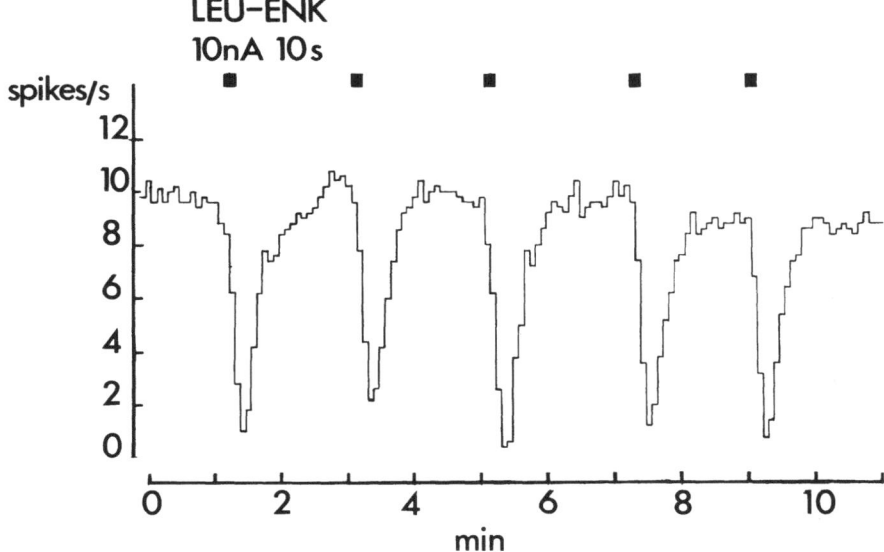

Figure 14.5 Effects of repeated applications of leu-enkephalin (LEU-ENK), (10 nA for 10 s) on the activity of a single neuron in the brainstem. Although the applications were repeated at 2 min intervals, the inhibitory response was fully maintained.

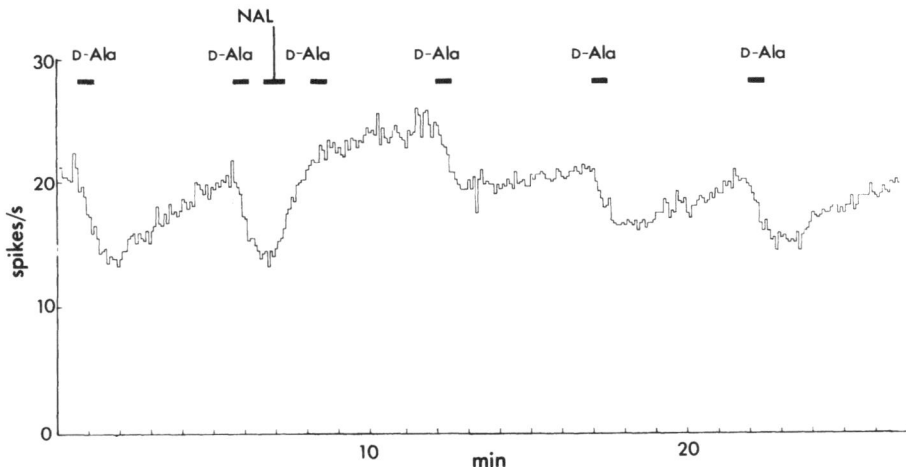

Figure 14.6 Antagonism by naloxone (NAL) of the response to [D-Ala^2]-enkephalin (D-Ala). Both substances were applied with currents of 30 nA.

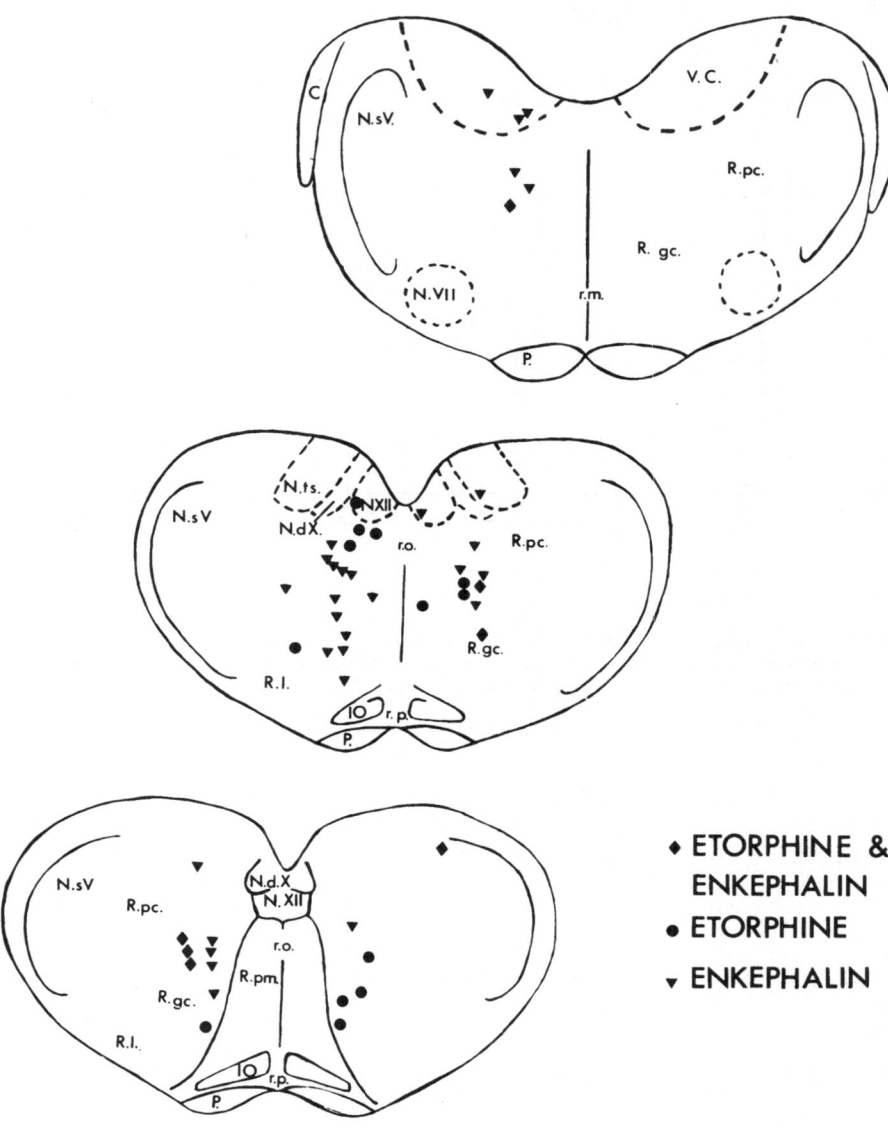

Figure 14.7 Transections of the rat brainstem showing the location of some of the neurons responding to iontophoretic application of enkephalins and/or etorphine. The nomenclature of Valverde (1961) for the divisions of the reticular formation has been used: N.dx., dorsal motor nucleus of the vagus; R.gc., nucleus reticularis gigantocellularis; N.XII, hypoglossal nucleus; R.l., nucleus reticularis lateralis; N.sV., nucleus of the spinal trigeminal tract; R.pm., nucleus reticularis paramedianus; N.ts., nucleus of the solitary tract; R.pc., nucleus reticularis parvocellularis; N.VII, facial nucleus; r.o., nucleus raphe obscurus; V.C., vestibulary complex; r.p., nucleus raphe pallidus; P., pyramidal tracts; r.m., nucleus raphe magnus; I.O., inferior olive; C., cortex.

Table 14.3   Effects of opioid peptides and etorphine on single
neurons in the brainstem

Compound	No. of cells inhibited	Percentage	Naloxone reversal
Met-enkephalin	70/89	79	8/8
Leu-enkephalin	78/95	82	8/10
D-Ala^2, Met^5]	75/85	88	13/14
BW 180C	28/41	68	9/10
Etorphine	55/62	89	20/21

EFFECTS ON SPINAL NEURONS

So far we have considered the actions of opiate drugs and opioid peptides on single neurons in the brain but the spinal cord is also recognised as an important site for analgesic effects (see Le Bars *et al.*, 1975, for references), although there appear to have been fewer studies in this region using iontophoresis. Renshaw cells are excited by morphine (Duggan and Curtis, 1972; Davies and Duggan, 1974; Davies, 1976; Davies and Dray, 1976*b*) and also by the two enkephalins, these effects being reversed by naloxone (Davies and Dray, 1976*b*). In the case of neurons in the dorsal horn, the situation is less clear. There are reports of excitation by morphine but these effects were often accompanied by changes in spike height and were not reversed by naloxone (Duggan *et al.*, 1977*a*). There are also reports that morphine was ineffective (Davies and Dray, personal communication) or produced weak depression (Duggan *et al.*, 1977*a*). There are two reports of clear-cut depression both of spontaneous activity of the dorsal horn cells by morphine (Zieglgänsberger and Bayerl, 1976) and of responses evoked by nociceptive stimuli with both morphine and the enkephalins (Duggan *et al.*, 1976; 1977*a, b*). In both cases, however, these positive results were obtained with modifications to the iontophoretic technique; in the one by using a microelectrode with the recording tip protruding beyond the end of the iontophoresis barrels (Zieglgänsberger and Bayerl, 1976) and in the other by using completely separate recording and iontophoresis electrodes to (a) record the activity of a single cell in the dorsal horn, and (b) apply drugs into the substantia gelatinosa (Duggan *et al.*, 1976; 1977*a*). Thus, depression of the activity of dorsal horn cells did not occur when the opiate or peptide was applied to the cell soma, but only when it was applied at a distance from the cell, possibly in the dendritic region. This observation may provide some important information about the localisation of opiate receptor sites in the spinal cord.

From this review of the data present available, we may conclude that the principal effect of morphine and related opiate drugs on single neurons in the brain, in all areas which have been studied so far with the exception of the hippocampus, is depression. This effect can be mimicked by levorphanol but not by dextrorphan and is reversed or antagonised by naloxone in the rat but not the cat. Similar effects are seen with the enkephalins.

EXCITATORY EFFECTS OF OPIATES AND OPIOID PEPTIDES

Depression is not the only action of opiate drugs and peptides since most workers also observed excitation of some neurons. In most cases the number of neurons excited was smaller than the number depressed although one group (Davies and

Dray, 1976*a*) consistently found excitation to be the predominant effect. They also found that excitation was naloxone sensitive, which does not accord with the lack of naloxone-sensitive excitation that we have found in the same region. At present it is not possible to account for these different findings on the basis of differences in techniques although there are still some possibilities to be explored. It is interesting to note in this context that etorphine, a more potent analgesic drug than morphine, has not in our hands ever produced an increase in neuronal firing. This lends some support to the hypothesis that depression of neuronal activity is the more important effect and is more likely to be related to analgesia.

It is difficult to account satisfactorily for the excitatory actions of morphine and the enkephalins, particularly since morphine, in all its known pharmacological actions, is a powerful depressant. As already indicated, excitation is the predominant effect in the hippocampus and is the only effect found on Renshaw cells. The hippocampus is reported to have a relatively low density of opiate receptors (Pert *et al.*, 1975) and also a low enkephalin content (Simantov *et al.*, 1976) furthermore, both the pyramidal cells of the dorsal hippocampus and Renshaw cells are cholinoceptive. In studies of the interactions between morphine and neurotransmitters (Bradley and Dray, 1974) a closer correlation was found between the effects of morphine and acetylcholine than with any other neurotransmitter and, since morphine is known to possess some acetylcholinesterase activity (Johanesson, 1962), it is possible that excitation might be related in some way to cholinergic receptors. Excitation at the neuronal level need not necessarily be associated with excitation at the 'integrated' tissue or whole animal level, that is in terms of behaviour. It is interesting that, in a recent study of respiratory neurons in the rat medulla, excitation was the main effect produced by iontophoretically applied morphine (Lucy, unpublished). Moreover, in the cat the behavioural excitation produced by morphine is well documented, although at the neuronal level the drug causes depression of activity (Gent and Wolstencroft, 1976*a, b*). The fact that this latter effect is not sensitive to naloxone reversal might reflect the atypical behavioural response in this species. There are as yet no reports about the nature of opiate receptors in the cat brain or whether the feline CNS even contains enkephalins.

OPIOID PEPTIDES: PRE- OR POSTSYNAPTIC NEUROTRANSMITTERS?

It has been suggested that the opioid peptides, such as the enkephalins, may be neurotransmitters, and that the opiate drugs and the enkephalins may have a common locus of action. The data from biochemical studies of the opiate receptor (Pert *et al.*, 1974) in general support this hypothesis and evidence is now accumulating from immunohistochemical studies (Elde *et al.*, 1976; Elde, this volume) showing that enkephalin is present in nerve fibres and terminals in certain prescribed regions of the brain (spinal cord, medulla, pons and mesencephalon, and also in the hypothalamus and basal ganglia, but not in the cerebral cortex). Thus, the histochemical localisation is consistent with a transmitter role.

The results reported above show that the enkephalins have marked effects on neuronal activity and that very small quantities are needed to produce these effects. In addition, the more biologically stable analogues ([D-Ala^2, Met^5]-enkephalin and BW 180C) were more potent and produced longer-lasting effects. Thus, it would appear that the mechanisms necessary for inactivation of the enkephalins are present in the vicinity of the neurons on which they produce their effects. These results therefore support the concept of the enkephalins as neurotransmitters.

The question remains as to whether the effects of the enkephalins are due to a pre- or a postsynaptic action. As yet there is no clear-cut answer to this question and further studies using intracellular recording will be required to provide a complete answer. There are, however, some tentative indications. Evidence for a presynaptic action of the opiates and opioid peptides is difficult to find and it appears that electrophysiological evidence is non-existent. However, it is known that morphine can inhibit the stimulation-evoked release of acetylcholine in the myenteric plexus (Cox and Weinstock, 1966) while naloxone can facilitate this release (Waterfield and Kosterlitz, 1975). In the brain, enkephalin may modulate the release of noradrenaline, since it has been reported (Taube *et al.*, 1976) that in slices of rat occipital cortex pre-loaded with [^{3}H] noradrenaline, met-enkephalin diminished the outflow of radioactivity evoked by field stimulation or high potassium in the medium. These data are most easily explained by postulating presynaptic receptors. In the whole animal, a significant reduction in the number of measurable opiate receptors was found following the cutting of dorsal roots in the spinal cord and consequent degeneration of sensory nerves (Lamotte *et al.*, 1976). From this it was concluded that some of the opiate receptors must be presynaptic.

On the other hand, met-enkephalin has been found to cause hyperpolarisation of neurons in the guinea pig myenteric plexus (North, 1977), a peripheral tissue which is thought to be a good model for studying the actions of opiate drugs. Morphine, however, was found to have no effect on resting membrane potential or on the membrane resistance of motoneurons, although it did block the glutamate-induced sodium influx associated with depolarisation (Zieglgänsberger and Bayerl, 1976). This blocking of sodium efflux elicited by excitatory neurotransmitters has been postulated as the mechanism for the depressant action of morphine on neurons. However, motoneurons may not be an entirely satisfactory model for studying the mechanism of action of opiates.

In support of the postsynaptic site of action is the finding that opiates and enkephalins block glutamate excitation of neuronal activity. This effect was first observed in the spinal cord (Dostrovsky and Pomeranz, 1973) and has since been found in many regions of the brain (Nicoll *et al.*, 1977; Zieglgänsberger *et al.*, 1976; Segal, 1977). It seems to be a fairly specific effect since it is not seen to the same extent with other excitatory amino acids such as D, L-homocysteic acid. As the excitatory action of glutamate is thought to be due to a postsynaptic action (Zieglgänsberger and Puil, 1973), the blocking of this effect must also be postsynaptic.

If the opioid peptides are in fact neurotransmitters operating through 'enkephalinergic' synapses, there may well be both pre- and postsynaptic receptors at these synapses, analogous to those proposed for adrenergic (Langer *et al.*, 1971) and cholinergic nerves (Szerb and Somogyi, 1973).

ACKNOWLEDGEMENTS

We thank the following for supplying some of the compounds tested: Reckitt and Colman for met- and leu-enkephalin; Burroughs Wellcome for BW 180C and met- and leu-enkephalin; and Endo Laboratories for naloxone.

REFERENCES

Baxter, M. G., Goff, D., Miller, A. A. and Saunders, I. A. (1977). *Br. J.Pharmac.*, 59, 455–56P
Benjamin, R. M. (1970). *Brain Res.*, 24, 525–29
Bird, S. J. and Kuhar, M. J. (1977). *Brain Res.*, 122, 523–33
Bradley, P. B. (1968). In *Pain* (ed. A. Soulairac, J. Cahn and J. Charpentier), Academic Press, New York, pp. 411–23
Bradley, P. B. and Bramwell, G. J. (1977). *Neuropharmacology*, 16, 519–26
Bradley, P. B., Briggs, I., Gayton, R. J. and Lambert, Lynn A. (1976). *Nature*, 261, 425–26
Bradley, P. B. and Dray, A. (1974). *Br.J.Pharmac.*, 50, 47–55
Bramwell, G. J. and Bradley, P. B. (1974). *Brain Res.*, 73, 167–70
Cox, B. M. and Weinstock, M. (1966). *Br.J.Pharmac.*, 27, 81–92
Davies, J. (1976). *Brain Res.*, 113, 311–26
Davies, J. and Dray, A. (1976a). *Br.J.Pharmac.*, 58, 458–59P
Davies, J. and Dray, A. (1976b). *Nature*, 262, 603–04
Davies, J. and Duggan, A. W. (1974). *Nature*, 250, 70–71
Dostrovsky, J. and Pomeranz, B. (1973). *Nature*, 246, 222–24
Duggan, A. W. and Curtis, D. R. (1972). *Neuropharmacology*, 11, 189–96
Duggan, A. W. and Hall, J. G. (1977). *Brain Res.*, 122, 49–57
Duggan, A. W., Hall, J. G. and Headley, P. M. (1976). *Nature*, 264, 456–58
Duggan, A. W., Hall, J. G. and Headley, P. M. (1977a). *Br.J.Pharmac.*, 61, 65–76
Duggan, A. W., Hall, J. G. and Headley, P. M. (1977b). *Proc.Aust.Physiol.Pharmac.Soc.*, 8, 49Pjs)
Eidelberg, E. and Bond, M. L. (1972). *Archs.int.Pharmacodyn.Thér.*, 196, 16–24
Elde, R., Hökfelt, T., Johansson, O. and Terenius, L. (1976). *Neuroscience*, 1, 349–51
Frederickson, R. C. A. and Norris, F. H. (1976). *Science*, 194, 440–42
Frederickson, R. C. A., Nickander, R., Smithwick, E. L., Shuman, R. and Norris, F. H. (1976). In *Opiates and Endogenous Opioid Peptides* (ed. H. W. Kosterlitz), Elsevier/North-Holland, Amsterdam, pp. 239–46
Gayton, R. J. and Bradley, P. B. (1976). In *Opiates and Endogenous Opioid Peptides* (ed. H. W. Kosterlitz), Elsevier/North-Holland, Amsterdam, pp. 213–16
Gent, J. P. and Wolstencroft, J. H. (1976a). *Nature*, 261, 426–27
Gent, J. P. and Wolstencroft, J. H. (1976b). In *Opiates and Endogenous Opioid Peptides* (ed. H. W. Kosterlitz), Elsevier/North-Holland, Amsterdam, pp. 217–24
Hill, R. G., Pepper, C. M. and Mitchell, J. F. (1976a). In *Opiates and Endogenous Opioid Peptides* (ed. H. W. Kosterlitz), Elsevier/North-Holland, Amsterdam, pp. 225–30
Hill, R. G., Pepper, C. M. and Mitchell, J. F. (1976b). *Nature*, 262, 604–06
Hughes, J. (1975). *Brain Res.*, 88, 295–308
Hughes, J., Smith, T. W., Kosterlitz, H. W., Fothergill, L. A., Morgan, B. A. and Morris, H. R. (1975). *Nature*, 258, 577–79
Johanesson, T. (1962). *Acta pharmac.tox.*, 19, 23–55
Korf, J., Bunney, B. S. and Aghajanian, G. K. (1974). *Eur.J.Pharmac.*, 25, 165–69
Lamotte, C., Pert, C. B. and Snyder, S. H. (1976). *Brain Res.*, 112, 407–12
Langer, S. Z., Adler, E., Enero, N. A. and Stefano, F. J. E. (1971). *Proc.XXVth.Int.Congr. Physiol.Sci.*, Munich, p. 335.
Le Bars, D., Menetrey, D., Conseiller, C. and Besson, J. M. (1975). *Brain Res.*, 98, 261–77
McCarthy, P. S., Walker, R. J. and Woodruff, G. N. (1977). *J.Physiol.,Lond.*, 267, 40P
Navarro, G. and Elliott, H. W. (1971). *Neuropharmacology*, 10, 367–77
Nicoll, R. A., Siggins, G. R., Ling, N., Bloom, F. E. and Guillemin, R. (1977). *Proc.natn.Acad. Sci.U.S.A.*, 74, 2584–88
North, R. A. (1977). *Br.J.Pharmac.*, 59, 504–05P
Pert, A. (1976). In *Opiates and Endogenous Opioid Peptides* (ed. H. W. Kosterlitz), Elsevier/North-Holland, Amsterdam, pp. 87–94
Pert, C. B., Bowie, D. L., Fong, B. T. W. and Chang, J.-K. (1976). In *Opiates and Endogenous Opioid Peptides* (ed. H. W. Kosterlitz), Elsevier/North-Holland, Amsterdam, pp. 79–86
Pert, B. C., Kuhar, M. J. and Snyder, S. H. (1975). *Life Sci.*, 16, 1849–54
Pert, C. B., Snowman, A. M. and Snyder, S. H. (1974). *Brain Res.*, 70, 184–88
Satoh, M., Zieglgänsberger, A. and Herz, A. (1976). *Brain Res.*, 115, 99–110
Segal, M. (1977). *Neuropharmacology*, 16, 587–92

Simantov, R., Kuhar, M. J., Pasternak, G. W. and Snyder, S. H. (1976). *Brain Res.*, 106, 189–97
Szerb, J. C. and Somogyi, G. T. (1973). *Nature new Biol.*, 241, 121–22
Takagi, H., Doi, T. and Akaike, A. (1976). In *Opiates and Endogenous Opioid Peptides* (ed. H. W. Kosterlitz), Elsevier/North-Holland, Amsterdam, pp. 191–98
Taube, H. D., Borowski, E., Endo, T. and Stark, K. (1976). *Eur.J.Pharmac.*, 38, 377–80
Valverde, F., (1961). *J.comp.Neurol.*, 116, 71–99
Waterfield, A. A. and Kosterlitz, H. W. (1975). *Life Sci.*, 16, 1787–92
Young, W. S., Bird, S. J. and Kuhar, M. J. (1977). *Brain Res.*, 129, 366–70
Zieglgänsberger, W. and Bayerl, H. (1976). *Brain Res.*, 115, 111–28
Zieglgänsberger, W. and Fry, J. P. (1976). In *Opiates and Endogenous Opioid Peptides* (ed. H. W. Kosterlitz), Elsevier/North-Holland, Amsterdam, pp. 231–38
Zieglgänsberger, W. and Puil, E. A. (1973). *Expl. Brain Res.*, 17, 35–49
Zieglgänsberger, W., Fry, J. P., Herz, A., Moroder, L. and Wünsch, E. (1976). *Brain Res.*, 115, 160–64

15
Biogenesis and metabolism of opiate active peptides

D. G. Smyth, B. M. Austen, A. F. Bradbury, M. J. Geisow and C. R. Snell
(Laboratory of Peptide Chemistry, National Institute for Medical Research,
The Ridgeway, London NW7 1AA, U.K.)

INTRODUCTION

Physiologically active peptides, elaborated intracellularly, are biosynthesised as single molecular species. In the case of peptide hormones, the functionally significant molecule is generated from the corresponding prohormone by enzymes with high specificity and the product preserved intact within secretory granules (Kemmler *et al.*, 1973). Thus a characteristic biological activity is not produced by a group of naturally occurring peptides with overlapping sequences. In the field of endogenous peptides with opiate activity, however, it has seemed as if this general rule does not apply.

A series of opiate active peptides has been reported to occur in the central nervous system, each representing some fraction of the C-fragment of lipotropin (figure 15.1). In this chapter we consider two alternatives: that each of the mor-

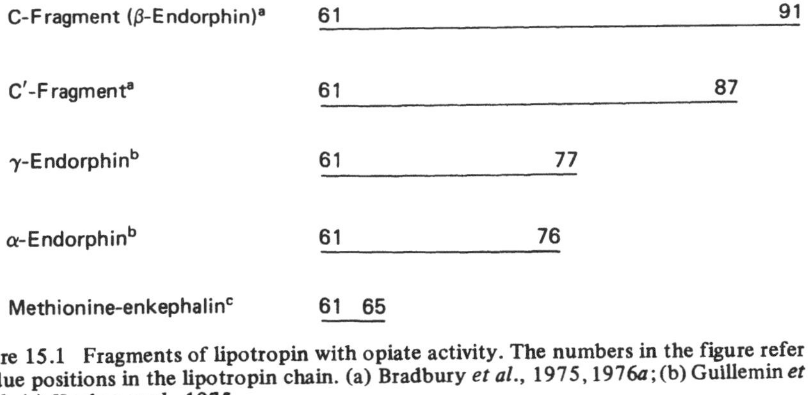

C-Fragment (β-Endorphin)[a]	61 ———————————————	91
C'-Fragment[a]	61 ————————————	87
γ-Endorphin[b]	61 ——————	77
α-Endorphin[b]	61 —————	76
Methionine-enkephalin[c]	61 65	

Figure 15.1 Fragments of lipotropin with opiate activity. The numbers in the figure refer to residue positions in the lipotropin chain. (a) Bradbury *et al.*, 1975, 1976*a*; (b) Guillemin *et al.*, 1976; (c) Hughes *et al.*, 1975.

231

phinomimetic peptides has a characteristic function or that the shorter peptides represent degradation products of C-fragment that retain a fraction of its central activity.

BIOGENESIS

That C-fragment, which has the sequence of residues 61–91 of lipotropin, is derived *in vivo* from lipotropin was indicated strongly when the contiguous peptides, N-fragment (residues 1–38); β-melanocyte stimulating hormone (β-MSH) (residues 41–58) and C-fragment (61–91) were isolated from porcine pituitary (Bradbury *et al.*, 1975, 1976) and this has been confirmed *in vitro* by pulse labelling experiments with pituitary cells (Chrétien *et al.*, 1976). It is notable that each of the peptides isolated from pituitary was formed by the action of a highly specific enzyme that cleaves the lipotropin chain at paired basic residues, the basic residues being removed by an enzyme with the specificity of carboxypeptidase B (figure 15.2). To obtain further understanding of the activation process, purification of the trypsin like enzyme was undertaken.

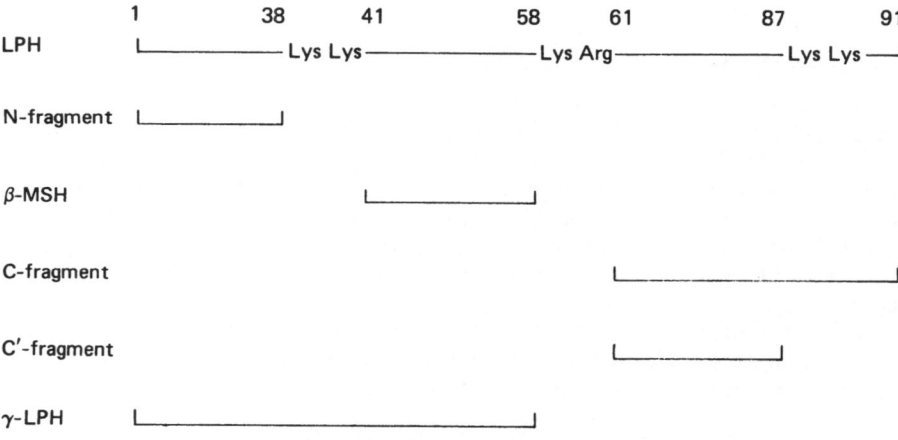

Figure 15.2 Peptide fragments of lipotropin isolated from porcine pituitary glands. The horizontal lines indicate the relationship of the isolated peptides to the intact lipotropin molecule.

Secretory granules were prepared from 100 porcine pituitary glands (approximately 50 g) by homogenisation in 5 mM Tris–HCl at pH 7.4 containing 0.25 M sucrose. Centrifugation was at 25 000*g* for 1 hour at 4 °C, to provide a pellet which was resuspended in the same buffer and centrifuged at 10 000*g*. The final pellet was lysed in hypotonic saline and the soluble enzymes were fractionated by gel filtration at pH 7 on a column (150 × 1.8 cm) of Sephadex G-100 with 0.2 M sodium phosphate as eluent (figure 15.3). The fractions were monitored for enzymic activity by incubation with the synthetic hexapeptide Lys–Asp–Lys–Arg–Tyr–Gly, which represents the sequence from residues 57–62 of lipotropin. Two activities were observed, an endopeptidase that released the dipeptide tyrosylglycine and an exopeptidase that released lysine.

Both activities emerged in the high molecular weight fractions excluded from the G-100 column. In contrast, pancreatic trypsin penetrated the gel. The active fractions, which were concentrated by vacuum dialysis, were resolved further by chromatography at pH 7 on a column (40 × 1 cm) of DEAE–Sephadex A25. Gradient elution was in 0.02 M Tris–HCl at pH 7 with a NaCl gradient from 0.1 to 1 M. The lysine releasing enzyme emerged in advance of the endopeptidase (figure 15.4).

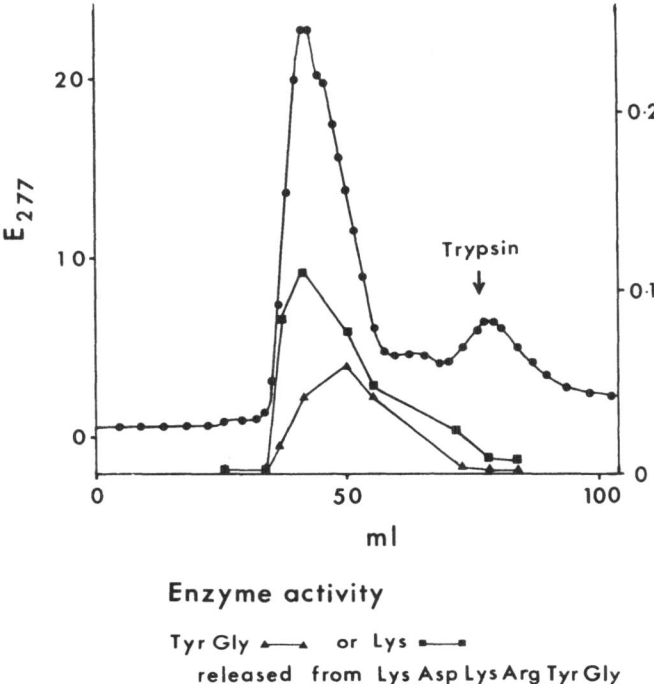

Enzyme activity

Tyr Gly ▲——▲   or Lys ■——■

released from Lys Asp Lys Arg Tyr Gly

Figure 15.3 Gel filtration of soluble enzymes obtained from the secretory granules of porcine pituitary on Sephadex G-100. Fractions (2 ml) were collected and portions (25 μl) were incubated with the hexapeptide Lys-Asp-Lys-Arg-Tyr-Gly (0.1 μmol) in 0.5 ml of 0.2 M sodium phosphate at pH 7.4 for 4 hours, 37 °C. The released dipeptide and lysine were determined by amino acid analysis.

The purified preparation of the trypsin-like enzyme exhibited a number of properties that distinguish it from trypsin. Its action was unaffected by soya bean trypsin inhibitor (1 mg/ml) and the optimum pH was in the region of 9.0. It was activated by reducing agents and partially inhibited by sulphydryl reagents. Its apparent molecular weight during gel filtration was greater than 100 000. Recent experiments have demonstrated the presence of a similar enzyme in rat brain; the brain enzyme, like that in pituitary, appears to be located intracellularly and is not membrane bound.

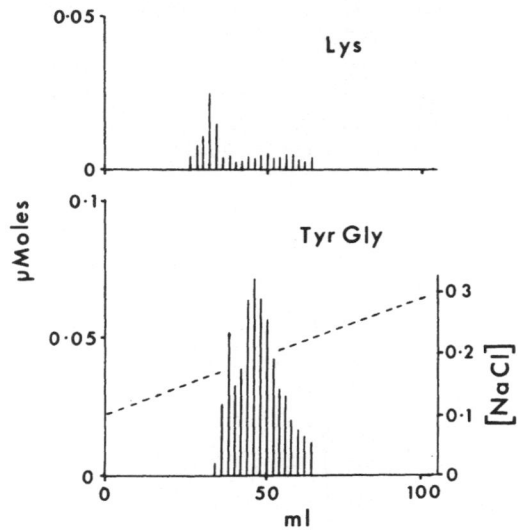

Lys or TyrGly released from Lys Asp Lys Arg Tyr Gly

Figure 15.4 Chromatography of soluble enzymes from the secretory granules of porcine pituitary on DEAE-Sephadex A25. Fractions (2 ml) were collected and incubation of portions (25 μl) was carried out as in the legend to figure 15.3.

CENTRAL ACTIONS OF C-FRAGMENT

The C-fragment exhibits a range of pharmacological activities when injected cerebroventricularly in the brain. Both in the cat and in the rat, the peptide acts as an analgesic agent with a potency about 100 times that of morphine (Feldberg and Smyth, 1976, 1977; Bradbury *et al.*, 1977). The effects persist for several hours and the action is reversed by subcutaneous naloxone. It is notable that the C′-fragment gave no detectable analgesia when 100 μg was infused in the cat; in the rat the C′-fragment had a long lasting analgesic action but the potency was at least two orders of magnitude less than that of C-fragment (Geisow *et al.*, 1977). These results call attention to the importance of the COOH-terminal tetrapeptide of C-fragment for the analgesic potency. Peptides representing portions of the C′-fragment sequence (γ- and α-endorphins and met-enkephalin) have a very low potency as analgesic agents and their actions are transient.

It has been suggested that met-enkephalin might be unable to express its potential as an analgesic because it is destroyed rapidly in the brain and an analogue resistant to degradation by exopeptidases was therefore prepared. *N*-methyltyrosine-1-methionine-enkephalin amide, despite having an affinity for brain opiate receptors at least equal to that of the unmodified pentapeptide, had a low analgesic potency (Feldberg and Smyth, 1977a; Bradbury *et al.*, 1977). Although the effects lasted several hours, confirming its stability, the protected pentapeptide was 50–100 times less active than C-fragment.

The intact C-fragment produces a range of additional activities (table 15.1). In each of these properties, the protected pentapeptide exhibited qualitatively the

Table 15.1   Central, opiate-like actions of C-fragment (lipotropin 61-91)

	Potency (relative to morphine = 1)	References
(1) Analgesia	100	Feldberg and Smyth, 1976, 1977*a*, *b*; Bradbury *et al.*,977; 1977;
(2) Inhibition of firing of isolated neurons	highly potent	Gent *et al.*, 1977
(3) Hyperglycaemia	100	Feldberg and Smyth, 1977
(4) Grooming activity	50	Gispen *et al.*, 1976
(5) Inhibition of prostaglandin E_1 stimulated adenylate cyclase	60	Collier, H.O.J., unpublished data

These pharmacological assays were carried out with C-fragment isolated from porcine pituitary by Bradbury *et al.* (1975).

same activity as C-fragment but the potency was no greater than that of morphine. The stabilised enkephalin was thus 50-100 times less active than C-fragment. It appears that the complete C-fragment sequence is important for the full expression of its central activities.

C-fragment is present in porcine brain (Bradbury *et al.*, 1976*b*) as well as in pituitary and it has high potency in a range of central actions. This suggests that the peptide could fulfil a functional role in brain and the long duration of its action would imply that it is involved in control mechanisms acting over a period of time. Evidence has now gathered that C-fragment inhibits the release of acetylcholine and noradrenaline (Lord *et al.*, 1977), dopamine (Loh *et al.*, 1976) and Substance P (Jessel and Iversen, 1977). These properties are consistent with a broad role as a neuromodulator.

DEGRADATION OF C-FRAGMENT

The contrast between the long lasting actions of C-fragment and the transient actions of the endorphins and enkephalins would suggest that the duration of action may be related to resistance to degradation by brain enzymes. It is known that the principal enzyme involved in the degradation of met-enkephalin in brain is an aminopeptidase (Hambrook *et al.*, 1976); the resistance of the series of opiate active peptides to proteolysis by aminopeptidase M was therefore investigated (Austen and Smyth, 1977*a*). A striking difference was seen between the high stability of C-fragment and C'-fragment on the one hand and the rapid degradation of the shorter peptides on the other (figure 15.5). As each peptide possesses the same NH_2-terminal sequence, the high stability exhibited by C-fragment was attributed to conformational properties which protect the NH_2-terminal residue. Peptides containing up to 13 residues were readily degraded, consistent with their inability to form stable elements of secondary structure. Corresponding degrees of stability were exhibited when the series of peptides was exposed to a membrane bound aminopeptidase prepared from a particulate fraction of rat brain (P2 Fraction of Whittaker *et al.*, 1964).

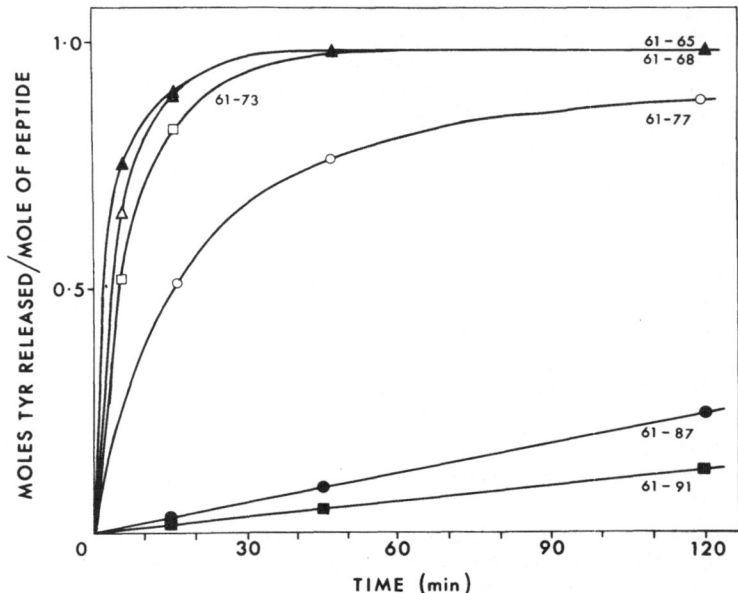

Figure 15.5 Rates of release of NH$_2$-terminal tyrosine from lipotropin C-fragment and a series of related peptides by aminopeptidase M. The peptides (0.34 mM) were incubated at 37 °C with the enzyme (0.02 mg/ml) in 0.1 M ammonium bicarbonate at pH 7.4. Aliquots of the solution were removed at intervals and the tyrosine determined by amino acid analysis.

Brain membranes from the rat possess carboxypeptidase activity in addition to the aminopeptidase (Geisow and Smyth, 1977). C-fragment incubated with carboxypeptidase A (Sigma) at an enzyme/substrate molar ratio of 1:200, pH 7.4, 37 °C, lost less than 1 per cent of the COOH-terminal glutamine residue in 2 hours. This was a somewhat surprising result since glutamine is normally released easily. When a synthetic hexapeptide Ala–Tyr–Lys–Lys–Gly–Gln with the COOH-terminal sequence of C-fragment was exposed to the same conditions, the same high stability was observed but the corresponding pentapeptide with a single lysine was readily attacked. Thus the unusual stability of C-fragment towards attack by carboxypeptidase is due to the presence of the paired lysine residues near the COOH-terminus and it is not due to conformational properties. As expected, C-fragment incubated with rat brain membranes which possess carboxypeptidase activity showed negligible release of COOH-terminal residues. The high resistance exhibited by C-fragment to the attack of exopeptidases implies that the inactivation of this peptide in brain can be attributed to attack by an endopeptidase.

Four endopeptidases were observed to cleave C-fragment specifically in the central section of the chain under mild conditions (figure 15.6). The NH$_2$-terminal region, though having a number of potentially vulnerable sites, remained intact and it was only when vigorous conditions of digestion were used that cleavage took place in this section (Austen and Smyth, 1977*b*). The results again point to the existence of preferred conformations in C-fragment which conceal the NH$_2$-terminal region of the peptide. When C-fragment was incubated at pH 7.4 with the P2 mem-

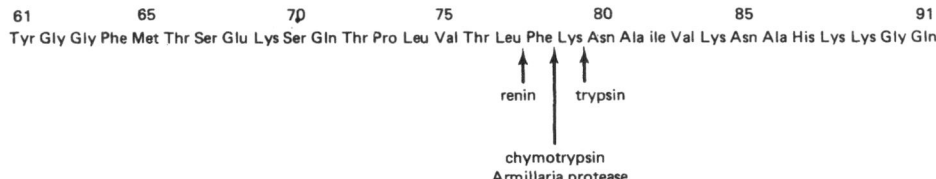

Figure 15.6 Sites of cleavage of C-fragment by endopeptidases.

brane fraction from rat brain, it was found that specific cleavage occurred with the formation of γ-endorphin together with a small amount of α-endorphin and met-enkephalin. Because of the rapid destruction of these products by amino peptidases, significant amounts were formed only when an aminopeptidase inhibitor, bacitracin, was included in the incubation medium. It was particularly notable that incubation of C-fragment with membranes at pH 5, a condition more appropriate to post-mortem acidity, led to rapid proteolysis by the brain enzymes and efficient conversion to γ-endorphin and met-enkephalin took place.

Incubation of [^{125}I] C-fragment with slices of rat striatum (Smyth and Snell, 1977) led to the formation of the same series of degradation products (figure 15.7), again bacitracin being included to prevent loss of the NH$_2$-terminal tyrosine. Evidence was obtained that the rate of degradation of C-fragment was not affected by

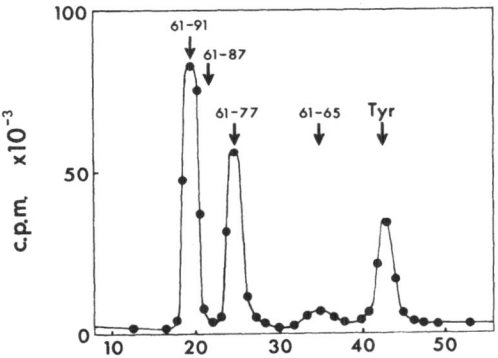

Figure 15.7 Gel filtration of [^{125}I] C-fragment after incubation with slices of rat striatum. Slices (approximately 200 mg wet weight) through the corpus striatum were prepared by section of fresh rat brains 3.5mm and 2.5mm rostral to the anterior part of the optic chiasma. The slices were preincubated (45 min, 37 °C) in 3 ml of saline (NaCl, 134 mM; KCl, 5 mM; KH$_2$PO$_4$, 1.25 mM; MgSO$_4$, 2.0 mM: CaCl, 1.0 mM; NaHCO$_3$, 16 mM; glucose, 10 mM) containing 10^{-4} M bacitracin. The slices were transferred and gently agitated at 37 °C in 3 ml of saline, or 3 ml of saline and 10^{-4} M bacitracin, containing 6 × 10 c.p.m. of [^{125}I] C-fragment (approximately 2 Ci/mmol). After 4 hours incubation, gel filtration was performed on Sephadex G-50 (150 × 0.9 cm column) in 50 per cent acetic acid. The eluate fractions (2.2 ml) were monitored for radioactivity. The arrows indicate the elution positions of authentic [^{131}I]- or [^{125}I]-labelled peptides.

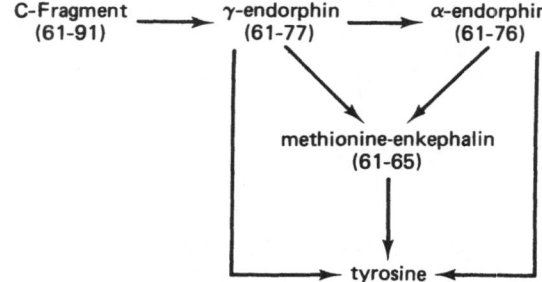

Figure 15.8 Degradation of C-fragment by membrane-bound enzymes from rat brain or by slices of rat striatum. The degradation of γ-endorphin proceeds principally by loss of NH_2-terminal tyrosine. In the presence of bacitracin ($10^{-4}M$) or at acid pH, the aminopeptidase is inhibited but the activity of the endopeptidases increases, leading to accumulation of tion of α-endorphin and met-enkephalin.

the presence of bacitracin, confirming that the first step in the degradative pathway involves attack by an endopeptidase and not an aminopeptidase. γ-Endorphin, in contrast, was markedly stabilised by the aminopeptidase inhibitor. The degradation appears to be mediated by extracellular enzymes because no enrichment of labelled peptides could be detected in the slices. The experiments do not rule out the possibility that formation of the endorphins and enkephalin might also take place by similar enzymes within the cell but there is as yet no firm evidence of stimulated release of γ- and α-endorphin or met-enkephalin from a cellular environment.

It is true that the ease of destruction and low receptor affinity of met-enkephalin are properties characteristic of a short lasting neurotransmitter (Kosterlitz and Hughes, 1975; Hughes *et al.*, 1975) and it has been suggested that γ- and α-endorphins may be elaborated to play a specific role in a behavioural pattern (Bloom *et al.*, 1976). Before engaging in further speculation, however, it is important that the concentrations and indeed presence of each of the 'opiate active' peptides in brain should be determined under conditions in which degradation is precluded. It is also important that the peptides should be resolved by chromatography before estimation by immunoassay, thereby reducing the demands made on the specificity of the antibodies. Such experiments are in progress.

REFERENCES

Austen, B. M. and Smyth, D. G. (1977a). *Biochem. biophys. Res. Commun.*, 76, 477–82
Austen, B. M. and Smyth, D. G. (1977b). *Biochem. biophys. Res. Commun.*, 77, 86–94
Bloom, F., Segal, D., Ling, N. and Guillemin, R. (1976). *Science*, 194, 630–31
Bradbury, A. F., Smyth, D. G. and Snell, C. R. (1975). In *Peptides: Chemistry, Structure and Biology. Proc. 4th Am. Peptide Sym.* (ed. J. Meienhofer), Ann Arbor Science Publishers, Michigan, pp. 609–15
Bradbury, A. F., Smyth, D. G. and Snell, C. R. (1976a). In *Polypeptide Hormones: Molecular and Cellular Aspects. CIBA Foundation Symposium 41 (new series)* (ed. R. Porter and D. W. Fitsimons), Elsevier Excerpta Medica North-Holland, Amsterdam, pp. 61–75
Bradbury, A. F., Feldberg, W. S., Smyth, D. G. and Snell, C. R. (1976b). In *Opiates and Endogenous Opioid Peptides* (ed. H. W. Kosterlitz), Elsevier North-Holland, Amsterdam, pp. 9–17

Bradbury, A. F., Smyth, D. G., Snell, C. R., Deakin, J. F. W. and Wendlandt, S. (1977). *Biochem. biophys. Res. Commun.*, 74, 748–54
Chrétien, M., Lis, M., Gilardeau, C. and Benjannet, S. (1976). *Can. J. Biochem.*, 54, 566–70
Feldberg, W. S. and Smyth, D. G. (1976). *J. Physiol., Lond.*, 260, 30–31P
Feldberg, W. S. and Smyth, D. G. (1977a). *J. Physiol., Lond.*, 265, 25–27P
Feldberg, W. S. and Smyth, D. G. (1977b). *Biochem. biophys. Res. Commun.*, 75, 625–29
Geisow, M. J. and Smyth, D. G. (1977). *Biochem. biophys. Res. Commun.*, 75, 625–29
Geisow, M. J., Deakin, J. F. W., Dostrovsky, J. O. and Smyth, D. G. (1977). *Nature*, 269, 167–68
Gent, J. P., Smyth, D. G., Snell, C. R. and Wolstencroft, J. H. (1977). *Br. J. Pharmac.*, 60, 272P
Gispen, W. H. Wiegant, V. M., Bradbury, A. F., Hulme, E. C., Smyth, D. G. and Snell, C. R. (1976). *Nature*, 264, 794–95
Guillemin, R., Ling, N. and Burgus. R. (1976). *C.r. hebd. Séanc. Acad. Sci., Paris, D*, 274, 783–85
Hambrook, J. M., Morgan, B. A., Rance, M. J. and Smith, C. F. C. (1976). *Nature*, 262, 782–83
Hughes, J., Smith, T., Kosterlitz, H. W., Fothergill, L., Morgan, B. A. and Morris, H. R. (1975). *Nature*, 258, 577–79
Jessel, T. and Iversen, L. L. (1977). *Nature*, 268, 549–51
Kemmler, W., Steiner, D. F. and Borg, J. (1973). *J. biol. Chem.*, 248, 4544–51
Kosterlitz, H. W. and Hughes, J. (1975). *Life Sci.*, 17, 91–96
Loh, H. H., Brase, D. A., Sampath-Khanna, S., Mar, J. B., Wei, E. L. and Li, C. H. (1976). *Nature*, 264, 794–95
Lord, J. A. H., Waterfield, A. A., Hughes, J. and Kosterlitz, H. W. (1977). *Nature*, 267, 495–99
Smyth, D. G. and Snell, C. R. (1977). *FEBS Lett.*, 78, 225–28
Whittaker, V. P. Michaelson, J. A. and Kirkland, R. J. A. (1964). *Biochem. J.*, 90, 293–303

16
Behavioural effects of
neuropeptides related to ß-LPH

D. de Wied (Rudolf Magnus Institute for Pharmacology Medical Faculty,
University of Utrecht, Vondellaan 6, Utrecht, The Netherlands)

INTRODUCTION

The lipotropic hormone β-lipotropin (β-LPH) was first isolated in 1964 from sheep
pituitaries (Li, 1964). It is located in discrete cells of the anterior and intermediate
lobes of the pituitary (Moon *et al.*, 1973). β-LPH consists of 90 or 91 amino acids
and the sequence does not vary much among the species (Li and Chung, 1976).
The residues 41–58 resemble those found in the melanocyte stimulating hormone
β-MSH (table 16.1). Thus, β-LPH may be a pro-hormone for this polypeptide.
Other β-LPH fragments were recently found in pituitary material and each of the
various peptides generated from β-LPH seems to be formed by cleavage of the
lipotropin chain at the carboxyl side of paired basic residues (Bradbury *et al.*,
1976*a*). Residues 47–53 are identical to residues 4–10 that are found in adrenocorticotropic hormone (ACTH) and in α- and β-MSH. This sequence has not been
found in pituitary extracts but this or a related peptide could well be generated
from β-LPH, ACTH, α- or β-MSH.

BEHAVIOURAL EFFECTS OF ACTH AND RELATED PEPTIDES

The implication of ACTH in central neuronal mechanisms was first suggested by
observations in hypophysectomised rats. Removal of the anterior lobe of the
pituitary (de Wied, 1964) or the whole gland (Applezweig and Baudry, 1955;
de Wied, 1969*a*, Weiss *et al.*, 1970) in rats impairs acquisition of shuttle-box avoidance behaviour. This behavioural impairment could be corrected with ACTH and
also with α-MSH and fragments of ACTH/MSH which are virtually devoid of
steroidogenic properties (de Wied, 1969*a*; Bohus *et al.*, 1973). These results suggested that the behavioural effect of ACTH is not mediated by the adrenal cortex.
Fragments of ACTH like $ACTH_{1-10}$, $ACTH_{4-10}$ but not $ACTH_{11-24}$ had a similar
beneficial effect on avoidance learning of hypophysectomised rats as the parent
molecule. Thus, the behaviourally active site appeared to be located in the N-terminal
part of ACTH, presumably in $ACTH_{4-10}$. From these experiments it was inferred

241

Table 16.1 β-Lipotropin (porcine)

```
H-Glu Leu Thr Gly Glu Glu Arg Leu Glu Gln Ala Arg Gly Pro Glu Ala Gln Ala (17)
                                                                    Glu
                                                                    Ser
                                                                    Ala (21)
                                                                              γ-LPH (1-58)

Ala Ala Glu Ala Ala Val Leu Gly Tyr Glu Leu Glu Ala Arg Ala Ala (21)
Ala
Glu
Lys
Lys
         β-MSH

(41) Asp Ser Gly Pro Tyr Lys Met Glu His Phe Arg Trp Gly Ser Pro Pro Lys Asp (58)
                          |_____ ACTH4-10 _____|                   Lys
                                                                         Arg

(76) Thr Val Leu Pro Thr Gln Ser Lys Glu Ser Thr Met Phe Gly Gly Tyr (61)
                                                 |___ met-enkephalin ___|
     |_____ α-endorphin _____|

Leu Phe Lys Asn Ala Ile Val Lys Asn Ala His Lys Lys Gly Glu-OH (91)
   β-endorphin
   C-fragment
```

that these pituitary hormones are precursor molecules for neuropeptides involved in acquisition and maintenance of new behaviour patterns.

In intact rats. ACTH and related peptides delay extinction of shuttle-box avo l-ance behaviour (de Wied, 1969b), pole-jumping avoidance behaviour (de Wied, 1966), improve maze performance (Flood *et al*., 1976), facilitate passive avoidance behaviour (Levine and Jones, 1965; Lissák and Bohus, 1972; Kastin *et al*., 1973a; de Wied, 1974; Flood *et al*., 1976), delay extinction of food-motivated behaviour in hungry rats (Leonard, 1969; Sandman *et al*., 1969; Gray, 1971; Guth *et al*., 1971; Garrud *et al*., 1974; Flood *et al*., 1976), delay extinction of conditioned taste aversion (Rigter and Popping, 1976) and sexually motivated approach behaviour (Bohus *et al*., 1975). On the basis of these observations we postulated that ACTH and related peptides are involved in motivational processes. However, these peptides affect learning and memory processes as well. They alleviate amnesia produced by CO_2 inhalation, electroconvulsive shock (Rigter *et al*., 1974; Rigter and van Riezen, 1975), by intracerebral administration of the protein synthesis inhibitor puromycin (Flexner and Flexner, 1971) and amnesia produced by anisomycin (Flood *et al*., 1976). Rigter *et al*. (1974) interpreted the effect of $ACTH_{4-10}$ on amnesia as an influence on retrieval processes. Gold and Van Buskirk (1976) found that post-trial injections of ACTH enhance or impair retention of passive avoidance behaviour, depending on the dose of ACTH used. These authors suggested that ACTH modulates memory storage processing of recent information. ACTH also improves correct performance rewarded by water through better use of environmental cues without a general effect on learning (Isaacson *et al*., 1976). These findings do not refute a motivational hypothesis since motivational effects operate in most of the paradigms used. Observations in man suggest that $ACTH_{4-10}$ facilitates selective visual attention (Kastin *et al*., 1975) but evidence for a motivational influence has been obtained as well (Gaillard and Sanders, 1975).

ACTH and related peptides do not materially affect spontaneous behaviour. Under various circumstances gross behaviour in a novel environment is not affected (Bohus and de Wied, 1966; Moyer, 1966; Weijnen and Slangen, 1970). Similarly, locomotor activity and food and water intake in the rat are not affected by α-MSH (Kastin *et al*., 1973b). A tendency towards increased activity was found in an open field situation in rats receiving α-MSH in addition to footshock prior to the test (Nockton *et al*., 1972). Facilitation of locomotor activity following α-MSH was reported in rats bearing lesions in the septal area (Brown *et al*., 1974). Increased locomotor activity was also found in mice after β-MSH but not after α-MSH or ACTH. In the rabbit, drowsiness, hyperpnoea and hypertension have been observed in reponse to β-MSH, or ACTH (Dyster-Aas and Krakau, 1965), while in the rat β-MSH induced drowsiness, hyperpnoea, hypertension and piloerection (Sakamoto, 1966). None of these actions is sufficiently pronounced as to affect conditioned avoidance behaviour in the rat.

BEHAVIOURAL EFFECTS OF PEPTIDES RELATED TO β-LPH

Several years ago, a number of peptides which possessed similar behaviour effects as ACTH and related peptides as assessed by inhibition of extinction of active, and facilitation of passive, avoidance behaviour, were isolated from hog pituitary material (Lande *et al*., 1973). These studies suggested the presence of potent behavioural peptides in the pituitary. One of these peptides which was obtained in pure form, yielded three oligopeptides after tryptic digestion. The amino acid

composition of two of these appeared to be similar to those of β-LPH$_{61-69}$ and
β-LPH$_{70-79}$ but the amount of material available at the time did not allow for
structure analysis. Since these peptides became available it was deemed of interest
to verify our previous findings on the behavioural effect of peptides related to
C-terminal β-LPH fragments.

Extinction of pole-jumping avoidance behaviour was used to assay the behaviour-
al effect of [Met^5]-enkephalin, β-LPH$_{61-69}$, α-endorphin, β-endorphin and a
number of related peptides. Following subcutaneous injection, α-endorphin
(β-LPH$_{61-76}$) appeared to be the most potent peptide. On a molar basis it was more

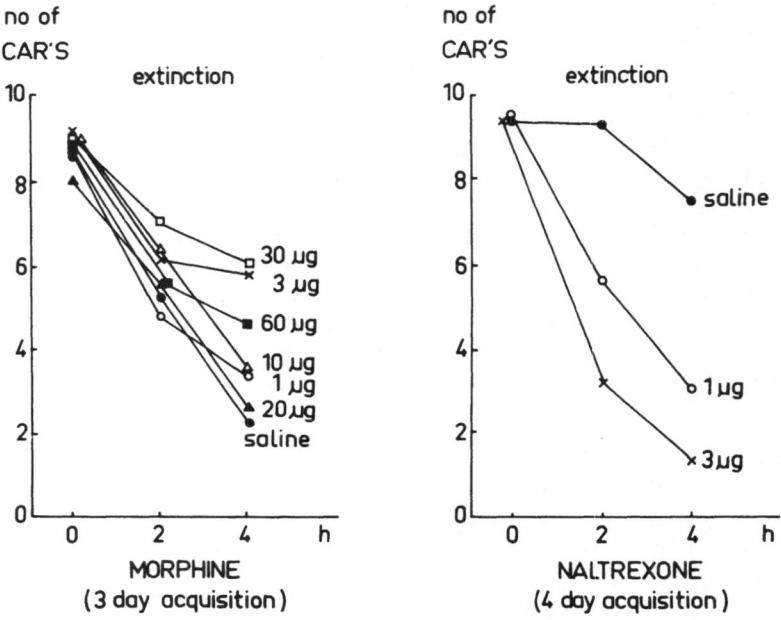

Figure 16.1   Effect of graded doses of morphine HCl and naltrexone HCl on extinction of pole-
jumping avoidance behaviour. Rats were trained in the pole-jumping test for 3 days. The next
day an extinction session of 10 trials was run. All rats that made 7 or more avoidances were
injected subcutaneously with the various doses of morphine. Extinction was studied again 2 and
4 hours later. In the case of naltrexone, training procedures lasted 4 days so as to make the rats
more resistant to extinction. The next day an extinction session of 10 trials was run. All rats
that made 8 or more avoidances were subcutaneously injected with naltrexone. Extinction was
studied again 2 and 4 hours later. CAR, condition avoidance response.

than 30 times more active than ACTH$_{4-10}$. Oxidation of methionine in [Met^5]-
enkephalin to the sulphoxide potentiated the effect on pole-jumping avoidance
behaviour (de Wied *et al.*, 1977). Intraventricular administration of the respective
peptides mimicked the effect of systemic administration, but much less peptide
was needed to elicit an equipotent behavioural effect. Interestingly, α-endorphin
and ACTH$_{4-10}$ were approximately equiactive following this route. Thus, the
difference in potency found following systemic administration seems to be related

to biotransformation and/or brain uptake mechanisms rather than to intrinsic behavioural activities. Passive avoidance behaviour was affected in a similar fashion by α-endorphin and $ACTH_{4-10}$. Both peptides facilitated passive avoidance behaviour. α-Endorphin was again more active than $ACTH_{4-10}$.

In view of the morphine-like activities of the endorphins, the effect of morphine and the specific opiate antagonist naltrexone was studied on extinction of pole-jumping avoidance behaviour. A tendency to delay extinction of the avoidance response was found following morphine but the effect was slight and not dose dependent. Naltrexone, however, facilitated extinction of the avoidance response, in relatively low doses (figure 16.1). The effect was clearly demonstrable in rats that were made more resistant to extinction by a 4 day training period. Naltrexone pretreatment, however, was unable to prevent the effect of α-endorphin, $ACTH_{4-10}$ or the potent $ACTH_{4-9}$ analogue (Org 2766) on extinction of pole-jumping avoidance behaviour. The same amount of naltrexone was sufficient to block β-endorphin-induced analgesia as measured on the hot plate (de Wied *et al.*, 1977).

As mentioned before, $ACTH_{4-10}$ possesses an anti-amnesic effect (Rigter *et al.*, 1974; Rigter and van Riezen, 1975; Flood *et al.*, 1976). Recently, Rigter (personal communication) found that [Met^5]-enkephalin, 1 hour prior to the retention test in doses ranging from 0.3–30 μg, markedly alleviated CO_2-induced amnesia for a passive avoidance response in rats. Also this effect is not blocked by naloxone. Kastin *et al.* (1976) showed that intraperitoneally administered [Met^5]-enkephalin and [D-Ala^2, Met^5]-enkephalin significantly improved maze performance of hungry rats for a food reward in a complex 12-choice Warden maze. Animals performed at faster speed and made less errors. [D-Phe^4, Met^5]-enkephalin had the same effect. This is interesting since this peptide has a low intrinsic opiate-like activity (Coy *et al.*, 1976). Morphine failed to facilitate maze performance.

ELECTROPHYSIOLOGICAL STUDIES

Electrophysiological studies with endorphins (de Wied *et al.*, 1977) revealed that relatively small amounts of these peptides affect theta activity in limbic midbrain structures. As found previously, systemic injection of $ACTH_{4-10}$ induces a shift from lower to higher frequencies in theta rhythm in hippocampus, septum and thalamus following stimulation of the reticular formation (Urban and de Wied, 1976). Intraventricular administration exhibits a similar shift in frequency distribution in the hippocampus during paradoxical sleep (de Wied *et al.*, 1977). The same effect was found with β-endorphin. We have stated that neuropeptides exert their effect on avoidance behaviour by facilitation of a selective arousal state in midbrain limbic structures (Urban and de Wied, 1976). This would increase the motivational value of environmental stimuli resulting in an increase in the probability that stimulus-specific behavioural responses occur. The endorphins exhibit similar effects on theta activity but much lower amounts are needed to affect active and passive avoidance behaviour, maze performance and to alleviate amnesia (Kastin *et al.*, 1976; de Wied *et al.*, 1977; Rigter, 1977 personal communication) than those which induce analgesia. This supports the hypothesis that the endorphins are more fundamentally involved in motivational, learning and memory processes than in pain perception.

EFFECT OF PEPTIDES RELATED TO ACTH AND TO β-LPH ON EXCESSIVE GROOMING

Peptides related to ACTH, MSH and β-LPH induce a stretching and yawning syndrome following intracerebral administration only (Ferrari *et al.*, 1963; Gessa *et al.*, 1967). In rodents, the onset of the syndrome is preceded by a display of excessive grooming (Ferrari *et al.*, 1963; Izumi *et al.*, 1973; Gispen *et al.*, 1975; Rees *et al.*, 1976). In rats, $ACTH_{1-24}$, α- and β-MSH are equipotent in this respect (Gispen *et al.*, 1975). $ACTH_{4-10}$ is less active in rabbits (Baldwin *et al.*, 1974) and inactive in rats (Gispen *et al.*, 1975) and mice (Rees *et al.*, 1976) but $[D\text{-Phe}^7]ACTH_{4-10}$ has appreciable activity. $ACTH_{4-10}$ could be activated by C-terminal chain elongation but also by shortening to $ACTH_{4-7}$. This peptide is as active as $[D\text{-Phe}^7]ACTH_{4-10}$ in this respect (Wiegant and Gispen, 1977). This is also the shortest sequence which has full activity in the pole-jumping avoidance test (Greven and de Wied, 1973). Excessive grooming induced by ACTH analogues can be completely suppressed by pretreatment with specific opiate antagonists. This suggests an interaction with opiate receptors in the brain (Gispen and Wiegant, 1976). Morphine also induces excessive grooming. Terenius (1975) found that $ACTH_{1-28}$ and $ACTH_{4-10}$ had an affinity for opiate receptors. Subsequent structure activity studies pointed to an active site within $ACTH_{4-10}$ with some indication for a second site distal to the C-terminus of this hexapeptide (Terenius *et al.*, 1975). Analysis of the binding revealed low selectivity of these peptides for agonist or antagonist binding sites, comparable to the partial agonist nalorphine (Terenius, 1976). In view of the relatively low affinity for the opiate receptor (IC_{50} of the order of $10^{-6}\text{-}10^{-5}$ M), $ACTH_{4-10}$ cannot be regarded as a physiological ligand for opiate receptors in the brain. This is in contrast to C-terminal β-LPH fragments which exhibit high affinity for brain opiate receptors (Bradbury *et al.*, 1976b). But the fact that ACTH interacts with opiate receptors in the brain may explain the interaction of these peptides with the analgesic effect of morphine (Gispen *et al.*, 1976a). Morphine induced spinal reflex activity is reduced by ACTH in preparations *in vivo* and *in vitro* (Zimmermann and Krivoy, 1973). ACTH counteracts the analgesic effect of morphine in rodents (Winter and Flataker, 1951). Gispen *et al.* (1976a) found that ACTH and ACTH fragments reduce the analgesic response of morphine by 50–60 per cent as measured on the hot plate test. The peptides themselves had no effect on the hot plate behaviour of the rates. Interestingly, this effect *in vivo* corresponds rather well with the affinity to brain opiate receptor sites *in vitro*. If excessive grooming is mediated by opiate receptors in the brain, the endorphins should be more potent than ACTH. Indeed, intraventricular administration of doses as low as 10 ng β-endorphin elicits excessive grooming (Gispen *et al.*, 1976b). α-Endorphin is much less active while $[Met^5]$-enkephalin is virtually inactive. β-Endorphin on a weight basis is ten times as active as $ACTH_{1-24}$. The grooming effect of β-endorphin is readily blocked by naloxone (Gispen *et al.*, 1976b).

OTHER BEHAVIOURAL EFFECTS OF ENDORPHINS

Other behavioural effects of endorphins have been reported. Bloom *et al.* (1976) demonstrated various behavioural effects of endorphins after injection into a lateral ventricle in rats using relatively high doses. β-Endorphin was the most potent peptide which caused the disappearance of corneal reflexes, a depression general motor activity and lack of responsiveness to noxious stimuli, resulting finally in a profound catatonic state with extreme muscle rigidity and loss of righting reflex. These effects were rapidly reversed by naloxone. The smaller β-LPH fragments did not induce the catatonic state. [Met⁵] -enkephalin, α-, β- and γ-endorphin produced acute episodes of 'wet-dog' shakes, which are regarded as typical withdrawal symptoms. These were surprisingly also prevented by naloxone. Such responses were not observed following intraventricular injection of morphine. This drug also failed to induce the catatonic state. According to Bloom *et al.* (1976) these observations suggest that the endorphins are involved in maintaining behavioural homeostasis. Similar effects were observed by Jacquet and Marks (1976). These authors found that when β-endorphin was injected into the periaqueductal grey of the rat profound sedation and catalepsia occurred. The smaller endorphins induced attenuated forms of this behaviour. These effects were interpreted as being neuroleptic in character and it was suggested that β-endorphin might be an important neuromodulator in the central nervous system (CNS). Sexual behaviour may also be affected by endorphins. β-Endorphin given intraventricularly in amounts of 1 and 3 μg interferes with normal sexual behaviour of male rats (Meyerson and Terenius, 1977). Mounting behaviour was reduced and became abnormal. No evidence for muscular rigidity was found at the 1 μg dose level. The effect could be prevented by pretreatment with naltrexone.

INFLUENCE OF SPECIFIC OPIATE ANTAGONISTS ON ANALGESIA, BODY TEMPERATURE AND CONDITIONED BEHAVIOUR

Specific opiate antagonists prevent access of narcotics to their binding sites. In themselves these antagonists have been regarded as being without apparent intrinsic effects. However, since the opiate antagonists interfere with the action of exogenous endorphins one would expect that they might exhibit intrinsic activities as a result of displacement of endogenous opiate like substances. Rats treated with amounts of naloxone sufficient to block morphine analgesia showed no change in the threshold for escape from footshock (Goldstein *et al.*, 1976). Morphine does increase the threshold for footshock and this effect is antagonised by naloxone. The threshold for pain detection may not therefore be controlled by endorphins. Perception of experimentally induced pain in man is not blocked by naloxone (El-Sobky *et al.*, 1976). However, naloxone blocks analgesia produced by electrical stimulation in the rat midbrain (Akil *et al.*, 1972) and that elicited by intraventricular injection of acetylcholine (Pedigo *et al.*, 1975). Analgesia produced by electroacupuncture in mice is also abolished by naloxone (Pomeranz and Chiu, 1976), and normal mice injected with naloxone exhibit hyperalgesia. Pomeranz and Chiu (1976) suggested that naloxone intereferes with the basal release of endorphins as well as the activated release following electroacupuncture. Analgesic effects in the cat produced by stimulation of the central inferior raphe nucleus are greatly reduced by naloxone (Oliveras *et al.*, 1977). This analgesic effect is ascribed to the release of endogenous opiate-like substances. Such substances are found in relatively high concentration in areas where stereospecific opiate receptors are found (Kuhar *et al.*, 1973).

Temperature regulation may also be influenced by endorphins. Goldstein and Lowery (1975), however, found a slight hypothermic effect of naloxone at doses that effectively block opiate receptors. Even under cold stress the effect of naloxone was small. Chronic morphine causes hyperthermia, and this hyperthermic response can be conditioned by an olfactory stimulus (Lal *et al.*, 1976). Such a conditioned hyperthermia is blocked by naloxone. This suggests that conditioned stimuli may also release endogenous opiate-like peptides. Naloxone reduces amphetamine-induced facilitation of avoidance behaviour (Holtzman, 1974) and our studies showed a marked effect of naltrexone on extinction of avoidance behaviour (de Wied *et al.*, 1977). If conditioned stimuli trigger the release of endorphins, naloxone could act by blocking the release or generation of endorphins from their respective precursor molecules. This would explain why the effect of exogenous endorphins is not blocked by naltrexone (de Wied *et al.*, 1977; Rigter, 1977 personal communication). Further studies are needed to clarify the issue, but it is clear that the primary physiological role of the endorphins is not on pain perception or temperature regulation but on complex behavioural mechanisms.

DIFFERENT RECEPTOR SITES FOR THE VARIOUS NEUROPEPTIDES

Irrespective of intrinsic behavioural activities, naloxone does not prevent the effect of neuropeptides on conditioned avoidance behaviour, and amnesia. It does, however, block excessive grooming, elicted by $ACTH_{1-24}$ or β-endorphin (Gispen *et al.*, 1976*a*, *b*; Gispen and Wiegant, 1976), or the various behavioural abnormalities induced by relatively high amounts of intraventricularly administered endorphins (Bloom *et al.*, 1976; Jacquet and Marks, 1976; Meyerson and Terenius, 1977). Thus the effect of ACTH and endorphins on motivational, learning and memory processes takes place independent of opiate receptors in the brain. This is in accord with the fact that the highly active $ACTH_{4-9}$ analogue does not induce excessive grooming (Gispen *et al.*, 1975), that oxidation of methionine in [Met^5]-enkephalin reduces its opiate-like effect but potentiates the effect on avoidance behaviour and that morphine does not materially affect avoidance behaviour or maze performance, but elicits excessive grooming (Gispen and Wiegant, 1976). It might be postulated, therefore, that at least two endorphin receptor sites are present in the brain.

In contrast to the abundant presence of putative receptor sites for opiate-like peptides (Kuhar *et al.*, 1973), the presence of receptor sites for ACTH fragments in the brain has as yet not been demonstrated. However, the tritium labelled $ACTH_{4-9}$ analogue (Org 2766) is taken up preferentially in the septal area following intraventricular injection. Hypophysectomy resulted in a significantly enhanced uptake of radioactivity in the septum as compared to that of control rats. The increased uptake in hypophysectomised rats can be prevented by chronic treatment of hypophysectomised rats with $ACTH_{1-24}$ or $ACTH_{4-10}$ given as a long acting zinc phosphate preparation (Verhoef *et al.*, 1977). Treatment with [D-Phe^7] $ACTH_{4-10}$, α-endorphin, or desglycinamide^9-lysine^8-vasopressin (DG-LVP) did not affect the facilitated uptake of the $ACTH_{4-9}$ analogue. These results are evidence for a specific uptake or binding of the $ACTH_{4-9}$ analogue in the septal region because competitive displacement occurred only with peptides that were both behaviourally and structurally closely related to N-terminal ACTH fragments. Such studies suggest the presence of putative receptor sites for these neuropeptides. The inability of α-endorphin to compete with the potentiated $ACTH_{4-9}$ analogue for receptor sites in the septal area suggests the existence of different receptors for ACTH and related peptides and C-terminal β-LPH fragments for the same behavioural effect.

CONCLUDING REMARKS

The sequence $ACTH_{4-10}$ is not only present in ACTH, α- and β-MSH, but also in β-LPH. Which of these pituitary hormones acts as the precursor molecule for this neuropeptide is not known. It is unclear from the identical effects of ACTH fragments and of endorphins on avoidance behaviour as to which part of lipotropin is involved in the formation of new behavioural patterns. The effect of ACTH on avoidance behaviour is independent of inherent corticotropic, MSH-, opiate-like and fat mobilising activities. Modification of $ACTH_{4-9}$ by oxidation of [methionine[4]] to the sulphoxide and substitution of [arginine[8]] by D-lysine and [tryptophan[9]] by phenylalanine markedly potentiates the effect on avoidance behaviour. This compound (Org 2766) is one thousand times stronger than $ACTH_{4-10}$ in delaying extinction of pole-jumping avoidance behaviour. It possesses 0.1 per cent of the MSH activity of $ACTH_{4-10}$ and it has in addition a markedly reduced steroidogenic effect as compared to $ACTH_{4-10}$ (Greven and de Wied, 1973). $ACTH_{4-9}$ has no affinity for brain opiate receptor sites, that is it does not induce excessive grooming (Gispen *et al.*, 1975) and the potentiated analogue has no opiate-like activity as assessed on the mouse vas deferens preparation (van Ree, 1976, unpublished observations) and negligible fat mobilising activities on rat epididymis cells *in vitro* as compared to $ACTH_{1-24}$ (Opmeer, van Ree and de Wied, 1977, unpublished observations). Oxidation of methionine in [Met[5]]-enkephalin potentiates the effect on avoidance behaviour but markedly decreases opiate-like activity (Ling and Guillemin, 1976). The analogue [D-Phe[4], Met[5]]-enkephalin is equipotent with [D-Ala[2], Met[5]]-enkephalin in facilitating maze performance while it has no appreciable opiate-like activity (Kastin *et al.*, 1976). Thus, the effect of peptides related to ACTH, α- and β-MSH and β-LPH on conditioned behaviour is dissociated from the classical endocrine effects of these pituitary principles and from opiate-like activity.

It is conceivable that β-LPH generates at least two classes of neuropeptides ($ACTH_{4-10}$/β-MSH and endorphins). In view of the similarities in CNS effect, that is affinity for opiate receptor sites; influences on motivation, learning and memory processes; effects on grooming, stretching and yawning; it is clear that overlapping information is encoded in pituitary peptides. Since the endorphins are at least as potent as ACTH fragments in affecting conditioned behaviour, one might postulate that the C-terminal β-LPH peptides contain the active and the N-terminal peptides the dormant sites in this respect (Gispen *et al.*, 1977). However, it is also possible that the organism generates hitherto unknown neuropeptides which may be more specifically involved in the formation of new behaviour patterns (de Wied, 1969*a*). The identical behavioural effects of peptides related to ACTH, MSH and β-LPH may then be a reflection of overlapping information in related peptides. The dissociation between endocrine, opiate-like and behavioural effects by substitution of $ACTH_{4-9}$ and the millionfold potentiation of the effect on avoidance behaviour by substitution of $ACTH_{4-16}$ (Greven and de Wied, 1977) indeed suggest the existence of highly specific behavioural molecules in pituitary and brain.

REFERENCES

Akil, H., Mayer, D. J. and Liebeskind, J. C. (1972). *C.r. hebd. Séanc. Acad. Sci., Paris*, **274**, 3603–05

Applezweig, M. H. and Baudry, F. D. (1955). *Psychol. Rep.*, **1**, 417–20

Baldwin, D. M., Haun, Ch. K. and Sawyer, Ch. H. (1974). *Brain Res.*, **80**, 291–301

Bloom, F., Segal, D., Ling, N. and Guillemin, R. (1976). *Science*, **194**, 630–32

Bohus, B. and de Wied, D. (1966). *Science*, **153**, 318–20

Bohus, B., Gispen, W. H. and de Wied, D. (1973). *Neuroendocrinology*, **11**, 137–43

Bohus, B., Hendrickx, H. H. L., van Kolfschoten, A. A. and Krediet, T. G. (1975). In *Sexual Behavior: Pharmacology and Biochemistry* (ed. M. Sandler and G. L. Gessa), Raven Press, New York, pp. 269–75

Bradbury, A. F., Smyth, D. G. and Snell, C. R. (1976a), *Biochem. biophys. Res. Commun.*, **69**, 950–56

Bradbury, A. F., Smyth, D. G., Snell, C. R., Birdsall, N. J. M. and Hulme, E. C. (1976b). *Nature*, **260**, 793–95

Brown, G. M., Uhlir, I. V., Seggie, J., Schally, A. V. and Kastin, A. J. (1974). *Endocrinology*, **94**, 583–87

Coy, D. H., Kastin, A. J., Schally, A. V., Morin, O., Caron, N. G., Labrie, F., Walker, J. M., Fertel, R., Berntson, G. G. and Sandman, C. A. (1976). *Biochem. biophys. Res. Commun.*, **73**, 632–38

Dyster-Aas, H. K. and Krakau, C. E. T. (1965). *Acta Endocr. (Kbh.)*, **48**, 409–19

El-Sobky, A., Dostrovsky, J. O. and Wall, P. D. (1976). *Nature*, **263**, 783–84

Ferrari, W., Gessa, G. L. and Vargiu, L. (1963). *Ann. N.Y. Acad. Sci.*, **104**, 330–45

Flexner, J. B. and Flexner, L. B. (1971). *Proc. natn. Acad. Sci. U.S.A.*, **68**, 2519–21

Flood, J. F., Jarvik, M. E., Bennett, E. L. and Orme, A. E. (1976). *Pharmac. Biochem. Behav.*, **5**, suppl.1, 41–51

Gaillard, A. W. K. and Sanders, A. F. (1975). *Psychopharmacologia*, **42**, 201–08

Garrud, P., Gray, J. A. and de Wied, D. (1974). *Physiol. Behav.*, **12**, 109–19

Gessa, G. L., Pisano, M., Vargiu, L., Crabai, F. and Ferrari, W. (1967). *Rev. Can. Biol.*, **26**, 229–36

Gispen, W. H. and Wiegant, V. M. (1976). *Neurosci. Lett.*, **2**, 159–64

Gispen, W. H., Wiegant, V. M., Greven, H. M. and de Wied, D. (1975). *Life Sci.*, **17**, 645–52

Gispen, W. H., Buitelaar, J., Wiegant, V. M., Terenius, L. and de Wied, D. (1976a). *Eur. J. Pharmac.*, **39**, 393–97

Gispen, W. H., Wiegant, V. M., Bradbury, A. F., Hulme, E. C., Smyth, D. G., Snell, C. R. and de Wied, D. (1976b). *Nature*, **264**, 794–95

Gispen, W. H., van Ree, J. M. and de Wied, D. (1977). *Int. Rev. Neurobiol.*, (in press) **20**, 209–50

Gold, P. E. and Van Buskirk, R. (1976). *Behav. Biol.*, **16**, 387–400

Goldstein, A. and Lowery, P. J. (1975). *Life Sci.*, **17**, 927–32

Goldstein, A., Pryor, G. T., Otis, L. S. and Larsen, F. (1976). *Life Sci.*, **18**, 599–604

Gray, J. A. (1971). *Nature*, **229**, 52–54

Greven, H. M. and de Wied, D. (1973). In *Drug Effects on Neuroendocrine Regulation* (ed. E. Zimmermann, W. H. Gispen, B. H. Marks and D. de Wied), Progress in Brain Research Vol. 39. Elsevier, Amsterdam, pp. 429–42

Greven, H. M. and de Wied, D. (1977). In *Melanocyte Stimulating Hormone: Control, Chemistry and Effects* (ed. F. J. H. Tilders, D. F. Swaab and Tj. B. van Wimersma Greidanus), Frontiers of Hormone Research Vol. 4. Karger, Basel, (in press)

Guth, S., Levine, S. and Seward, J. P. (1971). *Physiol. Behav.*, **7**, 195–200

Holtzman, S. G. (1974). *J. Pharmac. exp. Ther.*, **189**, 51–60

Isaacson, R. L., Dunn, A. J., Rees, H. D. and Waldock, B. (1976). *Physiol. Psychol.*, **4**, 159–62

Izumi, K., Donaldson, J. and Barbeau, A. (1973). *Life Sci.*, **12**, 203–10

Jacquet, Y. F. and Marks, N. (1976). *Science*, **194**, 632–34

Kastin, A. J., Miller, L. H., Nockton, R., Sandman, C. A., Schally, A. V. and Stratton, L. O. (1973a). In *Drug Effects on Neuroendocrine Regulation*. (ed. E. Zimmermann, W. H. Gispen, B. H. Marks and D. de Wied), Progress in Brain Research Vol. 39, Elsevier, Amsterdam, pp. 461–70

Kastin, A. J., Miller, M. C., Ferrell, L. and Schally, A. V. (1973b). *Physiol. Behav.*, **10**, 399–401

Kastin, A. J., Sandman, C. A., Stratton, L. O., Schally, A. V. and Miller, L. H. (1975). In *Hormones, Homeostasis and the Brain*. (ed. W. H. Gispen, Tj. B. van Wimersma Greidanus,

B. Bohus and D. de Wied), Progress in Brain Research Vol. 42, Elsevier, Amsterdam, pp. 143–50
Kastin, A. J., Scollan, E. L., King, M. G., Schally, A. V. and Coy, D. H. (1976). *Pharmac. Biochem. Behav.*, 5, 691–95
Kuhar, M. J., Pert, C. B. and Snyder, S. H. (1973). *Nature*, 245, 447–50
Lal, H., Miksic, S. and Smith, N. (1976). *Life Sci.*, 18, 971–76
Lande, S., de Wied, D. and Witter, A. (1973). In *Drug Effects on Neuroendocrine Regulation* (ed. E. Zimmermann, W. H. Gispen, B. H. Marks and D. de Wied), Progress in Brain Research Vol. 39, Elsevier, Amsterdam, pp. 421–27
Leonard, B. E. (1969). *Int. J. Neuropharmac.*, 8, 427–35
Levine, S. and Jones, L. E. (1965). *J. comp. physiol. Psychol.*, 59, 357–60
Li, C. H. (1964). *Nature*, 201, 924
Li, C. H. and Chung, D. (1976). *Proc. natn. Acad. Sci. U.S.A.*, 73, 1145–48
Ling, N. and Guillemin, R. (1976). *Proc. natn. Acad. Sci. U.S.A.*, 73, 3308–10
Lissák, K. and Bohus, B. (1972). *Int. J. Psychobiol.* 2, 103–15
Meyerson, B. J. and Terenius, L. (1977). *Eur. J. Pharmac.*, 42, 191–92
Moon, H. D., Li, C. H. and Jenning, B. M. (1973). *Anat. Rec.*, 175, 529–38
Moyer, K. E. (1966). *J. genet. Psychol.*, 108, 297–302
Nockton, R., Kastin, A. J., Elder, S. T. and Schally, A. V. (1972). *Horm. Behav.*, 3, 339–44
Oliveras, J. L., Hosobuchi, Y., Redjemi, R., Guilbaud, G. and Besson, J. M. (1977). *Brain Res.*, 120, 221–29
Pedigo, N. W., Dewey, W. L. and Harris, L. S. (1975). *J. Pharmac. exp. Ther.*, 193, 845–52
Pomeranz, B. and Chiu, D. (1976). *Life Sci.*, 19, 1757–62
Rees, H. D., Dunn, A. J. and Iuvone, P. M. (1976). *Life Sci.*, 18, 1333–40
Rigter, H. and Popping, A. (1976). *Psychopharmacologia*, 46, 255–61
Rigter, H. and van Riezen, H. (1975). *Physiol. Behav.*, 14, 563–66
Rigter, H., van Riezen, H. and de Wied, D. (1974). *Physiol. Behav.*, 13, 381–88
Sakamoto, A. (1966). *Nature*, 211, 1370–71
Sandman, C. A., Kastin, A. J. and Schally, A. V. (1969). *Experientia*, 25, 1001–02
Terenius, L. (1975). *J. Pharm. Pharmac.*, 27, 450–52
Terenius, L. (1976). *Eur. J. Pharmac.*, 38, 211–13
Terenius, L., Gispen, W. H. and de Wied, D. (1975). *Eur. J. Pharmac.*, 33, 395–99
Urban, I. and de Wied, D. (1976). *Expl. Brain Res.*, 24, 325–44
Verhoef, J., Witter, A. and de Wied, D. (1977). *Brain Res.*, 131, 117–28
Weiss, J. M., McEwen, B. S., Silva, M. and Kalkut, M. (1970). *Am. J. Physiol.*, 218, 864–68
Weijnen, J. A. W. M. and Slangen, J. L. (1970). In *Pituitary, Adrenal and the Brain* (ed. D. de Wied and J. A. W. M. Weijnen) Progress in Brain Research Vol. 32, Eslevier, Amsterdam, pp. 221–35
de Wied, D. (1964). *Am. J. Physiol.*, 207, 255–59
de Wied, D. (1966). *Proc. Soc. Exp. Biol.*, 122, 28–32
de Wied D. (1969a). In *Frontiers in Neuroendocrinology 1969* (ed. W. F. Ganong and L. Martini) University Press, London/New York, pp. 97–140
de Wied, D. (1969b). *Proc. 3rd Int. Congr. Endocrinology, Mexico*, June/July 1968. Excerpta Medica International Congress Series No. 184, Excerpta Medica, Amsterdam, pp. 310–16
de Wied, D. (1974). In *The Neurosciences, Third Study Program* (ed. F. O. Schmitt and F. G. Worden), MIT Press, Cambridge, Massachusetts, pp. 653–66
de Wied, D., Bohus, B., van Ree, J. M. and Urban, I. (1977). Submitted for publication
Wiegant, V. M. and Gispen, W. H. (1977). *Behav. Biol.*, 19, 554–58, abstr. no. 6241
Winter, C. A. and Flataker, L. (1951). *J. Pharmac. exp. Ther.*, 101, 93
Zimmermann, E. and Krivoy, W. A. (1973). In *Drug Effects on Neuroendocrine Regulation* (ed. E. Zimmermann, W. H. Gispen, B. H. Marks and D. de Wied) Progress in Brain Research Vol. 39, Elsevier, Amsterdam, pp. 383–94

Index